A Hole in the Ground with a Liar at the Top

A Hole in the Ground with a Liar at the Top

Fraud and Deceit in the Golden Age
of American Mining

by

DAN PLAZAK

THE UNIVERSITY OF UTAH PRESS
Salt Lake City

The Defiance House Man colophon is a registered trademark of the University of Utah Press. It is based upon a four-foot-tall, Ancient Puebloan pictograph (late PIII) near Glen Canyon, Utah.

18 17 16 15 14   2 3 4 5 6

LIBRARY OF CONGRESS CATALOGING-IN-PUBLICATION DATA

Plazak, Dan, 1951-
  A hole in the ground with a liar at the top : fraud and deceit in the golden age of American mining / by Dan Plazak.
       p. cm.
  Includes bibliographical references and index.
  ISBN: 978-1-60781-020-9 (alk. paper)
    1. Mines and mineral resources—Valuation—United States—History—19th century.
  2. Mines and mineral resources—Valuation—United States—History—20th century.
  3. Fraud—United States—History—19th century. 4. Fraud—United States—History—
  20th century..5. Swindlers and swindling—United States—History—19th century.
  6. Swindlers and swindling—United States—History—20th century. I. Title.
  TN272.P56 2005
  553.0973'09034—DC22

                                                                    2005029913

Preceding page: Illustration from Dan De Quille, *The Big Bonanza*, 1876

"A mine is a hole in the ground with a Cornishman at the bottom."
—Old British saying

"A mine is a hole in the ground with a liar at the top."
—Old American saying, usually attributed to Mark Twain, although there is no evidence that he ever said or wrote it

Table of Contents

Acknowledgments

Between classes at the Colorado School of Mines, I relaxed in the library by thumbing through oversize bound volumes of the *Engineering and Mining Journal,* whose back issues record more than 150 years of mining history. I was intrigued by the accounts of mining fraud, which the editors of *E&MJ* took special pains to expose and denounce.

When I wanted to learn about fraud in early American mining, I found that there was no book on the subject. The topic was largely ignored, and the information was scattered through dozens of books. Most of the facts rested comfortably forgotten in old newspapers and magazines. To find a book on the subject, I had to write it myself. Finishing it, however, took about eight years longer than I had foolishly estimated.

While researching, I came to know libraries across the country and spent uncounted hours searching the stacks and scrolling through microfilms of old newspapers. The best western history collection is no doubt in the Denver Public Library, where I became a semi-fixture for a year or two.

I need to thank those who took the time to read my efforts and offer comment, criticism, and encouragement: Emmy Booy, Joanne Burris, Sherma Erholm, Mark Feldman, Dale Jones, Kira Lapin, and Sherry Spitsnaugle. Greg Taylor helped with the research.

Introduction to Mining Fraud

"The ways of a man with a maid be strange, yet simple and tame,
To the ways of a man with a mine, when buying and selling the same."
—Bret Harte, quoted in Don Berkman, "Salting Solutions,"
ALS Chemex Minerals Division News 7, no. 4 (1998)

In the 1800s, fortune seekers pushed into the wilderness and discovered America's enormous mineral wealth. In just a few decades they made America the greatest mining nation in the world.

The booming high-tech industries of that era were mining and railroads. Railroads tied the country together with rapid travel, but the railroad manipulations and bribery of Fisk, Gould, and Vanderbilt are legendary. Mining was more on the up-and-up, but it, too, attracted some men with more ambition than scruples.

This book concentrates on fraud, deceit, and reckless finance in the American mining industry between the Mexican War and the First World War. This was a time when mining and its enormous wealth excited the public imagination as much as the high-tech start-ups of today and the barriers to fraud were much weaker than they are now. It was a golden era of expansion for American mining but also a dark era of mining fraud.

"Colorado to-day presents an example of a mining territory nearly
ruined by the villainous frauds of a class of men in whom there is
neither honor nor integrity."
—Salt Lake *Tribune*, May 10, 1873

Since antiquity, fraud and folly have visited every sort of business, and mining has been no exception. The history of mining fraud goes back centuries.

16th-century Germany

"It is complained that some sellers and buyers of the shares in mines are fraudulent. I concede it."

—Georg Bauer, 1556

In 1544 the Protestant reformer Martin Luther, himself the son of a miner, said about mining shares: "I am told that everything is full of fraud and injustice.... They have often tempted me with mining stocks, but I never wanted them!"[1]

Georg Bauer, in his *De Re Metallica*, wrote in 1556 that people tell lies about a mine to raise or lower the share price. The ploy of levying assessments to "freeze out" other shareholders also existed in Bauer's day, as did mine superintendents who diverted company money to their own pockets. Bauer said that "wicked men," mining small veins that branch off the main vein, try to claim ownership of the main vein. This is strikingly similar to the apex lawsuits that plagued nineteenth-century American mining, especially on Nevada's Comstock lode and at Butte, Montana.[2]

"All Is Not Gold That Shineth"

In 1576 Captain Martin Frobisher arrived back in London from a search for the Northwest Passage. From what is now arctic Canada, he brought back a captive Eskimo and a black rock. He had, without much thought, grabbed the black rock from a small bleak island off the coast of Baffin Island and gave it to one of his financial sponsors, Michael Lok.[3]

Frobisher's black rock, the humble souvenir of the New World, obsessed Lok, who decided that the stone must contain gold. The assayer of the Tower of London reported that it contained no gold. Lok took the stone to another assayer, then a third, each of whom told him the same thing: no gold.

Lok still believed that his stone contained gold, part of an enormous deposit in the New World. His behavior would be repeated through the years, as gold-obsessed men refused to believe that their rock was worthless. Lok finally found his assayer in Giovanni Agnello, an Italian living in London. Agnello told Lok that the stone indeed contained gold and as proof gave Lok a small quantity of gold powder supposedly derived from the rock. Agnello wrote of his success in finding gold: "It is necessary to know how to flatter nature." Agnello did not know how to flatter nature, but he knew how to flatter Lok.

Lok lobbied Francis Walsingham, Queen Elizabeth's secretary, for a charter to mine gold in the New World. Walsingham dismissed Agnello's result as "an alchemist matter," such as very many that had been brought before the queen. Lok gave Walsingham a piece of the black rock, but Walsingham's assayers found nothing, and he told Lok that Agnello "did but play the alchemist." Yet Lok finally obtained a charter for his Cathay Company to mine gold, and Queen Elizabeth became the largest investor.[4]

The gold-mining fleet sailed in May 1577 and returned to the finding place of the black rock. Frobisher's men searched nearby islands and found what the gold-tester Jonas Shultz, a German recommended by Agnello, assured them were gold and silver ore. Frobisher's men mined 158 tons of rock, stowed it aboard, and sailed away in August 1577, before the winter sea ice gripped the islands.

Frobisher landed two of his ships at Bristol, the crews marched through town to trumpets and cheering crowds, and their gold ore was stored in Bristol Castle for safekeeping. The third ship landed at London, and its ore was guarded in the Tower of London. Agnello and Shultz said that the ore was rich in gold, but reputable assayers disagreed.

Skeptical Francis Walsingham hired another foreign expert, Doctor Burchard, to examine Agnello's and Shultz's results. Burchard outdid even the schemers and found a gold concentration of fourteen ounces per ton of rock. But an observer filched some of the doctor's flux, a chemical added to help the rock melt, and found that it contained silver: Burchard was "salting" his assays.[5]

In May 1578 Frobisher sailed on his third arctic voyage, with a fifteen-ship fleet and orders to establish a permanent colony. The miners were recruited from Cornwall and Devonshire, places famous for skilled miners. The actual miners, however, stayed home, and Frobisher's recruits resembled real miners only in their West Country accents. Among the crew on the third voyage was a spy for Spain, a fact unknown until the Spanish archives released the secret correspondence in about 1900.[6]

One ship collided with an iceberg and sank with the pre-hewn lumber for the colonists' shelter. Frobisher decided to forego the permanent colony for that year and concentrate on mining.

Frobisher's lieutenant noted that "all the sands and cliffs did so glister and had so bright a marcasite that it all seemed to be gold" but recalled, "All is not gold that shineth." The gold finders found supposed gold ore on Countess of Warwick Island and along the coast of Baffin Island. Frobisher and his men mined 1,136 tons of rock, stowed it aboard, and sailed for home in August 1578.[7]

By the end of the summer, the investors realized that the rock was worthless. Frobisher's second load of ore did not receive the careful handling of the first. Frobisher dumped the rock into Dartford Harbor, and the remaining ore was "cast forth to mend the high-ways."[8]

Historians have hesitated to believe that Frobisher's expeditions were based on nothing more than fraud. Since rediscovery of the site, however, sample after sample of rock has been brought back to be tested for gold, and each test has reconfirmed that the mines are worthless. Agnello and Schultz were swindlers.[9]

Assay Frauds

"The assayer who finds a little something in every ore is always a popular assayer. The honest miner likes to be gulled and humbugged."

—Cheyenne *Daily Leader,* January 13, 1884

The assayer, who tests the metal content of the rock, is essential to mining. It is a skilled profession, but the techniques are well documented, and the results of one assayer can be duplicated by another—if both are honest. Dishonest assayers often say they have a secret assay process that allows them to detect gold that other assayers cannot see.

For thousands of years the most accepted test for gold and silver has been the fire assay, which takes about an ounce of rock and draws out the precious metal in the form of a little bead. No mind-altering drug is so powerful as the tiny gold doré bead from a fire assay. The client holds it in the palm of his hand, rolls it between thumb and forefinger, and feverishly multiplies the tiny bead by the many tons of rock in the deposit. Soon he is a millionaire in his mind, and nothing can dissuade him from the dream, for he has seen the gold and felt it himself. When being watched, fraudulent assayers resort to tricks of the ancient alchemists, often either sleight of hand or a doped flux, to slip some gold and silver into the sample.

Some assayers increased the gold and silver contents to encourage their clients, and prospectors preferred attractive lies to the dull truth. Mark Twain related this method in *Roughing It*:

> Assaying was a good business, and so some men engaged in it, occasionally, who were not strictly scientific or capable. One assayer got such rich results out of all specimens brought to him that in time he acquired almost a monopoly of the business. But like all men who achieve success, he became an object of envy and suspicion. The other assayers entered into a conspiracy against him, and let some prominent citizens into the secret in order to show that they meant fairly. Then they broke a little fragment of a carpenter's grindstone and got a stranger to take it to the popular scientist and get it assayed. In the course of an hour the result came—whereby it appeared that a ton of that rock would yield $1,284.40 in silver and $366.36 in gold!

Due publication of the whole matter was made in the paper, and the popular assayer left town "between two days."[10]

Twain's example may be fiction, but the practice was widely known. Mining history is full of high assays that could not be replicated by careful and honest assayers. If the prospector is wise, he gives a sample to a second assayer—and hopes that the second assayer is honest.[11]

Bucket Shops

A bucket shop was a low-class saloon that sold the dregs of other bars by the bucket. The term came to be applied to the dregs of the brokerage business: concerns that took buy orders from the public, usually on margin, but never bought the shares. Instead, the bucket broker acted like a gambling bookmaker, profiting from commissions and interest charges and figuring that at least as many customers would lose as win their bets.

Since bucket brokers profited from their customers' foolishness, they encouraged investments in stocks that they knew would fall. Many promoters of fraudulent mining and other companies were bucket brokers who incorporated guaranteed losers to sell to their customers.

Some seemingly legitimate brokers "bucketed" orders by failing to buy shares when their customers gave them a buy order that the broker thought to be a foolish investment.

Bulls and Bears

Bulls are those who believe that share prices will rise, and bears are those who expect prices to fall. Bulls and bears might actively campaign to drive prices up or down through manipulation. Both bulls and bears may spread false information. Bears may dump large blocks of stock on the market, hoping to start panic selling.

Watered Stock

Watered stock is a farm term for livestock fed salt and then given water to drink to increase their weight just before a sale. "Uncle" Daniel Drew, who gulled more than one Wall Street sharp with his country-boy manner, is said to have introduced the term to New York financial markets.

Wash Sales

In a wash sale the manipulator sells shares to himself. His broker makes the trade at the prearranged price, which is then a transaction of the stock market and is published in financial periodicals. Wash sales were often used to inflate reported prices of shares—shares for which there may actually be no market.

Match Sales

A match sale is much like a wash sale, except that two brokers are involved. The manipulator gives one broker an order to buy at a certain price, and another broker an order to sell at the same price. Neither broker may realize that their clients are the same.

Selling Short

A speculator *sells short* when he contracts to sell shares that he does not own for delivery at some future date. If the price declines between the contract and delivery dates, the short seller will pick up the shares at lower than the contract price and pocket the difference. Since short selling profits when the price declines, it is used by bears to profit from their pessimism.

Cornering the Market

A speculator corners the market in a stock when he controls enough shares to dictate the price. Attempts to create a corner are frequently aimed at markets where short selling is heavy, since short sellers are under legal obligations to deliver shares, whatever the price. The enormous profits of such a situation inspired Daniel Drew to versify:

> "He who sells what isn't his'n
> Must buy it back or go to pris'n."

Guinea Pig Directors

English companies often lured investors by hiring noblemen as directors. For promoters, the ideal director was greedy enough to sell his title but too lazy or ignorant to get in the way. Company directors were traditionally handed a guinea, an English gold coin, at each board meeting, but most received more than that token amount. The "guinea pig" director was a notorious joke in London's financial district, but English investors continued to entrust their money to figureheads from the nobility. For promoters on a budget, retired military officers, or even members of Parliament, might suffice as guinea pigs.[12]

When the British promoter Ernest Terah Hooley declared bankruptcy in 1898, his creditors sued. Although the trial did not extract from Hooley any of the money he had hidden away, it revealed the price of inducing members of the nobility to prostitute their titles for the adornment of Hooley company directorates. Hooley had paid the Earl of Wincheslea £10,000 to pose as chairman of one company. He paid another man £2,000 just for an introduction to Lord Ashburton. Lord Ashburton himself profited greatly from Hooley. In the British Embroidery and Machine Company, Ashburton was to enjoy half the promotion profits, but when the venture failed, Hooley absorbed all the losses.[13]

Americans congratulated themselves that their company directorates were free of noble patsies, but American companies had their own *ignoble* patsies to lure the unwary.

In the late 1800s former Union generals were sought-after guinea pigs. One ex-military guinea pig was Benjamin Butler, a Massachusetts politician whose vote getting won him a Union generalship during the Civil War. President Lincoln put him in charge of the occupation of New Orleans: far enough from the fighting that his lack of military skill would not give the army trouble, and far enough from Washington so that his political skill would not give Lincoln trouble. Butler's stay in New Orleans did nothing to endear him to Southerners, but his hard line against secessionists increased his popularity in the North. After the war Butler's populist politics made him an attractive figure for mine promoters to recruit as a director. Butler knew as little about mining as he had about military strategy. He served in so many rotten mining companies, such as the San Miguel Placer Company in Colorado, that the *Engineering & Mining Journal* noted that Butler's mere presence on a board of directors was warning enough to the wise.[14]

The San Pedro and Cañon del Agua Company, with mining property in New Mexico, announced in 1880 that they had persuaded the ex–Civil War general and ex-president U. S. Grant to become company president, causing the market value to jump $900,000 on the Boston exchange. The stock price sank right back down when Grant revealed that he had refused to have anything to do with the company, but by then the insiders had sold their stock.[15]

Also sought after as guinea pig directors were public officials. These served the dual purpose of inspiring share-buyer confidence and providing legal protection.[16]

In 1904 the champion guinea pig was former New York City mayor and US Senator Chauncy Depew, who served simultaneously on seventy-four company directorates. Perhaps it was just as well that Depew was too busy to give each company individual attention: he displayed his most famous bit of business acumen when he dissuaded a relative from investing in the automobile trade. "Nothing has come along," he assured his nephew, "that can beat the horse and buggy."

Promoters lacking the money to hire a celebrity director sometimes manufactured fake celebrities. A man named Guggenheim, but unrelated to the famous mining Guggenheims, offered to hire himself out to mining promotions, knowing that the public would assume the backing of his better-qualified namesakes. When Texas oil promoters needed to draw attention to their new swindle, they found an old janitor named Robert A. Lee, advertised him as the grandson of Confederate general Robert E. Lee, and named him figurehead of their new oil company. The General Lee Development Company took in almost $2 million on the strength of the phony family tree.[17]

Aylmer Vallance, biographer of the British promoter Ernest Terah Hooley, judged that Hooley's only lasting contribution was, through his employment of guinea pig directors, to reveal the vulgar greed of the British nobility and so hasten their decline in power.[18]

Why Are People Gullible?

That you cannot tell a book by its cover is a commonplace that everyone repeats but few really believe and almost nobody practices. We spend our lives judging books by covers and people by appearance; it would require tremendous effort to do otherwise. An old joke holds that the key to success is sincerity—once you can fake that, you have it made. Confidence men, or "con men," have mastered the art of faking sincerity to gain and then betray our trust. Those who judge others by steadiness of gaze and firmness of handshake are their victims.

> *"A man usually buys a mine not because it is worth the price he gives for it, but because he is justified in the expectation of finding some one who will pay more for it. The syndicate sells to the public, the public sell among themselves. The second man hopes to meet a third with more money and less sense. So the game proceeds. When the sequence has been exhausted, some one gets badly bitten."*
>
> —Thomas Rickard, 1898,
> *Mining and Scientific Press*, December 24, 1898, p. 630

During a stock boom, buyers often don't even care if shares are intrinsically worthless. Investors make foolish stock purchases, thinking they can find other fools to pay even more. This is the "greater fool theory." And it often works—for a while. Eventually, of course, the greatest fools are caught owning shares worth much less than they paid. The cycle has been repeated in every major stock boom in history.

The *Engineering & Mining Journal* once divided mine swindlers into four classes: the deliberately dishonest, the semi-dishonest, the honest but ignorant, and the visionary. The *Journal* noted that the deliberately dishonest make compelling villains for newspaper headlines, but that the other types were much more common and caused investors to lose the most money.[19]

On the cold printed page, most swindles look pathetic and silly. How can people be so stupid? But the written word cannot convey the seductive charm of con men, or the fervent wishful thinking of the victims. Be careful before you laugh at those who give their savings to a swindler who promises to make their dreams come true. It could happen to you.

The Imaginary Mine
of Doctor Gardiner

In 1851 Charles Davis was assigned the dullest of Washington bureaucratic duties: filing away the papers of the defunct Mexican Claims Commission. The Treaty of Guadalupe Hidalgo, which ended the Mexican War in 1848, stipulated that the United States would compensate American citizens with war-related claims against the Mexican government. The Mexican Claims Commission sat in Washington and heard from Americans who complained that they had suffered unjustly at Mexican hands during the war. Since the claims were being paid by the American government, Mexico did not bother to contest the allegations, and the commission ended its work in 1850.

But as Davis read and filed, two cases seemed not right. He was soon convinced that testimony was perjured and documents forged. Davis believed that Dr. George Gardiner had defrauded the United States government by inventing imaginary mines in Mexico.

Doctor George Gardiner

George Gardiner was a young American dentist working in Mexico City. He left the city, he told his friends, to enter the mining business.

The next time the American community heard from Gardiner was in 1847, during the Mexican War, when he arrived in American-occupied Veracruz, where his brother John Gardiner served as an interpreter for General Winfield Scott. George Gardiner told all who would listen of the wonderful silver mine he owned in the state of San Luis Potosi, and how the Mexican government had confiscated the mine when the war had started. The United States and Mexico had agreed in an 1831 treaty that, in case of war, law-abiding citizens of each country would be allowed to remain in the other country. Gardiner asserted that Mexican authorities had violated the treaty by ordering him out and had wrecked his machinery and buildings and filled up his mine shafts, resulting in the total loss of his investment.[1]

After the war, Gardiner went to Washington to press a claim for $700,000 for his silver mine, which he described as a great bonanza employing five hundred laborers, in which he had invested $300,000.

How to Buy Friends and Influence People

The truth was that after a couple of failed mining schemes, Gardiner had worked in San Luis Potosi as a dentist. There he met another American, John Mears. When the war began, Mears told Gardiner of his plan to claim compensation by making up a tale of confiscation and expulsion from a phony mercury mine. Actually, Mears spent the war unmolested, treating his Mexican patients in the town of Rio Verde. Mears's plan inspired Gardiner to create his own mining scam.[2]

Mears sold his claim to Gardiner and stayed safely in Mexico, while Gardiner returned to the United States to press both claims. After the 1848 election Gardiner wrote to Mears that Gardiner's influential friends in the new administration made approval of their claims more likely than ever. Gardiner bragged that he was a good friend of General Winfield Scott, whom he had served in the war. Following the influence-buying ways of Washington, Gardiner hired a former ambassador to Mexico to serve as lead counsel in return for fifteen percent of the award. The ambassador agreed to pay a US senator three percent of the award. Ohio congressman Edward Curtis, a friend of George Evans, president of the Claims Commission, joined the team. Gardiner hired another friend of Evans to help prepare the case, and for extra insurance also hired Evans's son-in-law.[3]

Gardiner's legal advocates told him what evidence he would need to convince the Claims Commission. He paid a Mexican diplomat in Washington to supply him with blank sheets bearing official stamps. He took the stamped pages to Mexico, where he and three Mexican friends manufactured documents. Gardiner returned with genuine-looking deeds, official documents, receipts for ore, and affidavits, and filed his claim with the commission in November 1849.[4]

The amount of the claim, $700,000, was a fabulous sum in 1849, greater than the value of any existing American mine. Commissioners questioned how a dentist in his twenties could have invested the enormous sum Gardiner claimed. Gardiner was an extremely persuasive witness and answered their questions convincingly. He told them that the Mexican mining financier Count Perez Galvez had been highly impressed by Gardiner's ability and had backed him financially. The Commission decided that Gardiner was telling the truth. They whittled down the amount, but Gardiner still walked away a wealthy man when they awarded him $428,750 in March 1850.

The Commission awarded the Mears claim $153,125, out of which Gardiner paid Mears $38,281, one-quarter of the award; gave another $28,711 to a helpful Mexican diplomat in Washington; and kept the remaining $86,133 for himself and a Mexican accomplice. Mears remained in Mexico. Gardiner celebrated by sailing for Europe, but not before bragging to a friend that he had won his fortune by fraud.[5]

The Exposure

When federal employee Charles Davis read Mears's and Gardiner's claims, he knew something was wrong. He had heard of Mears, a gambler and fugitive from justice who had narrowly escaped a lynching. That Mears owned a valuable mercury mine seemed incredible. More unbelievable was that Gardiner owned a rich silver mine. Davis had lived fifteen years in Mexico and knew the great silver mines of San Luis Potosi but had never heard of the bonanzas claimed by Mears and Gardiner. He concluded that they were frauds.[6]

Davis sent a letter to the attorney general, who ignored it. Davis forwarded a copy to a newspaper, and the matter became a scandal. Davis was summoned before a meeting of President Fillmore's cabinet, where the attorney general at first denied receiving the letter, then admitted that he may have seen it but not realized its importance.

A grand jury returned indictments for fraud against George Gardiner in July 1851. His brother John Gardiner, who was married into a socially prominent Washington, DC, family, had given the Claims Commission sworn affidavits supporting George Gardiner's claim, and was also indicted for fraud. John Mears was charged with submitting false documents but stayed in Mexico beyond the reach of American courts. The federal government froze Gardiner's assets in the United States.[7]

The accusations found Gardiner in Britain, which lacked an extradition treaty with the United States. Despite the risk, Gardiner returned to the United States in September 1851 to recover his assets. He posted bond of $40,000. It was an odd kind of bail, however, since it was paid—as were all of Gardiner's legal bills—out of his funds seized by the government. He claimed to be a victim of political persecution and demanded an immediate trial, but the district attorney asked for more time to gather evidence. Gardiner's eagerness to face the charges led many to believe him and to expect his vindication.[8]

While Gardiner was back in Britain awaiting trial, another federal grand jury indicted him for forgery and his bail was raised $20,000. By

June 1852 the district attorney had the evidence to convict, and now it was Gardiner who asked for delays.[9]

Politics as Usual

Democrats in Congress played the fraud for political advantage, since Gardiner had hired leading Whig politicians to further his case. Both the House and Senate held hearings to publicize the matter. The American minister to Mexico went to San Luis Potosi in November 1851 to search for Gardiner's and Mears's mines but could find no one who had even heard of the mines. Mexican officials contradicted Gardiner's and Mears's claims. The administration was gathering evidence only for the criminal trial, however, and refused to share it with publicity-hungry congressional committees.[10]

The Senate committee brought witnesses from Mexico who established that the Mears claim was a fraud. John Mears had taken the money Gardiner had paid him for his claim and become a druggist in Monterrey, Mexico. Since Mexico and the United States had no extradition treaty, Mears admitted that his claim had been a complete fraud.[11]

As to Gardiner's fabulous silver bonanza, José Barragan, the state comptroller of San Luis Potosi, came to Washington to tell senators that there were no silver mines in that part of the state and that Barragan would have seen the tax receipts if Gardiner had ever run a silver mine in San Luis Potosi, but there were no such tax receipts. Barragan testified that both the signatures and the state seal on Gardiner's mine title were forgeries. He identified numerous other forged documents of Gardiner's.[12]

At the conclusion of testimony unfavorable to his case, George Gardiner demanded the opportunity to testify and to cross-examine the witnesses against him. In fact, the committee had earlier invited Gardiner to participate, and Gardiner's lawyers had cross-examined witnesses. The committee agreed to accommodate Gardiner the following day, but Gardiner did not appear. The committee concluded that Gardiner's claim was "a naked fraud upon the treasury of the United States."[13]

Democratic senators used the scandal to blast the Whigs. Thomas Corwin, who had represented Gardiner while serving as a U.S. senator from Ohio, was now secretary of the treasury. The Whig lawyer–politicians countered that they were only following their professions as lawyers and pointed out that Democratic officials also lobbied the government for private clients.[14]

President Fillmore appointed five prominent people, including Major Abner Doubleday (later of baseball fame) to investigate, and they left Washington in late 1852. They went to the town of Lagunillas, the supposed location of the mine that had employed five hundred men, but found no mine. They offered $500 to anyone who could lead them to the famous mine, but no one in Lagunillas had heard of it. George Gardiner showed up in Lagunillas at the same time. The investigators identified and located Gardiner's principal Mexican accomplice, a wealthy landowner who was angry that Gardiner had not shared the award, and he provided the investigators with incriminating letters from Dr. George Gardiner and his brother John Carlos Gardiner.[15]

Gardiner's trial began in Washington in March 1853. His lawyers managed to keep from the jury some of the damning evidence gathered by the fact-finding commission. Gardiner even produced two witnesses who swore that they had seen the rich mines, with extensive equipment, masonry buildings, and surface facilities. Gardiner's lawyers were so confident that they let the case go to the jury without a closing argument. The jury deadlocked, with nine favoring acquittal and three conviction, and the judge dismissed them after a week.[16]

The government prepared for a new trial, but Gardiner had already beaten the government's best efforts to convince a jury that this sincere-sounding man was a swindler. George Gardiner was rich, handsome, and intelligent, and he and his fiancée enjoyed the Washington social life while waiting for the government to give up the prosecution.

But the prosecution pressed on. A second fact-finding mission to Mexico returned with more evidence. The new trial began in December 1853 with extravagant oratory. The prosecutor charged that the only holes Gardiner ever excavated were the ones he had drilled in the teeth of his patients. The defense counsel labeled the case a political vendetta, and with an eye to the jury, whose roster listed at least one identifiably Irish name, he likened the case against Gardiner to British persecution of Irish patriots.[17]

The speechifying ended, the spectators deserted the courtroom, and the trial inched along through the winter. This time, however, the jury heard all the damning evidence. One member after another of the government's fact-finding commission to Mexico told the jury that Gardiner's great bonanza was unknown in its supposed locality. They had searched for the mine and found nothing: the nearest mines were forty-six miles away in the next state. Gardiner's deeds were not in the records at the department capital of Rio Verde, and the copies of deeds Gardiner had provided to the Claims Commission differed suspiciously from the genuine deeds on file.

Gardiner's forged deed was last seen when it was left on a table for the defense counsel to examine. Dr. Gardiner walked out of the room with a sheaf of papers under his arm, and the document was never seen again. Gardiner appeared at his boardinghouse that evening "pale and agitated." He convinced his landlady to store a trunk and a carpetbag beneath her bed, telling her that the prosecuting attorney would obtain a search warrant for his room but would not think of searching the landlady's room.[18]

Despite the evidence, Gardiner and his lawyers seemed as confident as ever when the judge handed the case to the jury on a March afternoon. The jury split ten for conviction and two for acquittal when they retired for the night. Many expected a long deliberation, so only a few spectators were on hand the next morning when the jury informed the court that they had agreed on a verdict. Gardiner watched them closely as they filed in, perhaps hoping for a friendly glance, but he found none. The jurors pronounced him guilty on all three counts—two of forgery and one of perjury. The jury's quickness was so unexpected that the lead defense counsel was not present, but the second defense lawyer notified the court of his intent to appeal. The judge was ready with his punishment: he lectured Gardiner on the seriousness of perjury and sentenced him to ten years' prison and labor. The Baltimore *Sun* praised the "honest, intelligent, *uncorruptible* jury" (their italics perhaps expressing suspicion of the previous jury).[19]

A Dose of Strychnine

George Gardiner was taken to the jail, where he soon convulsed with spasms and vomited. The spasms stopped, but Gardiner remained very ill; some suspected attempted suicide. Gardiner conversed with his brother John in Spanish, which none of the others present understood. John Gardiner told the onlookers that his brother had not poisoned himself. Jailers sent for medical help, against the wishes of both Gardiner brothers, who said that George had suffered convulsions before and that they would pass.

Doctors could do nothing but watch the convulsions come and go. Between spasms George Gardiner proclaimed, "If I die, I die innocent." After his conversation in Spanish with the dying prisoner, John Gardiner went to George's boardinghouse and demanded his brother's belongings, including the secret trunk and carpetbag hidden beneath his landlady's bed. George Gardiner died that afternoon, an hour and a half after his convulsions began.[20]

Despite George Gardiner's protestations, his symptoms were those of a fatal dose of strychnine. In Gardiner's pocket was an envelope containing the poison. The coroner detected strychnine in Gardiner's stomach, along with some bits of paper of the type used by druggists to wrap chemicals.[21]

John Gardiner's trial for perjury was due to start in May, but he skipped out on his bail, disappeared, and was never found.

Dr. George Gardiner had coolly planned an elaborate swindle with dozens of forged and perjured documents. His obvious sincerity while lying under oath fooled many. So convinced was he that his power of persuasion could overawe poor fact that he dared to return twice to the United States to save his illegal fortune. Gardiner always showed extraordinary calm in the face of his own downfall, right down to his dying lies.

3

The Comstock: Mother Lode of American Mining Swindles

Some adventurous prospectors had staked their mining claim over a rock outcrop in the wilderness, when up rode Henry "Pancake" Comstock. Comstock told the prospectors that they were trespassing on his claim and demanded that they share their discovery. They knew that Comstock was lying, for there were no claim monuments and no signs of pick-and-shovel work anywhere about, but they gave Comstock a share of their claim to shut him up, and the resulting partnership was called Comstock and Company.

That is how the Comstock lode, the three-mile-long mineralized zone where silver veins pinched and swelled, was named after its first swindler, a blowhard who didn't discover it but demanded a part. Word spread, and in 1859 prospectors rushed to that remote hillside braided with silver. The towns of Gold Hill and Virginia City, the territory and then the state of Nevada, formed around the miners and mill men of the Comstock lode. As for Pancake Comstock, he squandered his share. He spent part of it buying a wife from her first husband, but the young woman didn't think much of the bargain and left with a third man.

Scoundrels also flocked to the Comstock lode. The Virginia City *Enterprise* complained in 1860 of "lazy, unscrupulous, rascally speculators." It reported: "By lying and misrepresentation they have succeeded in palming off an infinite amount of worthless property." The Comstock became the site of pervasive mine swindling. Other mining districts have seen frauds, but nothing else in the history of mining can compare to the orgy of swindling that plagued the Comstock.[1]

*Henry "Pancake" Comstock didn't discover the silver lode that
came to bear his name, but he lied and bullied his way into
a share. From Dan De Quille,* The Big Bonanza, *1876.*

The Forty Thieves

*"As at present organized, the Stock Exchange is simply a body of
men who, by a variety of means, some of which are in conflict with
the law, are enabled to fleece the gambling public."*

—*Engineering & Mining Journal*, July 30, 1892,
on the San Francisco Stock and Exchange Board

*"The Stock Exchange, the heart of San Francisco: a great pump we
might call it, continually pumping up the savings of the lower quarters into the pockets of the millionaires upon the hill."*

—Robert Louis Stevenson, *From Scotland to Silverado*

The history of Comstock swindling is inseparable from that of the San
Francisco stock exchanges. Early California mining and milling corpo-

rations were filled with fraud and failure, and by the mid-1850s few Californians would consider investing in a corporation. But the fantastic stream of silver from the Comstock lode swept away misgivings, and San Franciscans eagerly provided the large investments needed for shafts and ore mills.[2]

The Ophir was the first Comstock mine to form a stock corporation in 1860, and by year's end two or three were incorporating each week in San Francisco. About four hundred Comstock mining corporations sold shares, but most owned no property of any value. Share dealers at first bought and sold on San Francisco sidewalks. Forty share traders organized the San Francisco Stock and Exchange Board, California's first, in 1862, and the public dubbed them "the forty thieves." The number of San Francisco exchanges grew to seven by 1864. In 1877 there were four boards, with an estimated fifteen hundred brokers.[3]

Visitors were surprised by the get-rich-quick passion pervading all classes on the West Coast. Many laborers in San Francisco and the Comstock spent their wages on Comstock shares. Female servants in California and Nevada were particularly addicted to speculation, as were the Chinese in San Francisco. They did not invest but gambled for quick fortunes. A San Francisco railroad worker asserted, "I don't want your dividend-paying mines. Them stock what go up and down lively is the mines for me." All but a few speculators lost to the manipulators. The brokers themselves speculated heavily with their own and their clients' money, to the downfall of many.[4]

Most shares were bought on margin. Terms varied, but the speculator typically put up twenty to fifty percent of the share price, and the broker charged one and a half to two percent monthly on the balance. This highly leveraged market was dangerously unstable, as speculators were quick to sell shares not appreciating faster than their interest payments, and the slightest downturn could detonate panic selling.[5]

Share-price gyrations often had little to do with conditions down in the mines but everything to do with manipulation in San Francisco. Within a few months of the inauguration of the Stock and Exchange Board, a Virginia City correspondent asked: "Why does San Francisco allow all this? Have the people become insane, that they should unreflecting rush into a financial abyss?"[6]

Fraud was endemic to mining shares traded on the San Francisco exchanges. The *Mining & Scientific Press* noted that throughout California, Nevada, and Idaho mines owned by corporations were nearly without exception more corruptly and less profitably managed than those owned by partnerships. The incorporated mines of Eureka, Nevada, and Owyhee, Idaho, were said to be even more corrupt than those of the Comstock lode, but Comstock mines were always the most prominent and the most avidly manipulated. In 1876, eighty-five percent of the companies listed on the San Francisco Stock and Exchange Board were Nevada mining companies, predominantly Comstock mines.[7]

Stockbrokers, of course, profit from commissions and interest charges, whether their clients are enriched or impoverished. In a famous incident of the 1870s, a San Franciscan proudly showed a visitor the fine yachts moored along the waterfront, naming the stockbroker who owned each one. The puzzled visitor stammered, "W-where are the customers' yachts?"[8]

Public reaction to stock manipulation teeter-tottered with share prices. When share prices went up, happy investors ignored wrongdoing; it was only when prices went down that public indignation rose.

The "Friendly Custom" of Bribing Newspapers

On hand to chronicle the excesses of the first Comstock boom was Samuel Clemens, who arrived in Virginia City in 1861 and invented the pen name Mark Twain for his articles in the city's *Territorial Enterprise*. Clemens described the cozy arrangements by which mine promoters bribed reporters with shares (measured in "feet") of the company.

New claims were taken up daily, and it was the friendly custom to run straight to the newspaper offices, and give the reporter forty or fifty "feet," and get them to go and examine the mine and publish a notice of it. They did not care a fig what you said about the property so long as you said something. Consequently we generally said a word or two to the effect that "indications" were good, or that the ledge was "six feet wide," or that the rock "resembled the Comstock" (and so it did—but as a general thing the resemblance was not startling enough to knock you down). If the rock was moderately promising, we followed the custom of the country, used strong adjectives and frothed at the mouth as if a very marvel in silver discoveries had transpired. If the mine was a "developed" one, and had no ore to show (and of course it hadn't), we praised the tunnel; said it was one of the most infatuating tunnels in the land; driveled and driveled about the tunnel till we ran entirely out of ecstasies—but never said a word about the rock. We would squander half a column of adulation on a shaft, or a new wire rope, or a dressed-pine windlass, or a fascinating force pump, and close with a burst of admiration of the "gentlemanly and efficient superintendent" of the mine—but never a word about the rock. And those people were always pleased, always satisfied. Occasionally we patched up and varnished our reputation for discrimination and stern, undeviating accuracy by giving some old abandoned claim a blast that ought to have made its bones rattle—and then somebody would seize it and sell it on the fleeting notoriety thus conferred on it.[9]

Samuel Clemens Promotes the North Ophir Fraud

"If I don't know how to blackmail the mining companies, who does, I should like to know?"

— Samuel Clemens, May 18, 1863, *Mark Twain's Letters*

In his book *Roughing It*, Clemens described swapping favorable newspaper coverage for shares as a "friendly custom," but he named it more honestly in private as "blackmail." William Wright, who wrote mining news under the name Dan DeQuille for the *Territorial Enterprise*, accepted shares in return for favorable write-ups. When Wright left for nine months in late 1862, his temporary replacement Clemens continued the practice, wrote cheerful letters home of his lies on behalf of the mining companies that paid him bribes, and said that if he had more business sense, he could make his position pay $20,000 per year.[10]

An example of Clemens's promotion of worthless properties was the North Ophir, deceptively named after the Comstock's rich Ophir mine. The owners became dissatisfied with the rock's meager silver content and supplemented it by pouring melted silver half-dollars down the shaft.[11]

Pure silver clinging to the rock in the North Ophir shaft began a stock sensation in March 1863. Among those profiting from the crude salting was Clemens. In a letter to his sister in Missouri, he bragged that his booming had boosted North Ophir shares from $13 to $45 per foot in a single day. There is no indication that he was privy to the fraud, but it made no difference to Clemens, as long as he received his five feet of the mine. A careful observer punctured the boom after noticing that some of the silver in the "ore" was stamped with dates and eagle heads.[12]

Justice for Sale

When thirty armed men jumped the Savage claim in 1860 and forced off the original discoverers, the Savage men went to court. Surprisingly, a majority of the jurors ignored fact and law to award the claim to the jumpers. But even while the claim jumpers and their financial backers in San Francisco celebrated, a juror swore an affidavit that his decision had been bought for $250 and a share of the claim. The courts had became a means of theft on the Comstock lode.[13]

Samuel Clemens used his position as a writer for the Virginia City, Nevada, Territorial Enterprise *to exact blackmail from Comstock mine promoters in exchange for favorable write-ups. Courtesy Library of Congress.*

Mine ownership in the American West was on a finders-keepers basis. Americans in the California gold rush of 1848 adopted Mexican law that gave the discoverer of gold or silver the right to mine the deposit. The procedure was at the time illegal under US law, but Washington was far away, and prospectors in each western mining camp voted their own rules of self-government, including makeshift mining laws. The federal government finally legalized the practice in 1866.

To secure legal title to a gold or silver deposit, the prospector staked his claim by marking the claim boundaries with piles of rocks or by driving wooden stakes into the ground and posting a written notice. The maximum size allowed for a claim varied but was generally a rectangle some hundreds of feet in length, running parallel to the vein for hardrock claims, or along the streambed for gravel (placer) claims. The democracy imposed by jealous latecomers usually allowed a claim to be only large enough to be worked by a small group. When the miners voted rules, however, they failed to anticipate the complications of geology. It was only after they followed the veins underground that they learned that veins twisted, split, joined, and crossed one another. Miners following what they thought were separate veins fought underground gun battles when their tunnels intersected.

Lawsuits descended on the Comstock like a biblical plague. Veins twisted and split down in the ground, and lawyers twisted the law and split hairs in the courtroom. In 1863 one out of every forty-seven Comstock residents was a lawyer. Many suits were spurious, hoping to steal title to a mine or blackmail owners into an out-of-court settlement. By 1862 every valuable mining property on the Comstock was mired in litigation. The Ophir company alone was involved in twenty-eight lawsuits before 1867, and other companies were not far behind.[14]

Reliable facts are scarce, but all observers agreed that judicial corruption was rampant on the Comstock in the early 1860s. Witnesses unwilling to perjure themselves were replaced with others less finicky. Under influence of liquor or consulting fees, experts described veins where none existed. Inconvenient witnesses left town under death threats, and one witness was shot at as he rode away from the courthouse. Evidence was manufactured or destroyed. According to rumor, judges and juries were corrupted with numbing regularity.[15]

The struggle over the Nevada judicial system reached a fever pitch in 1863 and 1864. With the fabulous silver mines as the prize, owners and would-be owners did not stop at the edge of the law. Even today it is not certain who was corrupt and who was not. The fight was complicated by political rivalry, and the facts obscured by the bitter partisanship of the newspapers.

Towering physically and politically—but not morally—above other Comstock lawyers was William Stewart. He had a domineering personality, a remarkable memory for the facts of a case, and, according to his enemies, no ethics. The Virginia City *Union* wrote: "He was successful as a lawyer on account of his *peculiar* talents in getting control of the judges, juries and witnesses by means not approved by honest men." Stewart later admitted that he "fought fire with fire," but the historian Grant Smith noted, "In fact, he appears to have furnished most of the fire."[16]

The cost of litigation was heavy. William Stewart estimated that litigation cost Comstock companies $10 million through 1865, a sum greater than their dividends during that time. He himself pocketed $500,000 in legal fees in four years.[17]

The Remarkable Floating Yellow Jacket Claim

When miners on the Union and the Princess claims found ore, the owners of the neighboring Yellow Jacket claim moved their boundary markers three hundred feet downhill to steal the new discoveries—or so testified the owners of the Union and Princess. Such a move violated mining law, but the Yellow Jacket owners insisted that their claim had always been at the new location.

The Union and Princess brought twenty-eight witnesses to testify; the lawyers for the Yellow Jacket, William Stewart and Sandy Baldwin, found fifty-one witnesses to swear to the contrary. One witness swore that he had read a Yellow Jacket boundary marker notice many times in its present location, but when the opposing lawyer handed him the notice itself and asked him to read it, the witness was forced to admit that he could not read at all.[18]

After lawyers argued for two days over the location of a tree stump used as a landmark, the judge and jury decided to go see the famous stump for themselves. It was too late. The stump had been removed overnight and its location so well disguised that no one could tell where it had been just one day earlier. Despite the evident weakness of their case, the Yellow Jacket company won the verdict and became one of the great mines of the Comstock. The Union and Princess owners were left with nothing.[19]

North Potosi v. Savage: Who Bribes the Most

If William Stewart's account can be credited, a man came to him during the trial of the Savage mine versus the North Potosi mine, and offered to testify in favor of the Savage, Stewart's client, for $500. Stewart claims he refused the offer, but the man went to the superintendent of the Savage, who gave him his $500. When called as a witness, however, he disclosed how he had been bribed by the Savage, and turned the $500 over to the court.

Stewart believed that the opposition had bribed even more and had paid off a majority of the jury. Stewart paid the deputy in charge $14,000 and a fast horse to recount in detail how the bribed jurors had been paid. The deputy, said Stewart, told him everything then took the money and the horse and was never seen again. Stewart exposed the bribed jurors in open court, and the jury deadlocked and was dismissed. The outcome sent North Potosi stock sharply down, and the Savage bought a controlling interest in the North Potosi to prevent further lawsuits.[20]

Chollar v. Potosi

The Chollar Mining Company sued the Potosi Mining Company in 1861 over a disputed silver vein. After two trials a juror revealed that the Chollar company had bribed him and others, and when the Potosi company began investigating rumors of the bribery, the Chollar company paid him and another juror to sail back to the East Coast. The juror also accused the Chollar people of forcing him at gunpoint to sign an affidavit that accused the Potosi company of paying him $1,500 to make false charges of bribery against Chollar.[21]

Burning Moscow v. Ophir

In the spring of 1863, the Ophir mine tunneled into the Burning Moscow mine. The Ophir claimed ownership of the Burning Moscow vein and obtained an injunction preventing the Burning Moscow from removing any ore. Judge John North dissolved the injunction in December 1863 and ruled that the vein belonged to the Burning Moscow.[22]

William Stewart, counsel for the losing Ophir, charged that North had been bribed. He said that Burning Moscow's lawyer James Hardy, while drunk, had told him that his client had prearranged the outcome by donating one hundred feet of company stock to Judge North. North denied the accusation, as did his supposed accuser, James Hardy. Stewart sent a written apology to the newspapers but continued to spread the story against his rival.[23]

The Burning Moscow and Ophir companies fought on, spending about $1 million on litigation. The legal contest ended in the depressed stock market of 1865 when the Ophir bought the Burning Moscow outright for $70,000.[24]

Judge North and His Minnesota Mill

At the time of his appointment, Judge John North was building an ore-treatment mill. When finished, his Minnesota Mill contracted to treat ore from the same mining companies appearing before John North in court.

In November 1863 the Grass Valley Company applied to Judge North for an injunction against the Potosi Company. North granted a temporary injunction. Soon after, one of the owners of the Potosi lent $15,000 to Judge North. On December 15, 1863, North sided with his creditor by ruling in favor of Potosi.[25]

The Gould & Curry mine, which had cases pending before Judge North, sent ore to his Minnesota Mill. According to William Stewart, while the Gould & Curry had an injunction pending against the El Dorado mine, North complained that his mill was not receiving enough good

Gould & Curry ore. The Gould & Curry promised to fix the matter, and North directed one of his employees to see that the ore selected was adequate. When he returned from the Gould & Curry, North's man reported that the ore was satisfactory, and North signed the injunction, which idled the El Dorado mine. Judge North also granted the Gould & Curry injunctions against two other neighbors, the North Potosi and the Sinaloa companies.[26]

Back to the *Chollar v. Potosi* Case

When North's decision favoring the Potosi came before the Nevada supreme court, which consisted of the three territorial judges sitting together, Justice North supported the Potosi, Justice Turner favored the Chollar, and Justice Powhatan Locke vacillated with his deciding vote. After much dubious behavior on the part of the lawyers on both sides, Locke joined North in ruling for the Potosi. However, the Chollar Company persuaded Locke—according to a version reported in newspapers at the time, by putting a gun to his head—to write an addendum preventing Potosi from using the decision as precedent in future litigation. But the Potosi Company then offered Locke armed protection against threats of violence from the Chollar people and convinced him to remove the addendum from the records.[27]

Cleaning Out the Judiciary

The Chollar–Potosi case boiled over into public accusations of corruption, and the Gold Hill *News* and Virginia City *Territorial Enterprise* were soon printing column after column accusing all three judges of corruption. At the forefront of the denouncers was William Stewart, even though he had been in the thick of the Chollar's efforts to unduly influence judges.[28]

Judge Turner was accused of becoming suddenly wealthy while on the bench. According to an often-told story, some litigants showed up at Turner's house late one night to deliver a bribe. Turner's wife opened the door, told them that she would accept the bribe, and held out the skirt of her nightgown to catch the money. But when they tossed her the sack with $10,000 in gold, the weight tore the nightgown from her shoulders and left her standing naked in the doorway.[29]

The three judges announced their intention to disbar Stewart at the next meeting of the court. However, Judge North resigned on August 22, 1864, because of ill health. Judge Turner opened his court that day, but after a private meeting with William Stewart, he announced his resignation. Stewart later wrote that he had blackmailed Turner into resigning by threatening to show a signed receipt and a cancelled check to Turner for

$5,000 in bribe money from the Hale & Norcross mine, one of Stewart's clients.[30]

Stewart and his allies celebrated in the back room of a saloon. In their euphoria the lawyers decided to make a clean sweep, and two of them went to fetch Judge Locke. "If he is locked in his room," Stewart punned, "locks can be broken." Powhatan Locke arrived and was told to resign. Weak-willed as ever, Locke asked William Stewart what he should do. Stewart told Locke to write a letter of resignation, which he did on the spot.[31]

Judge North sued Stewart for slander and demanded $100,000. Rather than wait out the delays of a court case, North agreed to waive damages, and he and Stewart submitted the case to three referees. The referees declared, "The evidence fails to show any act of corruption, on the part of the Plaintiff, and we therefore pronounce the character of John North free from each and every imputation cast upon it by the accusations of the Defendant W. H. Stewart." They found Stewart guilty of slander and ordered him to pay court costs. However, they faulted North for his conduct toward Stewart on one occasion and for his conduct toward Judge Locke's position in the Chollar and Potosi case.[32]

North won his lawsuit but lost in the court of public opinion, when Stewart's newspaper allies twisted the panel's mild criticism of North as a proof of the corruption charges. Stewart had succeeded in destroying his political rival, and the new state legislature named Stewart a United States senator from Nevada in 1864.[33]

The Chollar and Potosi companies continued the lawsuits, which cost both companies about $500,000. They finally ended the legal battles in April 1865 by merging into the Chollar-Potosi Company.[34]

Telegraphs, Codes, and Code Breakers

While the Chollar and Potosi companies fought in court, their share prices seesawed with each new legal twist. Some speculators realized that a fortune depended on the long telegraph wire connecting the Comstock mines to the San Francisco exchanges.

D. C. Williams had been fired from his job as telegrapher in Virginia City when his employers discovered that he was divulging the contents of the telegrams sent and received by stock speculators. Williams became a stockbroker himself but recklessly speculated for his own account and went bankrupt. During the Savage versus North Potosi case, he first tried to send his own very long message to delay the telegraph at Virginia City to give his partners in San Francisco time to act on his coded message. But the telegraph company refused to send his telegram ahead of the news about the lawsuit.[35]

In July 1864 Williams began hanging around the telegraph station at Sportsman's Hall, near Placerville, California. All the news from Virginia City came by wire over the Sierra Nevada to Sportsman's Hall, where it was relayed to San Francisco. Williams had arranged with a confederate in Virginia City to send news of the Chollar-Potosi decision in a coded message. He approached the telegrapher at Sportsman's Hall and offered him $1,300 to cooperate in a scheme to cut the wire from Virginia City as soon as the coded message was received. Williams's employers in San Francisco would then have time to profit from the inside information.

However, Williams's loitering attracted the suspicion of law officers looking for bandits. The telegraph operator informed his boss, who had Williams arrested. Letters in Williams's possession implicated a number of stockbrokers and speculators, but all denied any connection, and when Williams asked a prominent San Francisco brokerage firm to post his bail, they refused and said that they knew nothing of him.[38]

Insiders at the mines relied on coded telegrams to communicate to their San Francisco brokers, but others intercepted and decoded them. William Sharon, who represented the powerful Bank of California at Virginia City, sent his men to discover who was deciphering his secret messages. By the time his men forcibly brought in the code breaker, Sharon was already using a new and more complicated code. Sharon told the smart young man that his code breaking had come to an end with the elaborate cipher, whereupon the young man instantly read a message encoded with the new system. Sharon convinced the code breaker that the Comstock was unsafe for him.[37]

Shutting In the Mines

Mine management went to extraordinary lengths to keep information secret, to better manipulate share prices. A practice unique to the Comstock was that of holding miners incommunicado at the mine for days at a time, while rumors and share prices ran wild.[38]

In January 1868, mine managers halted miners coming to the surface from their shift in the Hale & Norcross mine. Everyone expected that the miners would cut into an ore body that day, and management wanted exclusive knowledge of the results. The miners consented to be sequestered in the company offices next to the shaft house while they were off shift, and were paid a handsome $12 per day, plus cigars and whiskey. Two more shifts of miners joined the first, and the company posted guards to prevent miners from communicating with outsiders.[39]

Speculation in Virginia City and San Francisco exploded. Share prices jumped from $1,400 to more than $2,200 as company insiders traded on their inside information. The miners returned to their homes after three days, and the company announced that they had struck a rich vein.[40]

In February 1869 the managers of the Imperial-Empire mine kept the miners underground at the end of their shift and sent down food and bedding. The surface crew was replaced to prevent their communicating with the miners by prearranged signals. This time, the public suspected Imperial-Empire management of trying to orchestrate a rise in share prices. The ploy backfired, and share prices dropped by a third. The miners were released after a week underground, and the walls of the mystery drift revealed that they had found no ore whatsoever.[41]

The failure of the tactic in the Imperial-Empire caused mine owners to wait three years before another shut-in. John Jones and Alvinza Hayward had bought a controlling interest in the Savage mine in 1871, and they were eager to profit by stock manipulation. Savage shares rose in February 1872 on rumors of a big find, and management encouraged the rumors by shutting in the miners. Quoted share prices in the Savage quickly rose to five times their former level of $62 amid charges that the quoted prices were from fake wash sales staged by insiders.[42]

After the miners came out, selected outsiders were permitted to view the mine but not take samples. One visitor was offered $250 for one of his mud-covered boots by a stock dealer hoping to gain inside information by assaying the mud. Publicity brought by the shut-in caused shares to continue their rise even after the miners were let out, until prices reached a high of $725 on April 25. The price rise ruined a Gold Hill stockbroker who had gambled by going short on Savage shares.

The rumored discovery was a fraud, and the Savage mine lost money for the rest of its life, although Jones and Hayward kept themselves wealthy by sending Savage ore to their mill. The operating losses of the Savage were paid by assessments levied on its ever-hopeful shareholders. Despite the shady deal, the Nevada legislature sent John P. Jones to Washington as a senator in 1873.[43]

The Savage's success with the tactic encouraged the Ophir management to shut in four miners driving an exploratory drift only a few weeks later. This time, however, the shut-in backfired as rumors spread that the miners were held prisoner. Ophir management offered sworn affidavits by the miners that they remained underground of their own free will, but investor confidence remained low. Share prices fell to less than half their former level, as insiders sold out their positions. Management released the miners after three days but built a bulkhead across the drift and posted a guard at the mine shaft. When visitors finally got a peek at the drift, they described a streak of silver too small to create excitement.[44]

Managers of the Imperial-Empire shut in their miners again in October 1872, and once again spread rumors of a great strike. The tactic succeeded better than it had three years earlier. The price rose from $7 to $12 then declined to $10 in a skeptical backlash.[45]

The practice of shutting miners in fell into disuse after 1872 because drilling with the new diamond-tipped bits was an easier and cheaper way to keep new explorations secret. Before diamond drilling, the only way to take a look at the rock was to dig a tunnel into it. Once the tunnel was dug, anyone could wander in and take a look. With drilling, however, the results (either rock fragments or long cores of about one-inch diameter) could be removed to a secure place and viewed only by select insiders. All that was left to see in the mine were small holes in the rock face, giving no clue as to what lay a few feet or a few hundred feet into the rock. Diamond drilling was an efficient innovation adopted in mines worldwide, but because it allowed mining companies to keep the results secret, one Comstock critic charged that it had been invented "either by Jim Fair or the devil."[46]

Assessment Mines

"Men are often elected Trustees who do not own a single share of stock, or who have merely hired it for election day, in order that they might have a fair chance to plunder the mine for twelve months."

—*Mining & Scientific Press*, March 15, 1873

Since most Comstock shares were held in trust by San Francisco brokers, sharp operators learned that they did not have to own shares to control a mining company but only had to pay the brokers to vote proxies their way. That this often worked against the interests of their clients seemed not to bother most brokers.[47]

Some "assessment mines" worked at a loss for years because the managers knew that assessments, also called "Irish dividends," were a steadier source of income than profits. Fading shareholder hopes could be revived by occasional bonanza rumors, or even an occasional small dividend. Such a "teaser" dividend was the Ophir company's $1.00-per-share dividend in January 1880, quickly followed by four assessments that same year, totaling $4.50 per share.[48]

The Belcher and Crown Point mines sometimes canceled dividends, even though they had ample funds, for purposes of inside manipulation. Assessments were used to force the price down so that insiders could buy the stock cheaply. The San Francisco *Bulletin* noted of Nevada: "There is scarce a leading mine here but the poorer stockholders have been sub-jected to this coercive, or as it is termed in financial slang, 'freezing out' process."[49]

Insiders Fleece Other Insiders

Drilling with the new diamond-tipped drill bits gave insiders a potent method of gaining secret information. But there were layers and layers of insiders, and they sometimes used the diamond drill to swindle one another.

The Prospect Mining Company was a property near the Comstock. When the company president learned that a diamond drill had cored a fabulous 180 feet of good ore, he kept it secret, began buying all the stock he could, and advised his friends to do the same. Buying drove the stock from $6 to $8 a share, after which the price faltered.

The price eventually fell to $0.15, and company officers learned that the Prospect mine manager had salted the diamond drill core and enriched himself by selling the stock short to his own employers. The out-swindled insiders had no legal recourse, although about a month later a defrauded San Francisco stockbroker saw the ex-manager at the train station in Vir-ginia City and beat him so severely that the manager was hospitalized. The broker then boarded the first train out of Nevada to avoid arrest.[50]

An often-told tale had "Slippery Jim" Fair of the Bonanza firm tell-ing his wife—in strict secrecy—that a certain stock was due to rise. Mrs. Fair confided the insider "point" to a few close friends, and soon all of San Francisco—point-hungry as it was—was buying the stock, except Jim Fair, who was selling his shares at a fat profit. In fact the stock was over-priced and instead of rising, its price fell. When his wife tearfully told him that his point had cost her $7,000, he reimbursed her out of a small part of his profit.[51]

Some miners in the Bullion mine learned too well from their employ-ers. In 1877, miners salted the cuttings from a drill hole by adding pow-dered ore. After the Bullion mine managers assayed the salted cuttings, they began buying shares. In reality they were buying the shares from their employees.[52]

When the Bullion tunnel reached the supposed bonanza, there was no vein and the managers realized they had been out-swindled. Manag-ers fired all the drillers in that heading, but after each man swore before a justice of the peace that he had not salted the cuttings, the management relented and rehired the men. The public enjoyed seeing the Bullion mine insiders swindle themselves by their own greed.[53]

Salting the Lady Bryan Mine

The Lady Bryan mine assessed shareholders heavily in the 1870s but only sank deeper into debt. When the management announced a thirteenth assessment, shareholders demanded to inspect the books. Company officers refused for months but finally allowed one shareholder, George Reynolds, to see the accounts. Reynolds did not like what he saw: treasury shares given free to trustees, money spent in suspect ways, notes of indebtedness issued without consideration, and trustees illegally evading assessments. While Reynolds inspected the books, the company secretary and president distracted and harassed him until closing time. Reynolds returned on Monday to find that 116 pages had been removed from the minutes, 128 pages from the ledger, 36 from the journal, and 118 from the cashbook. Reynolds sued the officers; they paid him $8,000 to settle out of court—and charged the amount against the company.[54]

Shareholders revolted and petitioned the court to appoint a receiver. The showdown came at the annual company meeting in December 1876. The secretary had little to report because the books and papers of the company were missing. When a shareholder objected that the meeting had been called contrary to the bylaws, the secretary said that no one could prove the point one way or another because the bylaws were missing. As opposing sides squabbled, a process server walked to the front and handed each trustee a summons to appear in court, and the meeting ended in chaos.[55]

The court declined to appoint a receiver, and the shareholders fell out, with mutual accusations of selling out to the corrupt management. The Lady Bryan mine was sold at a sheriff's auction in 1878 for $25,000, but the company was soon back in business, although no more successful.[56]

The Bullion mine salting of 1877 was closely repeated three years later in the Lady Bryan mine, when a drill hole ahead of the workings gave cuttings with $900 per ton in gold and silver. Word leaked, and share prices rose from $0.50 to as high as $4.00. A week later, when the mine tunnel caught up with the drill, the mine captain John Kelly led a crowd of experts to inspect the vein. They pronounced it fine-looking ore and collected rocks glistening with silvery metal.[57]

One of the experts tested his specimen immediately upon returning to his office. When he heated the rock, the "silver" immediately melted into a glob. "Solder!" he exclaimed, and ran down the street to order his broker to sell out immediately. The other experts obtained the same results: the rock was worthless. The following day Lady Bryan shares sunk as low as $0.15.[58]

There was no law against salting, but the Lady Bryan fired the five miners on the crew where the salting had taken place—a sixth miner had been fired the week before. The six convinced the management to allow them to testify to their innocence, and they also insisted that Foreman John Kelly testify. Kelly swore on a Bible and answered sharp questions for more than a day. Each miner then likewise swore on the Bible, solemnly testified to his own innocence, and returned to work. A later story had it that when Kelly stepped up to the Bible to take his oath, he realized that someone had removed the pages of scripture from their binding and replaced them with a patent-office report—in the language of the Comstock: someone had "salted" the Bible.[59]

The Bonanza Firm

The greatest success story of the Comstock was the Irish quartet of John Mackay, James Fair, James Flood, and William O'Brien. Mackay and Fair, as mine superintendents, had accurate information about the mines. Flood and O'Brien ran a popular San Francisco saloon near the stock exchange and were well placed to know the pulse of stock operations. The four-way partnership was formally Flood and O'Brien but better known as the "Bonanza firm," after their 1872 discovery of the great ore body known as the "big bonanza" in the Consolidated Virginia mine.

The first success of the partnership was their takeover of the Hale & Norcross mine in 1869. Flood and O'Brien picked up a controlling interest so quietly that the Bank of California, which had misrun the mine for years, did not wake to the threat until too late. The Bonanza partners rescinded an unneeded assessment that the old management had levied solely to drive share prices down, and the mine began to pay regular dividends under the expert direction of Fair and Mackay.[60]

The Bonanza firm at first had a record of treating shareholders better than did the other Comstock barons. Fair and Mackay were skilled miners, and their properties were well run. Even the *Mining & Scientific Press*, which disapproved of their ways of diverting profits to themselves, defended the Bonanza firm in 1877 against the worst attacks of the San Francisco *Chronicle* and noted that the Bonanza firm mines paid the highest portion of proceeds as dividends of any on the Comstock. Year by year, however, the Bonanza firm came more and more to resemble the other respectable robbers who swindled stockholders in Comstock mines.[61]

False Annual Report for the Consolidated Virginia Mine

Share prices in the Consolidated Virginia mine suffered in the crash of 1875, and the Bonanza firm was eager to regain its losses by showing that there was no limit in sight to the big bonanza. Unfortunately, as miners followed the ore body deeper into the earth, it grew smaller.

The big bonanza ore body was a sheet-like mass oriented north-south, and nearly vertical. The Comstock lode, the broad zone of disturbed rock that enclosed the bonanza, also ran north-south, but sloped down to the east. The big bonanza could not extend deeper than its intersection with the lower boundary—the footwall—of the Comstock lode. Knowledge-able observers knew this, but the Bonanza firm hoped that the public would not learn it.

Superintendent James Fair inserted into the 1875 Consolidated Virginia annual report a false description of ore in a winze (a vertical tunnel inside a mine) from the 1,550-foot level that more than doubled the apparent size of the deep ore body. The winze was in a different place than described, and the report said that the winze was entirely in ore, but in fact it had encountered the southern edge and had to be deviated northward to stay in ore. Even the historian Grant Smith, a stout defender of the Bonanza firm, called the report "inexcusable in its misrepresentations."[62]

Lies From the Bonanza Firm

"I tell you sir, that these mines will pay dividends right straight along, and will continue paying them when those who are now maligning them are cold in death."

—James C. Flood, April 22, 1876

Stock manipulators agreed that James R. Keene was the best. He had come to America from London as a teenager then moved west. After he made a small fortune speculating in Virginia City, he took it to San Francisco, where he bought the *Examiner*. He edited the *Examiner*, and studied and practiced law for a couple of years, but was drawn back to stock speculation. He found himself on the wrong side of a big price movement in Comstocks and went bankrupt. But Keene learned from his mistake and after a few years was again speculating, and winning big, in mining shares.[63]

Keene helped manipulate Ophir shares upward but was among the first to sense the market turning. He always seemed to know when insiders were buying shares and when they were selling at the same time as they puffed up shares for the suckers. Later, on Wall Street, he was legendary for his ability to "read" the ticker tape to discern what was behind trading activity—and therefore when to buy or sell ahead of the public.

Keene discovered in 1875 that the Bonanza firm was secretly unloading shares, so he sold Comstock shares short in anticipation of a decline.

Keene's short selling earned him the enmity of James Flood of the Bonanza firm, who was eager to maintain share prices until he could liquidate his holdings. Flood accused Keene of bearing shares out of spite after losing money in a bad deal. Flood told newspaper reporters that the mines were as good as ever and advised the public to "buy as much as they [could] afford to."[64]

Keene accused Flood of hypocrisy, secretly selling shares while advising others to buy. Keene was no mining expert, but he truthfully noted that the famous bonanza ore body was growing poorer with depth and told the public in April 1876 that based on the known geology of the Comstock lode, the ore body would terminate by the 1,640-foot level. For telling the truth—in his own interest, of course—he was vilified as a "bear" trying to wreck the market.[65]

Despite his unpopularity Keene knew that the facts down in the ground were on his side: the big bonanza was being rapidly mined out, and no amount of false Bonanza firm indignation could stretch the ore body one inch. Keene's bear raids continued through July 1876, after which he reversed himself and expressed guarded optimism about the California and the Con Virginia mines. James Keene left San Francisco a few months later to test his prowess on Wall Street.[66]

The Last of the Big Bonanza at the 1,650-Foot Level

The Bonanza firm engineered one last publicity coup in March 1877, when it deepened the Con Virginia to the 1,650-foot level and led a group of boosters through the new working. George D. Roberts, who within a few years would move to New York City to give lessons in San Francisco-style mine swindling, enthused about the ore body, although the spoil-sport San Francisco *Chronicle* called his description "several percent clearer than black sulphreted mud."[67]

In fact, the 1,650-foot level was the disappointing tail end of the big bonanza. Nevertheless, the Bonanza firm's newspaper allies shamelessly exaggerated it as a rich proof that the bonanza ore body was inexhaustible.[68]

The Bonanza firm kept up the pretense until November 1878, when the Con Virginia laid off a large part of its workforce, and John Mackay admitted that the bear rumors were fact: the mine was running out of ore, and what was left was decidedly poorer. The news panicked the exchanges, and Comstock shares lost more than forty percent of their value in one day. Mackay, who just months before had lied that the mine looked as good as ever, suffered well-deserved criticism.[69]

Burke versus Flood

The Consolidated Virginia mine was still smoldering and repairing from an underground fire in 1875 when a shareholder, Squire P. Dewey, worried that the company would stop paying dividends. Like any good San Francisco gambler, he went looking for an inside "point." Dewey asked James C. Flood, the treasurer of the Consolidated Virginia Mining Company, how much was in the company treasury. Flood told him that there was less than $500,000, which Dewey knew was insufficient to pay the upcoming dividend.

Here the tales diverge. Dewey maintained that there had been more than $2 million in the treasury and that Flood had lied. Flood insisted that his statement to Dewey had been true, but the cash in the treasury did not include gold and silver bullion on hand yet to be sold. Dewey sold his shares—Flood said that Dewey also shorted the stock—and lost $50,000 when the Con Virginia paid its regular dividend after all. Whatever the facts, Dewey dedicated himself to attacking the Bonanza firm.

Dewey disturbed the annual meeting in January 1877 by opposing the reelection of the Bonanza firm and their allies as officers and directors. Only six percent of the shares voted with Dewey, so the old officers were easily returned. The Bonanza firm consented to a pair of management reforms that Dewey proposed, but then Dewey proposed that the company buy its own ore mill. The Con Virginia, like all mines the Bonanza firm controlled, sent its ore to the Nevada Mill Company, owned not coincidentally by Mackay, Fair, Flood, and O'Brien. The arrangement guaranteed enormous profits for the Bonanza firm at the expense of the mine shareholders. The Bonanza partners and their supporters were willing to make cosmetic reforms, but this threatened their income, and they defeated the resolution.[70]

Dewey offered peace to the Bonanza firm—for a price—but they called him a blackmailer and would not pay him a cent. Dewey's lawyer demanded that the Con Virginia directors file a lawsuit against the Bonanza firm for mismanagement, a highly unlikely event, since the Bonanza firm had named the directors. Each side blasted the other through their respective newspaper allies.[71]

For some reason, Dewey did not want to institute a lawsuit himself, so he transferred one hundred shares to the San Francisco *Chronicle* reporter John H. Burke. Dewey's lawyer filed suits on behalf of Burke in May 1878 against Flood, Fair, Mackay, and others associated with the Con Virginia, for more than $35 million. The complaint charged that the Bonanza firm and others had defrauded shareholders by giving sweetheart contracts to companies they owned that supplied water, lumber, and milling services to the Con Virginia. The courts threw out the suit for misjoinder, so Burke split it into three narrower suits, totaling over $40 million, which he filed in October 1878.[72]

The defendants fought and delayed, but the first suit—for $10 million—came to trial in December 1880. Only Squire Dewey and one other shareholder joined Burke in the lawsuit. The complaint maintained that the members of the Bonanza firm had bought claims adjacent to the Con Virginia ground for $36,000 then sold them to the company for stock worth $313,000, shares that later became worth $10 million. The facts were never disputed, although the defense took issue with some of the dollar figures.[73]

The first suit was decided in March 1881 when the judge found that the firm of Flood and O'Brien had defrauded Consolidated Virginia shareholders of $926,000. He found, however, that the defendants had made no effort to hide their actions and so had not regarded their actions as fraudulent. The judge ordered that the money be given directly to the shareholders and gave shareholders sixty days to come forward to register to receive their money.[74]

The defendants paid the three plaintiffs an undisclosed sum to dismiss all the cases. Nonparticipating shareholders, now unable to register to collect their money, tried to overturn the dismissal, but the petition to reinstate the case was denied, and all but Burke, Dewey, and Noyes were left with nothing.[75]

Letting Light into the Mines

The common practice of barring visiting stockholders from the mines changed in 1877 when the California legislature passed a law allowing anyone representing ten percent of shares to inspect mines on the first Monday of each month. Since all the major mining companies except the Yellow Jacket were incorporated in California, they obeyed the law and allowed shareholder visits.

The Nevada legislature passed its own law in 1879, mandating two visiting days each month, but some mining companies used the conflicting laws as an excuse to bar all visitors. On the first Monday of August 1879, the Sierra Nevada and the Hale & Norcross mines refused entrance to visitors. A stockholder came to blows with John Mackay himself when the bonanza king barred him from the Sierra Nevada mine. A few days later the Hale & Norcross superintendent was arrested in San Francisco for the no-admittance policy. Predictably, Hale & Norcross shares rose on rumors about a new discovery that the company was supposedly keeping secret.[76]

Stealing the Woodville Mine

When John W. Pearson won control of the Woodville Mine, he announced a reorganization. He advertised for the return of the Woodville Gold & Silver Mining Company shares, two of which would be exchanged for one share of the new Woodville *Consolidated* Gold & Silver Mining Company. Almost all shareholders exchanged their shares and became owners of the Woodville Consolidated. Unfortunately for the shareholders, they had unwittingly given their original Woodville shares to John Pearson himself. The Woodville Consolidated Mining Company was worthless, since Pearson had never transferred the assets of the old Woodville company to the Woodville Consolidated. The swindle went undiscovered for three years, after which the Woodville Consolidated shareholders sued to get their mine back.[77]

No Justice for Shareholders in the Justice Mine

The Justice mine was one of those off the main lode that eternally hoped to find another Comstock. Stockholder assessments seemed to be rewarded when miners found a moderate-sized ore body that promised finally to make a profit. The company continued to lose money, however, and in 1876 the shareholders replaced the management with one that promised reform.

The new reform management filled the May 1877 annual meeting with self-congratulation. The mine was well run, they said, and the old debt was rapidly being paid off. Some shareholders complained, however, that management was inflating mill charges and performing other hocus-pocus to make company monies disappear.[78]

E. J. "Lucky" Baldwin, the largest shareholder, tried to engineer a "deal" in Justice by spreading rumors of a secret ore find in the lower levels. Schultz, the company president, refused to abet the rumors and so earned Baldwin's enmity.[79]

The Justice management arranged in November 1877 to sell a portion of their ground to the Alta mine. The acquisition rocketed Alta shares from $1 to $24. The lawyer for both sides—the same person—arranged for the Alta Company to pay $250,000 for the property, an expenditure duly recorded in the Alta cashbook. However, by the time the $250,000 arrived in the Justice treasury, it had inexplicably shrunk to $1, that amount faithfully written as cash received in the Justice cashbook. The sale of what many regarded as the best part of the Justice ground enraged shareholders, who once again revolted.[80]

The Alta sale gave Baldwin an excuse to force Schultz out. The struggle climaxed on December 1, 1877, when a special meeting called by Baldwin and other shareholders ejected the officers and the entire board of directors. However, some of the reform men also suffered from shady reputations. The new president, E. J. Baldwin, had recently spread false rumors of an ore discovery. Sam Curtis, his handpicked superintendent, had four months previously been fired from the Ophir mine for defrauding that company. The new board rescinded the sale of part of the Justice ground to the Alta and appointed a committee to examine the company books.[81]

The committee arrived at Virginia City to find that forty-six pages—seven months of transactions—had been torn from the cashbook. A company letter book was partially burned. The investigators patched together the financial picture with some difficulty and discovered that despite rosy official statements and constant shareholder assessments, the Justice was more than $350,000 in debt. The San Francisco *Chronicle* observed, "A plainer case of fraud and collusion could not well appear."[82]

Besides the company's being financially mismanaged, the mine itself had been badly and inefficiently worked. The committee discovered that the Justice had sent its ore to the California mill, owned by the company's ex-president George Schultz. Schultz shipped a great deal of low-grade ore to his mill, which kept the mill busy and profitable, but the metal recovered was worth less than the inflated mill charges.[83]

One of the Justice company assets was a controlling interest in the Woodville mine. Schultz had ordered the Woodville Consolidated company to levy an assessment, and the Justice Company disbursed $60,001 to pay its portion, but the money was pocketed by ex-president Schultz and ex-treasurer Von Bargen. In addition, Schultz ordered the Woodville Consolidated officers to turn over the assessment monies paid by other parties to himself for safekeeping—money that also disappeared into his dark pockets.[84]

The story became even more curious when investigators discovered that Woodville Consolidated shares actually owned by the Justice Company were worthless, since the Woodville Consolidated had been cheated out of its property several years earlier—the mine was still technically owned by the original Woodville Mining Company. New management went to the courts to recover their money, but Daniel and Seth Cook, lately officers of the Alta, dodged subpoenas and had their lawyer stall. The Justice Company attached the property of ex-officers Schultz and Von Bargen and sued the Alta Company to reverse Schultz's fraudulent sale of Justice property.[85]

The 1886 Deal

Many San Francisco stockbrokers bucketed client orders during stock run-ups that the brokers believed unjustified. They were betting that they could buy the stock later at a lower price and pocket the difference. The bucket brokers made money only if their clients lost, but that seemed to bother brokers not at all.

James L. Flood, son of James C. Flood of the Bonanza firm, recognized in 1886 that he could trap the brokers in their own buckets. The younger Flood was a leading force in the Consolidated California and Virginia mine, which had not paid a dividend in six years and whose shares were selling around $2.00. When the company announced that it had discovered a new ore body, shares rose modestly to $2.50 in the face of cynical disbelief.[86]

When the Savage mine announced that it would stop paying its share of the cost of pumping out the lode's underground water, there was a general expectation that the mines—all connected underground—would flood, prompting brokers to sell all Comstock shares short. Instead of falling, however, Con Virginia shares continued to rise. Flood had proved himself a true son of his father in stock manipulation. While bucketing brokers and short sellers listened to their own bear rumors, Flood and associates had quietly bought enough shares so that there were not enough Con Virginia shares left to satisfy the short commitments.[87]

The younger Flood had achieved the rare and profitable "corner." Brokers had to buy Con Virginia shares from Flood, who was determined not to sell cheaply. Con Virginia shares jumped to $35 on November 27, 1886, as worried brokers bought to cover their short positions, forcing prices up.[88]

Rising prices attracted more buyers, which fed the rise, and other Comstock shares rose in sympathy with Con Virginia. Wildcat shares inactive for years now found eager buyers. Stock movements on the San Francisco exchanges were often like gangland wars: much activity and rumors, but the principals rarely explained to the public. The public knew only that Comstock share prices were headed skyward and thought they could get rich by going along. Laborers, clerks, and washerwomen took their savings out of hiding to buy Comstock shares and crowded the exchange galleries and nearby streets to witness the result of their gamble. Between $3 and $4 million were withdrawn from savings banks in ten days to plunge into Comstock shares. Buying by the shorts to cover their commitments drove prices up even further, and Con Virginia shares hit $55 on December 4.[89]

A great surge in share purchases was a boon to honest brokers who stuck to executing orders. However, the San Francisco exchanges were full of crooks gambling with client money, and the juggernaut of high share prices threatened to crush the brokers who had bucketed orders. Three brokerages failed on December 2 and sent the shorts into panic. The president of the Pacific Stock Exchange went bankrupt and hid from the law. The Frankel Brothers, lately brokers in Virginia City, cleaned out their cash drawer and fled to the hills, fearing for their lives lest their cheated customers find them. The sheriff discovered Sol Frankel's hideout a week later, locked him in the Virginia City jail, and kept visitors away, fearing violence to the prisoner.[90]

Five more brokerages failed on December 3, and there were more failures each day. Investors discovered too late that the bankrupt brokers had illegally sold off portfolios held in trust. Customers demanded that their brokers hand over the stock certificates. The San Francisco and Pacific exchanges closed for the afternoon of December 6 to give the bucket brokers more time. Trading moved out into nearby streets and alleys, where Con Virginia shares rose to $66. A total of fifteen brokerage houses failed, leaving them more than $5 million in debt to their cheated clients. The Bank of Gold Hill, Nevada, had also been bankrupted by the panic and its owner arrested for embezzlement.[91]

Since bankrupt brokers cannot pay their debts, it profited Flood to allow his prey to remain solvent. The remaining shorts had seen fellow brokers bankrupt and hounded and were willing to pay nearly anything to settle. The terms were never made public, but it is said that the elder Flood, James C., returned from the East Coast to San Francisco on the evening of December 6 and dictated the settlement.

At first, none but the insiders realized that the 1886 deal was over. On the morning of December 7, when Con Virginia was first called on the San Francisco Mining Exchange, a broker confidently offered to sell his shares at $100 a piece, but no one bought. An asking price of $75 also found no takers. When the first sale was made at $42, it was clear that prices had no support. The bears took over and pounded all share prices down. Con Virginia closed at $32, and the whole list suffered. All the small-time plungers who had taken money out of banks and mattresses and put it into the stock exchanges once again lost their savings.[92]

The Consolidated California & Virginia declared a dividend in early 1887 and continued to pay modest dividends totaling $15.60 per share over the next five years before going back to levying assessments.[93]

The Mill Ring

"Never in the history of mining have such outrageous frauds been perpetrated on shareholders as those by the Comstock mill ring."

—*Engineering & Mining Journal*, February 20, 1892

Comstock insiders discovered early on that while public attention was fixed on the mines, the bulk of the profits could be diverted to the ore-milling companies—owned by the insiders themselves. By paying inflated mill charges, managers could siphon off the profits, leaving only a token profit to its stockholders—or even a loss to be made up through assessments.

The first mill combination formed in 1867, when the Bank of California acquired seven mills by foreclosure. Bank insiders formed the Union Mill and Mining Company and bought the mills from the bank. Their "fortified monopoly system" forced the mines owing money to the Bank of California to send their ore to the Union company mills under terms highly advantageous to the mills. Within two years the Union company controlled seventeen mills and treated most Comstock ore.[94]

When John P. Jones and Alvinza Hayward took control of the Crown Point mine in 1871, they started the Sierra Nevada Milling Company to direct most of the Crown Point profits to themselves. The Crown Point mine kept the Jones and Hayward mills busy, even when the company-owned mill was idle for lack of ore. The shareholders revolted in 1877, but since the great majority of shares were held in trust by stockbrokers, who rarely hesitated to betray that trust, Jones and Hayward secured re-election of their dummy directors by buying proxy votes from the brokers for between $0.80 and $1.95 per share.

Likewise the Bonanza firm of Mackay, Flood, Fair, and O'Brien set up the Pacific Mill Company to work the ore of the mines under their control. The names of the milling companies changed as business alliances shifted, but they all followed the same principle: cheat the shareholders of the mines.[95]

The Comstock Mill Company was formed in 1886 by Senator John P. Jones, John Mackay, and James L. Flood. In 1885 Jones had leased the Consolidated Virginia Mine and found some ore in the upper levels from which he made a profit. The Con Virginia insiders were eager to join Jones in his profitable venture.

Jones relinquished his lease to the company and in return became a partner in the new Comstock Mill Company, set up to mill the Con Virginia ore. Mining the Con Virginia ore was profitable only for the owners of the Comstock Mill. In fact, the Jones lease—which the company managers were so eager to terminate—was more advantageous to the Con Virginia shareholders. Jones had paid $0.50 per ton to the company. In the first quarter of 1886, by contrast, the company mined its own ore for a profit of only $0.03 per ton. The bulk of the profits were absorbed by the mills of Jones and the Bonanza firm.[96]

The Alta Silver Mining Company levied an assessment upon shareholders so that the Alta mine could build its own mill rather than suffer at the hands of the mill ring. However the board of directors then leased the new Alta mill to themselves and plundered the shareholders with the usual mill ring overcharges, questionable assays, and disappearing-bullion tricks.[97]

Since most Comstock mining companies were incorporated in California, they were subject to California law. The mill companies, however, incorporated in Nevada, where the laws and jurors were not overly concerned with California stockholders' rights.[98]

Some mines mixed ore with waste rock to inflate milling charges. The mills themselves were trusted to assay the ore—and thus determine how much gold and silver they had to return to the mining company. Shareholders alleged that mills not only overcharged but also simply stole bullion that should have been turned over to the mining companies.

"Little Joker" Mills

The early ore mills were notoriously inefficient and sent considerable gold and silver down the streambeds with the waste tailings. The waste was rich enough to support small downstream mills carefully reprocessing the tailings. By 1867 the downstream operators were recovering $164,000 of gold and silver per year that the big mills threw away. The mills wised up and improved their recovery, but they did it in such a way as to keep the additional values for themselves.

Mills in early years often worked the ore for free and guaranteed to pay a minimum of sixty-five percent of the mine assay values. In practice, the "minimum" actually meant that they returned no more than sixty-five percent. In 1868 the mills recovered seventy-five to eighty percent of the gold and silver and pocketed the difference. Extraction became more efficient, recovering eighty-five to ninety percent, but the percentage returned to the mining companies remained low. Originally the mills had based percentages on assays made at the mine, but the basis shifted to the assays made at the mills, which for technical reasons were somewhat lower. Once the mines were under the control of the mill owners, the mines also paid milling fees. In addition, mill companies erected "little joker" mills adjacent to the main mills to further treat the "waste" from their primary mills. Under the fiction that they were only obligated to pay the mining companies the recoveries of the primary mills, the mill companies kept all the metals recovered by their little jokers. This system encouraged mill companies to do a poor job of treating ore in the primary mills—all the more precious metal left to be recovered in the little joker.[99]

Some milling companies were not content with the existing inefficiencies of their processes and had mill men deliberately divert silver concentrates downstream into the little joker. An investigator for the Mining Stock Association even photographed an employee of the Nevada mill shoveling Hale & Norcross Company concentrates into the little joker attached to the Nevada mill.[100]

The mill companies refused to disclose how much gold and silver they retained. Some idea of the amounts of bullion diverted to the milling companies may be seen by the fact that in 1888 all the mines on the Comstock reported to shareholders that they produced $5.5 million in gold and silver, but the Wells Fargo Company alone reported that it had shipped a total of $7.5 million worth of bullion out of the district.[101]

The Bonanza firm owned not only the mills but also the companies that sold water, lumber, and supplies to their captive mines. The mines paid a Bonanza firm transportation company to haul ore to a Bonanza firm mill. Gold and silver bullion from the mines then went to banks owned by the ring, which collected commissions on bullion sales.

The Governor of Nevada Swindles
the Kentuck Mine Shareholders

Charles C. Stevenson gained control of the Kentuck Mining Company and engineered a sweetheart deal in 1882 by which the company allowed him to work the mine and send the ore to his own mill in exchange for a per-ton royalty and sixty-five percent of the assay values. Another director realized that Stevenson was getting a steal and offered to pay the company a cash bonus and double Stevenson's per-ton royalty, but Stevenson had the other directors in his pocket and awarded himself the contract.[102]

Stevenson then quietly changed the terms of the contract from payments based on mine assays to payments based on mill assays. Mill assays were lower than mine assays—usually by eight to ten percent—but Stevenson's mill assays ran thirty to thirty-five percent lower. Since Stevenson ran the mining company, none in the management questioned whatever he chose to pay the Kentuck company.

It took shareholders a few years to realize how thoroughly Stevenson was cheating them, and they organized to stop him. By that time, Stevenson had been elected governor of Nevada, and he refused even to make an accounting. Shareholders sued to force him to disclose how much he had taken from the mine, but the governor destroyed the records of his mill assays. In 1890 a San Francisco court appointed a referee to investigate the Kentuck mine financial dealings, but the matter became moot when Governor Stevenson died in office later that year.[103]

The Hale & Norcross Lawsuit

Alvinza Hayward and W. S. Hobart gained control of the Hale & Norcross mine in 1887. They arranged for Hale & Norcross ore to be treated in the Nevada mill, of which they each owned forty percent, and of which US Senator John P. Jones owned twenty percent. In addition, the Nevada Milling Company brazenly bribed the officers of the Hale & Norcross by paying them as individuals a percentage of the mill profits.[104]

Stockholders formed the Mining Stock Association to fight the mill ring. In August 1890 the association's chairman, M. W. Fox, demanded that stockholders be reimbursed for more than $2 million cheated from them under the Hayward and Hobart management. When the company refused, Fox started a class action suit for $2,225,000, charging a "fraudulent conspiracy" between company directors and the mills that treated the ore.[105]

The Hale & Norcross president dodged a subpoena and hurried out of California so that he would not have to explain why he received large sums of cash from the Nevada Milling Company. The Hale & Norcross superintendent was not so agile and was served. He left the state anyway, on the advice of his lawyer, and the court issued a warrant for his arrest. The company officers took refuge in Nevada, and the trial started without them.[106]

Hale & Norcross directors—those who didn't flee subpoenas—testified that they knew little about their company. They owned only token shares and professed complete ignorance of how they came to be directors. One dummy director—whose services were in such demand that in addition to his being a director of Hale & Norcross, he was a director of three other Comstock companies and the president of four more—told the court that he had no knowledge of the basic business dealings of the Hale & Norcross. Another director testified that he believed that he used to be a director, having once received a notice to that effect, but he had never attended a meeting or paid any attention to the Hale & Norcross. Perhaps as a reward for his perfect ignorance, Alvinza Hayward had given him a gift of stock in the Nevada Milling Company. The amnesia epidemic also struck stockbrokers called as witnesses. One admitted that he had sold proxy votes to the Hale & Norcross management, but others were at a loss to explain how shares entrusted to them wound up being voted in elections for company directors.[107]

Alvinza Hayward and W. S. Hobart, the two sharpers who controlled the Hale & Norcross and owned most of the Nevada Milling Company, claimed that they knew nothing about the little joker mills they owned. US Senator John P. Jones testified in a deposition from Washington that he knew virtually nothing of the milling company of which he owned a large interest. Evan Williams, manager of the Nevada Milling Company, could hardly plead such ignorance as his partners; he described the little joker mills and admitted their product was kept entirely by the mill company.[108]

Williams admitted making regular payments out of the Nevada mill accounts to President Levy of the Hale & Norcross Company, but had no idea why he paid the money. Neither could W. S. Hobart recall why he had written checks to President Levy. Alvinza Hayward also admitted making payments to Levy but had no idea for what. No one seemed to know. Records showed that Hale & Norcross president Levy regularly received one-eighth of the profits of the Nevada Milling Company.[109]

The trial highlighted the close relation between the Nevada Milling Company and the United States mint at Carson City, where the Comstock mining and milling companies took their bullion. Evan Williams, superintendent of the Nevada Milling Company, was also vice president of the Bullion Exchange Bank, which handled all the Hale & Norcross bullion. The cashier of the Bullion Exchange Bank was also—thanks to the patronage power of Senator Jones—chief clerk of the Carson mint. The assayer for the Carson mint was Evan Williams's brother-in-law. Many suspected that Haywood, Hobart, and Jones had installed their own employees at the mint to keep secret the amount of bullion deposited by the milling company. The *Engineering & Mining Journal* saw no way to reform the situation other than shut down the Carson mint entirely.[110]

Testimony revealed that, based on mill assays, the milling company gave the Hale & Norcross an average of only fifty-two percent of the gold and silver from their ore and kept the rest; for some lots the return was as low as thirty-five percent—and the directors and managers of the Hale & Norcross had never objected. Based on the mine assays, returns were even worse. Because of the sampling method, mill assays were about eight to ten percent lower than mine assays, but for some Hale & Norcross ore, mill assays were a miserable eighty-eight percent lower than the mine assays—and the milling company paid based on their own mill assays.[111]

The plaintiffs called John Mackay as an expert witness. Mackay, who was not connected with the Hale & Norcross, said that he would not stand for such low bullion returns from a mill as those that the Nevada Milling Company gave Hale & Norcross. Mackay's testimony greatly helped the case against Hale & Norcross. He did not realize that he had also helped share owners in a future lawsuit—against Mackay himself.[112]

A Million-Dollar Judgment

"The head of the Comstock mill ring, which has at last been convicted of conspiracy and fraud, is U.S. Senator John P. Jones. How long will United States Senators be willing to have a common thief as an associate?"

—*Engineering & Mining Journal*, June 11, 1892

In May 1892 the judge found the defendants guilty of "an unlawful combination and conspiracy" and ordered them to refund $1,011,835 to the stockholders. Hayward, Hobart, and the company president were judged to have defrauded the shareholders. The other ex-directors of the company were found guilty of gross negligence. The judge held the ex-directors liable not only to civil suits for their actions but also to future criminal prosecution. He ruled that the recoveries of the little joker mills belonged to the Hale & Norcross and not to the Nevada Mill Company.[113]

The *Engineering & Mining Journal* hailed the decision as "the greatest victory that legitimate mining has ever obtained in this country over the thieves who are its worst enemies." The *Mining & Scientific Press* called it "a revelation" and praised the judge for rare honesty and courage "in the face of the strongest financial, political, and social influences." In an era when money ruled, the decision was a shock to Comstock manipulators accustomed to legal immunity.[114]

Editorials in East Coast newspapers used the Hale & Norcross case to warn investors of corrupt Comstock management. A Frenchman circulated trial testimony to his fellow Comstock investors in France, Germany, and Britain. The publicity caused greater reluctance to invest in Nevada mines and a consequent slump in share prices. Not everyone blamed the wrongdoers, however. The San Francisco *Post* blamed the Stock Association and said that the "remedy in this instance has certainly proved worse than the disease."[115]

All defendants appealed. Judgments against dummy directors were eventually reversed by friendlier courts on the grounds that they were guilty of negligence rather than fraud. The company officers filed appeal after appeal and succeeded in reducing the award but not in overturning the decision.[116]

The $1.50 per share finally returned to shareholders as part of the judgment in 1898 was the last money that the Hale & Norcross paid its shareholders.

Everyone but the defendants cheered the Hale & Norcross victory, but all realized that the Hale & Norcross swindles were not unique or even unusual among Comstock mining companies. A spate of smaller lawsuits hit the courts to force directors to reveal the mine assays needed to keep the mill returns honest. The Mining Stock Association looked for a bigger target.[117]

The Savage Lawsuit

Simultaneously with the Hale & Norcross lawsuit, the Mining Stock Association sued the management of the Savage mine for an accounting and to prevent the management from giving the ore to the Nevada mill. The Savage mine was adjacent to the Hale & Norcross mine, which was also run by Hayward, Hobart, and Jones, and for the same purpose: to fraudulently transfer the profits to the Nevada Milling Company. The facts were much the same, but the legal strategy was faulty and the plaintiffs were unable to call witnesses before the judge dismissed the case in January 1892.[118]

The Mining Stock Association also started two lawsuits against Hayward, Hobart, and Jones over their management of the Chollar mine and the Potosi mine. As in the Savage suit, the facts were the same, but they failed to replicate their legal success.[119]

New Savage and Hale & Norcross Directors

The revelations at the trial had harmed business at the exchanges so much that the San Francisco Stock Exchange created a committee to evict the offending directors of the Hale & Norcross and the Savage at the next election. However, they defeated a more sweeping reform that would have stopped the lucrative practice of selling proxies.[120]

On March 9, 1892, Hale & Norcross shareholders elected a "reform" slate of directors, controlled by James L. Flood. Flood beat another "reform" slate proposed by the stockbrokers, although if history was any guide, the brokers' idea of reform was to substitute themselves as beneficiaries of the swindle. The pretense of reform wore thin when Flood's directors cooperated with those they replaced to fight the Mining Stock Association lawsuit. The new directors took business away from the Nevada mill and sent their ore to their own Brunswick mill. The Brunswick mill, however, also had a little joker, and its returns to the mine were no better than those of the Nevada mill. The money still disappeared but into different pockets. Perhaps as a temporary sop to shareholders, the new management began negotiations for the Hale & Norcross to buy its own mill.[121]

The "reform" Hale & Norcross directors also failed to make the mine profitable, and after years of steady assessments they too were thrown out. The mine shut down during a legal struggle for control in 1896.[122]

The annual meeting of the Savage company in July 1892 also put in new management, although Flood was able to name only two dummy directors out of five. Hayward and Hobart retained two dummy directorships, and the brokers gained one. Like the Hale & Norcross, the new Savage management made some reform noises about buying their own mill, but that appeared to be only a stalling tactic to thwart a Mining Stock Association proposal to put the ore-milling contract out for competitive bid. The Savage mine never returned to profitability, continuing to live on a steady diet of assessments.[123]

The Consolidated California & Virginia Lawsuit

The legal attack now turned to the Consolidated California & Virginia Company. In December 1891, shortly after John Mackay's testimony in the Hale & Norcross suit, a shareholder filed class action suits eventually totaling $4.5 million against Mackay, Senator Jones, and some of their dummy directors. He charged that Mackay, Jones, and company had conspired to defraud stockholders by sending Con Virginia ore to the Comstock Milling and Mining Company, which had overcharged the Con Virginia by $1.8 million.[124]

According to Mackay's own testimony in the Hale & Norcross case, Comstock ore containing as little as $14 per ton could be mined and milled at a profit. Now Mackay was accused of mismanaging a mine to the extent that the Consolidated California & Virginia's $23 ore lost money.

After four years of delays, the trial began in August 1895. Senator Jones again stayed safely in Washington but testified in a deposition that he did not even know whether or not he was an officer of the Comstock Milling Company.[125]

The Bonanza firm had learned its lesson in the case of *Burke v. Flood* and had stopped robbing shareholders quite so openly. In 1877 O'Brien, Mackay, and James C. Flood—three out of four of the Bonanza firm—were on the board of the Consolidated Virginia. The year after the Dewey–Burke lawsuit, they replaced themselves with handpicked dummy directors. Now none of the Bonanza firm served as directors of the mines they controlled.[126]

The Hale & Norcross trial had laid the groundwork, as in these questions to D. B. Lyman, superintendent of the Consolidated Virginia mine, in February 1892:[127]

Q. From whom do you receive your instructions?

Lyman. Mackay and Flood.

Q. They are heavy stockholders, are they not?

Lyman. Yes, Sir.

Q. Do you receive instructions from any other shareholders of the Company?

Lyman. No, Sir.

Q. If you did, would you carry them out?

Lyman. I would not. I look only to Mackay and Flood for orders concerning the operations of the mine.

In the same trial Charles Fish, president of the Consolidated Virginia Mining Company and a longtime officer of mines run by the Bonanza firm, swore that he had sent ore to Mackay's own mill on orders from Mackay himself.[128]

Q. Who compose the Comstock Mill and Mining Company?

Fish. J. P. Jones, John W. Mackay, and J. L. Flood.

Q. Who signed the contract for the mill company?

Fish. Senator J. P. Jones.

Q. And who signed for the mining company?

Fish. I did.

Q. By whose orders?

Fish. Mr. Mackay's.

Contradicting sworn testimony by Superintendent Lyman and President Fish, Mackay now denied under oath that he gave orders to company officers, denied that he had placed them in office, and told the court that since 1883 he had had little to do with the Consolidated California & Virginia. Judge Seawell found in favor of the defendants in December 1895 and ruled that the plaintiffs had not demonstrated what was the common understanding in financial circles and the sworn testimony of company officials: that Mackay and Jones ran the Con Cal & Virginia Mining Company.[129]

The Bonanza firm's dummy directors not only avoided legal consequences but stayed on to continue the swindle with the help of proxy votes from brokers. Shareholders challenged the election by brokers' proxies but were rebuffed by the court in 1896.[130]

The Clique of Six Crooked Brokers

"Never has a more shameful conspiracy existed against great and promising mines than that of the 'clique of six crooked brokers,' who have for some years controlled the middle and north-end Comstock mines and the United Comstock Pumping Association."

—*Mining & Engineering World*, April 12, 1913

The California constitution of 1879 made individual stockholders liable for corporate debt. This was intended as a reform, but stockbrokers learned that they could turn it to advantage. Owners of mining shares did not want to risk liability and so allowed shares to remain in the name of the brokers. The brokers thus became the shareholders of record and controlled the mines.[131]

The mill ring declined under legal assault, and death and retirement removed the original Comstock manipulators. The brokers now held sway and named the company directors. They thwarted any outsiders trying to acquire controlling interests. When brokers detected attempts to buy controlling blocks of stock, they forced share prices prohibitively high.

Comstock manipulation was so notorious that other mining stock exchanges, such as the Colorado Mining Stock Exchange in 1890, refused to list Comstock shares. Mere listing on the San Francisco exchanges became disreputable. The San Francisco exchanges tried in 1892 and 1893 to lure non-Comstock companies onto their boards, but nearly all California mining companies spurned the offer, to the applause of the *Engineering & Mining Journal* and the *Mining & Scientific Press*, the leading mining periodicals of the East and West Coasts. A spokesman for the California Mining Association boasted in 1899 that California gold mines were so honestly run that not one of them was listed on the San Francisco stock exchanges.[132]

Ore became scarcer in the Comstock mines, and dividends became small and sporadic. After 1888 there was no year in which Comstock shareholders received in dividends as much as they paid in assessments. The San Francisco exchanges declined along with the Comstock, until the San Francisco Stock Exchange could shut down for a week in the summer of 1891 and not be missed. An amendment to the California constitution prohibiting stock sales on margin also hindered trading. In 1892 the exchange sold its prestigious building on Pine Street and moved to humbler quarters.[133]

By 1897 interest in Comstocks had declined to the point that the *Engineering & Mining Journal* happily predicted the end of the San Francisco stock exchanges, a deliverance that the magazine foresaw as resulting in the businesslike and profitable operation of the Comstock.[134]

Comstock companies announced that the only way to keep the mines open was to reduce wages. The miners knew how the game worked, however, and pointed out that the inflated expense of salaried hangers-on was greater than the cost of miners. A. C. Hamilton, for example, was paid $2,500 to be superintendent of six different Comstock mines yet found the work so undemanding that he spent most of his time elsewhere. A committee of miners listed company after company with phony expenses, and the companies dropped the matter rather than suffer more embarrassing disclosures.[135]

Most Gold Hill mines, forming the southern part of the Comstock lode, fell under the control of R. F. Morrow and W. E. Sharon, both experienced manipulators. In 1907 they collected $153,234 in assessments for the companies under their thumbs and spent ninety percent of it on overhead. An investor gained control of the Yellow Jacket, which under Morrow and Sharon had not been producing any ore, and began hoisting out 180 tons per day of ore, at the same time cutting expenses in half.[136]

Six San Francisco stockbrokers controlled companies of the northern Comstock lode. They cared even less for shareholder rights than had the Comstock barons they replaced. At the 1908 Chollar annual meeting, the puppet directors all together owned $3.10 worth of stock in the company. The old mill ring had at least profited from keeping miners and mill men at work producing bullion; the broker ring, however, profited most from shutting mines down and levying assessments.[137]

San Francisco brokers maintained political influence in Nevada by putting politicians on salary. In 1908, in addition to his duties as a public servant, the Nevada state treasurer found time to collect salaries as the superintendent of six Comstock mines. W. G. Douglas, while serving as Nevada secretary of state, was also the superintendent of five Comstock mines.[138]

Zadig & Company brokers became dealers in dynamite so that they could charge captive mines $0.18 for dynamite sticks costing $0.13 elsewhere. Shareholders complained of the graft in dynamite and other supplies, but Herman Zadig refused to stop.[139]

The reputation of Comstock shares fell so low that two-thirds of the shares in the Exchequer mine were forfeited by shareholders who refused to pay a $0.10 assessment in 1893. When the famous Gould & Curry levied $0.03 per share in 1914, more than half the shares fell delinquent and reverted to the company treasury. Other companies were no better at coaxing assessments from stockholders, and the majority of shares reverted to the Alpha, Bullion, Best & Belcher, Chollar, Hale & Norcross, and Savage companies. As investors shunned the Comstock, the brokers lost interest in their captive companies and promoted other Nevada mining booms at Tonopah, Goldfield, Bullfrog, and Rawhide.[140]

The resourceful brokers tried to revive shareholder hopes. In 1898 they announced the formation of the Comstock Pumping Association to lower water levels and permit exploration in the deep parts of the northern Comstock. The results gratified the brokers but not the shareholders. Gould & Curry shares, as low as $0.03 in July, rose to $0.29 in December. Consolidated California & Virginia shares meanwhile rose from $0.07 to $1.40. The mines collected enough assessments to build their own electrical generators to run the pumps but instead gave the $100,000 to a puppet company owned by the brokers, which built a hydroelectric generating plant then charged the mines $100,000 per year for electricity.[141]

Companies supported the pumping associations unequally. When the Ophir mine found a good ore body in 1908, its shareholders found themselves paying not only forty percent of the expenses of the Pumping Association but also twenty-five percent of the costs of the Ward Shaft Association, a drainage project for the southern Comstock mines that was of no benefit whatever to the Ophir mine.[142]

Reformers gained directorships in the Union Consolidated and Sierra Nevada companies, but the broker-ring directors still held power, and they fired Whitman Symmes, the mine manager who was cooperating with the reformers. Shareholders wanting to know the truth about their mine hired Symmes to inspect the Sierra Nevada mine and collect ore samples, but the new manager refused to let Symmes collect ore samples even though the law gave him the right to do so.[143]

Eastern shareholders tried in 1910 to merge the Comstock companies together for rational and economical development, but San Francisco brokers blocked the plan.[144]

Assessment Frauds

By holding client shares in trust, the brokers were able to sell the same share, usually on margin, to several different clients. When assessments were levied, each "owner" of the share would forward the assessment to his broker, who would pay only one assessment to the company and would pocket the rest. Obviously, this was only profitable for assessments, and dividends would be disastrous for the brokers. Likewise, if share prices rose and clients decided to cash in, the broker would lose heavily. This gave the brokers a financial interest in keeping the Comstock mines unprofitable and share prices low.[145]

The Comstock lode died slowly and quietly. After World War I, manipulators lost interest in the lode, and the properties were consolidated and operated with the purpose of mining at a profit. By that time, however, there wasn't much ore left in the mines.[146]

By the Standards of Their Own Time

"It is safe to say that more money is lost in rich mines than in poor prospects."

—*Mining & Scientific Press*, December 15, 1906

Comstock investors lost money in worthless properties but were also swindled out of their investments in some of the richest mines in the world. The small-time frauds were always justly condemned, but both contemporaries and later historians have been ambivalent about the great Comstock barons who lied and manipulated, to the impoverishment of the small shareholder.

The Comstock barons were a tough lot who thrived in the corrupt environment and rarely gave a sucker an even break. Today they would quickly run afoul of securities laws. However, the two most complete and influential histories of the Comstock, by Elliot Lord in 1882 and Grant Smith in 1943, argued that the fault lay not with the Comstock barons, but with the small investors.[147]

Elliot Lord wrote, "When stockholders are both greedy and careless, they must expect that their agents will be equally selfish." He justified the insiders who hogged nearly all the profits by asserting that they bore most of the risk and work. They certainly worked assiduously, but they worked to put all the risk onto the shareholders. Far from being risky, the insider-owned supply and ore-milling companies always profited greatly because they were guaranteed favorable contracts at the expense of the captive mining companies.[148]

Dishonest management tried every means to defeat attempted reforms. Lord criticized ineffective shareholder efforts to end the frauds, but since shareholders were mostly working men and women of limited means, it is unreasonable to blame them for losing unequal contests with those possessing the greatest economic and political influence on the Pacific Coast.[149]

Grant Smith maintained that unfavorable judgments of the Comstock capitalists were unfair retroactive applications of modern ethical standards. There certainly were practices that, while unethical and illegal by modern standards, were legal, tolerated, and even applauded at the time. The San Francisco *Chronicle* noted that insiders Mackay, Flood, Fair, and

O'Brien had enriched their friends by advising them to buy stock before news of the Consolidated Virginia bonanza became public. The use of inside information that would send them to jail today was approved by the *Chronicle* as "a bright spot of good, honest charity."[150]

John Mackay in particular is difficult to dislike. He was an unpretentious man who rose by intelligence and hard work. Most Comstock millionaires quit the mines for the easy life as soon as they could, but Mackay stayed in Virginia City, and for years the millionaire pulled on his boots every morning to run his mines. Although he indulged—at great expense—his wife's international social climbing, he himself had no enthusiasm for gilded society. He remained a devoted husband untainted by the scandals of many of his wealthy contemporaries. Neither did he share the lust for political power that propelled fellow mine owners to buy their way into the United States Senate—though Mackay was more popular than any of them and could well afford the bribes. The miners respected his prowess at bare-knuckle boxing, a skill he retained well beyond the age when most men grow sedentary. He was known on the Comstock for fair dealings with his employees and refused to join other mine owners in cutting wages when the price of silver dropped. But for all his attractive qualities, John Mackay also used reprehensible business practices that impoverished the shareholders of the mines he controlled.

Of the illegal practice of giving favorable contracts to their own ore mills, Smith wrote that the Bonanza firm charged reasonable rates for milling. In fact, the milling contracts were outrageously unfair. The recovery that the Bonanza firm mill returned to the Con Cal & Virginia mine in 1891 hovered about sixty-five percent of pulp assays, scandalously low compared to the Occidental mill, which during the same period recovered ninety percent. The *Mining & Scientific Press* editorialized that "nothing but fraud or the most reckless of management could produce such results." The *Engineering & Mining Journal* asserted that the Bonanza firm directors, "either arrant knaves or positive dummies," were cheating the Con Cal & Virginia mine through overpriced supplies, overpriced transportation, and milling charges three times too high. The magazine accused the Bonanza firm milling company of "fraudulently working the ore of the Consolidated California & Virginia Mine and *stealing* the proceeds."[151]

Much of the lying, trickery, and secret deals were legal at the time but were still widely condemned. Grant Smith defended the Comstock operators as no worse than Rockefeller, Fisk, or Gould.[152] It is a mistake, however, to accept the conduct of the era's most vilified businessmen as the ethical standard of the time. A better standard is to judge the Comstock barons as they were judged in their own time:

"There is but little question that many of the silver mines of Nevada, which are now used to defraud our citizens, if under honest and competent management would be a source of great wealth, instead of continually running shareholders into debt."

—San Francisco *Bulletin*, June 8, 1864

"It is the general opinion that there must be a change in the system of managing mining companies....The public have been plundered so unmercifully that their patience and confidence are nearly exhausted."

—San Francisco *Alta California*, May 28, 1865

"In the first place, ninety-nine out of one hundred of the mines owned by these companies are worthless. The mines under the best management would never be made to pay dividends, and such as could are usually manipulated by the managers."

—San Francisco *Call*, March 13, 1870

"No mines in the world are more subject to the interested money making manipulations of the Trustees and Superintendents having them in charge than those of the Pacific coast."

—Gold Hill *News*, February 22, 1872

"As a rule, with here and there an honorable exception, we believe that the management of our Nevada mines has been dishonest and criminal."

—San Francisco *Chronicle*, January 4, 1877

"It has become the general custom for the trustees to lay all sorts of plots to steal the proceeds of the mine, sometimes the mine itself."

—Oakland *Transcript*, May 16, 1877

"Mining conditions on the lode are barnacled with 30 years of gambling, of mismanagement, of more or less corruption, of experiments and non-savey."

—Virginia City *Enterprise*, May 1890

Some criticism, especially that from the San Francisco *Chronicle*, was allegedly motivated by blackmail, and certainly the *Chronicle* owners, the de Young brothers, were not overburdened with ethics—but neither were most of their rivals. Newspapers defending the Comstock barons had at least equally interested motives, since Comstock manipulators could buy, bully, or persuade favorable press. William Sharon lent money to the Gold

Hill *News* in 1872, with strings attached. Two years later he bought the hostile Virginia City *Enterprise* to support his senatorial bid. John Mackay later bought the *Enterprise*, and his partner James Fair bought the San Francisco *Alta California*. In 1893 the *Engineering & Mining Journal* noted that the *Enterprise*, the captive organ of stock manipulators, had for years cooperated in misinforming investors.[153]

The most prominent American mining periodicals also harshly criticized Comstock management. These were not muckrakers but conservative advocates of mining and capitalism. The criticism of respected mining spokesmen of their own day shows that the Comstock manipulators were considered unethical by their contemporaries, and we today have no reason to revise their reputations upward.

> "There is a vast difference between legitimate mining and the system of stock jobbing so prevalent at the present time in this city and elsewhere."
>
> —*Mining & Scientific Press*, July 2, 1864

> "The effect of this wild and speculative fever was the ruin of hundreds or thousands of those who indulged in its excesses, and it brought out again into clear relief the disreputable management which has so long obtained in Comstock mines and which is a shame and disgrace to San Francisco."
>
> —*Engineering & Mining Journal*, January 11, 1879

> "Can it be considered a startling coincidence that among these gentlemen, owners of the mills, are found the millionaires of the Comstock? The paupers are found among the stockholders of the mining companies, who intrust their interests to those who, from personal interest or criminal neglect, sacrifice them and their poor earnings to the mill owners."
>
> —*Mining & Scientific Press*, June 14, 1890

> "Thus is mining carried on at the famous Comstock lode to the disgrace of the whole mining industry of this country."
>
> —*Engineering & Mining Journal*, May 23, 1891

The Effects of Fraud

Comstock mismanagement not only robbed investors but harmed mining itself. Much investment was absorbed in the great gambling casinos posing as stock exchanges in San Francisco. An additional and incalculable amount of capital was repelled by the bad reputation of the management.

Comstock mining engineer Almarin Paul estimated in 1877 that production of western mines would have been fifty percent higher but for frauds and inside manipulation. He complained: "Men are sent to the penitentiary for stealing trifles, while Simon and his cohorts steal thousands, yes, millions, rob poor men, women and orphans and yet still go scot free."[154]

The San Francisco Mining Exchange, the descendant of the original San Francisco Stock and Exchange Board, wandered into irrelevance as its listings shrank in number and value. The federal Securities and Exchange Commission finally closed the exchange in 1967, after repeated violations of securities laws by members and officials of the exchange. Fittingly, the deal that finally prompted the SEC to close the exchange was a manipulation of an old Comstock mining company, the Best & Belcher, the shares of which had become just pieces of paper without any assets behind them. At the closing bell, the exchange had only twelve members and twenty-two listed stocks, and a day's transactions might total less than $800. Nevertheless, the exchange president defended the institution, saying, "Everyone has the gambling instinct—why, to buy at a nickel and sell at thirty cents, you can't beat that." By this time, the small-scale mine-stock manipulation seemed even quaint, so that a commentator complained that the SEC lacked a sense of humor.[155]

Snow Job at the Emma Mine

"The operations of unscrupulous speculators in mining property, in Utah, are doing, and have done, more harm to the mining interests of the Coast than can well be computed in dollars and cents."

—*Mining & Scientific Press*, September 30, 1873

The mountain rises so steeply next to the ski-resort village of Alta, Utah, that you have to crane your neck to see all the old mine openings. It attracts little attention now, for there are many old mines in the canyon, and the skiers care far more about snow conditions. But one of the old holes is the Emma mine, once the center of high finance, international lawsuits, congressional hearings in Washington, and parliamentary debate in London. Many men each in turn tried to swindle the others, until English and Scottish stockholders were left with a worthless hole in the ground for which they had paid millions.

Soldiers discovered silver in Little Cottonwood Canyon in 1864. They dug out and smelted a load of ore but found that the bullion was mostly lead. The closest metal refinery that could extract the silver was in San Francisco, and the cost of shipping by wagon to the refinery was more than the bullion was worth. The miners abandoned Cottonwood Canyon to winter storms in December 1867.[1]

Robert Chisholm was a New Yorker who went west and worked in the lead mines of Illinois. In 1851 he left his family in Illinois and continued west. He and prospector Thomas Woodman located a mining claim they named the Monitor in Little Cottonwood Canyon in August 1868.

Cheating Lyon Out of His Interest

The transcontinental railroad revolutionized Utah mining by providing cheap transportation to distant smelters. New York businessman James Lyon arrived in Utah in October 1868 to invest in mines. He agreed to work the Monitor claim temporarily in exchange for a one-third interest.

Two prospectors, believed to be the Emma discoverers Woodman and Chisholm, pose with a donkey. Courtesy Utah State Historical Society. All rights reserved.

Lyon produced seventy-five tons of ore in a few weeks then hired Woodman to deepen the shaft and returned to New York. He heard no more of the Monitor claim and assumed that it had turned out worthless.[2]

Chisholm and Woodman continued to mine, and in 1869 they dug into a body of rich ore. The two decided to erase Lyon's interest by relocating the claim. By law, the long axis of a mining claim parallels the vein. Chisholm and Woodman relocated the claim in February 1870, swiveling it ninety degrees to run northeast. This they did in their own names, maintaining that Lyon had owned part of the original Monitor claim but not the repositioned claim. To help sever Lyon's interest, they changed the name of the relocated claim to the Emma, after Chisholm's ten-year-old daughter.[3]

Chisholm and Woodman further muddied the ownership by selling part interests to two investors. The four partners then sold four hundred feet of the three-thousand-foot-long claim to Salt Lake City businessmen. Production began in earnest in July 1870, when the mine produced thirty-one carloads of ore so rich that even after it was hauled by wagon to Salt Lake City, then taken by rail to a smelter in New Jersey, it still returned a great profit. Daily shipments of five thousand dollars worth of ore made the Emma the foremost mine in Utah.[4]

In early 1870 James Lyon in New York learned that Chisholm and Woodman had struck a rich ore body. Lyon returned to Utah and found that their Emma mine was none other than the Monitor, of which he owned a third. He sued to recover his property but received little satisfaction from the courts.

The other owners continued to try to sell the mine. The asking price ran up and down depending on the latest developments in the mine and the courts. A San Francisco syndicate examined the mine in October 1870 but thought the four-hundred-thousand-dollar asking price too steep. The next following month the owners tripled the asking price to $1.2 million and offered the Emma to San Francisco capitalist William Lent. Lent sent a mining engineer, who advised that the mine was not worth it. The owners finally sold a half interest to Trenor Park and Henry Baxter for $375,000 in March 1871.[5]

Trenor Park was a Vermont lawyer who had made his fortune in California. After Park and his partner, General Henry Baxter, bought a controlling half-interest in the Emma, they claimed to have spent $1.5 million developing the mine. This was an exorbitant sum, of which Park and Baxter demanded their partners pay their proportionate shares. It appeared to be a classic case of freezing out.

Enter Senator William Stewart

"Dishonest speculators lie in wait for the mine owners first; then come the incompetent managers to complete his ruin; and what these two classes cannot succeed in bagging, the lawyers are sure to get, if they have a chance."

—*Engineering & Mining Journal*, January 3, 1871,
on the Emma mine lawsuits

James Lyon decided that the mining law was blind only to justice and that he needed politically powerful counsel. Lyon obtained an introduction to US Senator from Nevada William Stewart, who agreed to help Lyon recover his property in return for a one-quarter interest. Stewart had a reputation for highly effective dishonesty as a lawyer on Nevada's Comstock lode. Stewart's high-spending ways in Washington, including building a mansion on Dupont Circle, made him eager for lawyerly plunder. Now that Lyon had made his Faustian pact with Stewart, there would be the devil to pay.[6]

After legal skirmishes, all sides decided that the only way to settle would be to incorporate and sell the Emma mine. Park and Baxter formed the Emma Mining Company of New York in April 1871 and negotiated a division of shares. Stewart made a deal whereby Lyon would receive at least five hundred thousand dollars. "You will, without doubt, get that," wrote Stewart, and Trenor Park told Lyon that he would likely receive seven hundred thousand dollars once the mine was sold. Stewart advised Lyon to take the offer, so they would " not be working for a worked-out mine."[7]

Selling a Worked-Out Mine

"At the present rate of extraction the mine will soon be stripped, and for the present practically exhausted."

— W. P. Blake's report on the Emma mine, July 27, 1871

The great ore chamber of the Emma mine was a cavity seventy feet long, seventy feet high, and twenty-five to thirty feet wide, and miners were removing ore as rapidly as possible. Trenor Park and Henry Baxter hired mining expert W. P. Blake to assess the value of the Emma. Blake reported bluntly in letters to Park and Baxter dated July 26 and 27, 1871, predicting the imminent exhaustion of the Emma ore body and warning, "Deposits in limestone are always regarded by miners as treacherous; we cannot depend on them." Park and Baxter knew what they had to do: keep Blake's report secret and sell the Emma mine to the British.[8]

Trenor Park brought in a new mine manager, Silas Williams, whom he described as "the best man to prepare a mine for the inspection of engineers I ever saw." He laid off about one hundred miners and kept only six working in the mine.

Stewart Switches Sides

Trenor Park and William Stewart arrived in London in September 1871 to sell the Emma. They came in the middle of one of the London financial district's periodic enthusiasms for mining shares. Previous London mining-share booms had been based on Cornish tin mines and Indian gold mines, but this boom would include a new twist: overpriced American silver mines.

Stewart was still representing his client Lyon, but he conceived of bigger profits. He wrote Lyon on October 24 to assure him that Lyon would receive his $500,000, but then on November 3 he warned Lyon that the payment might be only $400,000. On November 9 Stewart wired for Lyon to come to London immediately. Lyon arrived in London on November 20 and received a sad story from Stewart about how the expenses of the flotation were enormous and Lyon might receive nothing at all. This Lyon refused to accept, and Stewart finally offered him $200,000 to assign all rights to Stewart himself. This was a blatant conflict of interest and a sham performance, for shares had already been sold on the basis of $5 million capitalization, and there was plenty of profit. But Lyon accepted his attorney's word and took the $200,000, out of which Stewart had the nerve to deduct his $50,000 fee, giving $150,000 to his cheated client. The agreement gave Stewart five thousand Emma shares, worth $500,000.[9]

Trenor Park, in trying to pay Stewart a compliment, told Lyon: "He is a very honest man; I tried to buy him. I offered him fifty thousand dollars if he would throw you in your suit." Stewart and Park continued their partnership by investing together in silver mines at Panamint, California.[10]

Stewart returned to the United States, where, now openly acting for Trenor Park, he bought the interests of the prospectors Chisholm and Woodman, paying less than half what the shares were worth in London; Park's profit in the matter was more than $1 million. After Stewart bought Lyon's interest for $150,000, he earned another $275,000 in fees from Park.[11]

Albert Grant Takes His Cut

In London, Park, Baxter, and Stewart needed someone to lead them through the intricate dance of an initial public offering on the London market. They found their guide in Baron Albert Grant. Grant, born Albert Gottheimer in Dublin, gained his title in 1868 from King Emmanuel of Italy in exchange for a charitable contribution to a Milan gallery named after the king. Grant became one of the most prominent financiers in London, but his title and the wealth that bought it were intensely resented by the British upper class.[12]

Grant had figured prominently in floating the Lisbon Steam Tramways Company, for which the Portuguese minister to Britain served as chairman. The prospectus emerged in July 1871 to favorable notice in the *Economist,* and also in the *Times,* whose financial writer was taking bribes from Grant. Stock sold largely on the minister's prestige until shareholders discovered that Grant, the Portuguese minister, and other promoters had pocketed nearly half the proceeds, not leaving enough to build the streetcar lines.[13]

Grant continued trying to buy respectability. In 1872 he was made a commander of the Portuguese Order of Christ. In 1874 he bought run-down Leicester Square in London, turned it into a public garden adorned with a statue of himself, and donated it to the city. When he ran for Parliament from Kidderminster that same year, he ordered forty thousand pounds worth of the district's principal product, carpets. Grant's patronage assured his election, but parliamentarians decided that his methods violated the Corrupt Practices Act and refused Grant a seat in the House of Commons.[14]

In September 1871 Grant agreed to organize the Emma mine stock offering for a fee of one hundred thousand pounds. Since Baron Grant's failed companies were earning him a bad name, he hid his involvement. Grant knew how the thing was worked in "the City," London's financial district. He paid a member of Parliament to become a director of the Emma Silver Mining Company and to persuade three other members of Parliament to do likewise. Beyond the regular fees, each of Grant's sty of guinea pigs received five hundred pounds. Grant spread money liberally: a hefty finders fee to the share brokers who introduced him to Park, £25,000 to a respected banking house to be the company bankers, and £40,000 ($200,000) to Senator William Stewart to become a director.[15]

The Optimist: Professor Benjamin Silliman Jr.

"These several facts, in my opinion, establish beyond all reasonable doubt the conclusion that the Emma is a true mineral vein of great power, and place it in the category of the great mines of the world."

—Benjamin Silliman Jr., October 16, 1871

The Emma Mine had already produced some $3 million worth of ore in its brief burst of glory, but the sellers needed to convince investors that it would continue to issue silver. The promoters turned to Benjamin Silliman Jr., a professor of chemistry at Yale College. Silliman had great academic prestige, but as a historian noted, "He saw millions in everything." Park in London wired instructions to hire Silliman to examine the Emma mine in September 1871. Silliman traveled to the mine and was dazzled.[16]

Professor Silliman called the Emma ore body a "true mineral vein," believing that the ore body would extend as far down into the earth as anyone would care to mine, and concluded that the fact was "beyond all reasonable doubt." Editor Rossiter Raymond of the *Engineering & Mining Journal* doubted it. He had examined the Emma ore body a few months earlier and noted that little advance development had been done, so that in-sight ore reserves extended little beyond the miners' picks. Other mining experts, the magazine pointed out, classed the Emma as a lenticular deposit, meaning that however rich, it might be very limited in size. In fact, rumors already circulated that the Emma ore body was exhausted and that the investors were buying an empty shell.[17] As the *Engineering & Mining Journal* noted on December 12, 1871: "As it is our impression that no single mine in the country is worth $5,000,000, we feel of course sincere regret that the proprietors of the Emma Mine have obtained so much for the property."

Silliman completed his work on October 16, 1871, and Baxter spent about $4,000 telegraphing the long report to London, where it had considerable success in convincing British investors that the Emma mine was worth £1 million.[18]

The Prize Guinea Pig Director: Minister Robert Schenck

Robert Schenck arrived in London in 1871 as the new American minister. His term in that sought-after diplomatic post should have been the final laurel of a career as congressman, minister to Brazil, and general in the Union army. Instead, he would soon see his honesty attacked on two continents.

He found that ministership was a costly honor and that the salary would not support himself and his family at a level befitting his title. Schenck dined with Park and Stewart in London and heard how they were there to float a rich mine in Utah. The reputation of American companies was low in the City, because of the wild manipulations of Erie Railroad stock, and Stewart saw Schenck as a way to create confidence in the Emma Company.

Schenck agreed to serve as a director. Park gave Schenck an interest-free loan of £10,000 to buy Emma stock, with the guarantee that Schenck would receive at least twenty-four-percent yearly return on the shares; alternatively, he could sell back the stock at the original price at any time.[19]

His directorship was immediately criticized on both sides of the Atlantic. London's *Economist* magazine, which had given favorable notice earlier in the year to Lisbon Steam Tramways with the Portuguese minister prominently at its head, now decided that a diplomat should not use his name to float a stock offering. No mention was made of the United States senator or of the members of Parliament also serving as Emma Mining Company directors. Despite the controversy, Schenck's directorship favorably impressed British investors.[20]

Schenck's appointment had been a reward for loyal Republican Party service after his fickle constituents failed to return him to Congress. But now Schenck was adding another embarrassment to the scandal-plagued Grant administration. Secretary of State Hamilton Fish telegraphed Schenck from Washington on November 28, 1871, telling him that his association with the Emma was "ill-advised and unfortunate." Schenck understood that he was being told to choose either the Emma or his diplomatic post, and he chose diplomacy. Buyers paid only £2/share as down payment for the £20 Emma shares, with an agreed-upon schedule of future payments, however, and Park did not want Schenck to resign before the buyers had committed more money. After procrastinating for a week,

Schenck wired back that he would quit the Emma Company. The announcement was not made until January 12, 1872, however; in a public letter ghostwritten by Trenor Park, General Schenck emphasized that his resignation was for personal reasons and that he continued to have great faith in the Emma Silver Mining Company. By this time the entire public offering had been subscribed, subscribers had paid half the £20 price, and the next £5 payment was only three days away.[21]

Even after Schenck's withdrawal from the board, Trenor Park continued to give him inside stock tips and in one instance he gave Schenck £1,894, which Park said he had gained by stock speculations in Schenck's name. Schenck had given Park no investment money and no authority to speculate on his behalf; he had no idea which stocks had been invested in but cheerfully accepted the windfall.[22]

The promoters issued the prospectus on November 9, 1871. They even included part of Professor Blake's highly adverse report—the part describing past Emma production—but they carefully edited it to suggest a prediction of future richness.[23]

The Emma Silver Mining Company of Utah, Ltd., had a capitalized value of £1 million, approximately $5 million, divided into fifty thousand shares of £20 each. It was a hefty sum, but it was only the previous summer that an Englishman had calculated that the Emma was making more than $13,000 in profit every day, which would amount to almost $5 million per year—if the mine could continue the dizzying pace. The American owners were to sell half the shares and retain the other half, which they were unable to sell without the company's permission. The Emma mine meanwhile had increased production to a level that, if continued, would make the mine worth every bit of the £1 million and more.[24]

To induce share sales the company would pay its first dividend on December 1, 1871. The twenty-five thousand shares were oversubscribed two hours after the prospectus came out. Share prices immediately rose above the par value of £20 and sold for £25.[25]

Ugly Rumors I

Despite the monthly dividends and Silliman's glowing description, reports reached London that the Emma mine was exhausted, and share prices slumped. The promoters blamed the rumors on manipulators. The company sent one of its directors, the member of Parliament E. Bridges Willyams, to Utah. Accompanying Mr. Willyams was his friend, the novelist George Lawrence. They rode, along with Professor Silliman and their host Senator Stewart, in a private railcar with fine whiskey and fine company. Willyams and Lawrence arrived at the village of Alta City to the sight of only chimneys sticking out of the thick blanket of snow.[26]

Willyams discovered that a fellow Cornishman was in charge of the underground workings, and together the two sampled the mine, with Willyams's new friend guarding the samples until they were turned over to Silliman for assay. Silliman's assays revealed rich ore, and Willyams wired back to London instructions to buy more shares. He and Silliman sent a joint telegram back to London: "Very great improvements since last report." Share prices rose.[27]

Silliman's February 29, 1872, report confirmed that the renowned professor was an incompetent judge of mining property. Based on the rock samples from Willyams, he wrote that development since his October report had revealed at least two years of additional ore reserves, giving the mine enough ore in total to mine for the next three years without further exploration. The professor followed with another wide-eyed report on April 2, reporting that another eight thousand tons of ore had been discovered in the month of March alone. He later admitted that he had based his April report solely on information received from the mine manager.[28]

The Emma directors tried to gain a quotation on the London Stock Exchange, but the exchange was wary of companies of "irregular origin and antecedents" and imposed a rule against listing companies held more than one-third by their promoters. Since half the Emma shares were held by the vendors, the Stock Exchange refused to give them a quotation. The Emma Company removed the obstacle by allowing the vendors to cash in their remaining shares at the market price of £23. The new shareholders had no misgivings, for Professor Silliman had assured them that the latest discoveries provided "a guarantee for the future of regular and satisfactory dividends."[29]

Buying the Opinion of the *Mining Journal*

The high price of Emma Silver Mining shares, called Emmas, was not left to chance. The promoters wrote laudatory articles and arranged to have them published as editorials in the London *Mining Journal* on April 13 and 20, 1872, for a bribe of £500 and a call—a right to buy at the current price of £23—on 250 shares. The April 13 commentary called the deposit "inexhaustible" and predicted that the monthly dividends in May would double and that "consequent high dividends for 40 or 50 years to come are [were] already assured facts." The April 20 piece was more of the same nonsense and suggested £50 as an appropriate price for Emma shares.[30]

Emma shares rose, and the *Mining Journal* editors exercised their call to buy them at £23. Rather than wait for the price to rise to their predicted £50, however, they immediately sold at £30 per share, making a nice profit on their own publicity. The Stock Exchange still declined to list the Emma, a stand for which the *Mining Journal* exhausted its vocabulary, labeling it "an intolerable injustice," "oppression," and "not only invidious, but unfair, untenable, and callous."[31]

Willyams was back in London in May 1872, where to the repeated cheers of a special meeting of shareholders, he reported that the Emma was worth twice what they had paid for it. Trenor Park paid Willyams £5,000 plus a call on two thousand shares for his trip. The call on shares gave Willyams a quick profit of £14,000. His novelist friend Lawrence wrote *Silverland*, a travelogue with only high praise for the Emma mine, Senator Stewart, and Professor Silliman. Lawrence wrote that rich silver ore seemed to be everywhere, and mocked the rumors that it was "an exhausted 'shell.'"[32]

The company continued sending the rich ore all the way to Swansea, Wales, for smelting. The *Economist* criticized the wastefulness of the high shipping charges, but it was good publicity, the monthly dividends of one and half percent of the par value continued, and the stockholders were happy.[33]

Rubery and Harpending Butt In

As he so often did, Asbury Harpending involved himself in controversy. In his autobiography Harpending painted himself as an upright businessman who refused to handle the Emma mine, and exposed the exaggerated claims of the Emma promoters. Harpending claimed that he had warned Baron Grant that the Emma was not worth the overblown promoted price and then exposed the promotion to the public when Grant persisted in selling stock.[34]

The truth is that Harpending and his English friend Rubery—the two had become fast friends when they were arrested together in San Francisco—claimed that Grant owed them money connected with promoting the Mineral Hill mine in Nevada, and they threatened to publicize Grant's role in the Emma unless he paid them. Grant balked at the blackmail, so Rubery bought one Emma share and stood up at the March 1872 shareholders meeting and exposed Grant's part in forming the company.[35]

Trenor Park admitted that Grant had helped promote the Emma, but lied to the shareholders that Grant's fees had been less than £2,000, when in fact they had been £120,000. The shareholders believed Park. The Emma had paid regular dividends, and Albert Grant's role was past. The chairman told shareholders that annual dividends would probably mount to thirty or forty percent of par value rather than the promised eighteen percent.

Cursed by Snow

Mining in Little Cottonwood Canyon was always plagued by heavy snowfall. The broad Salt Lake Valley a few miles away could be bathed in warm sunshine even as winter storms piled snow up along the Little Cottonwood, immobilizing the transportation of men, ore, and supplies to and from Alta City, the settlement that had sprung up near the Emma. Miners were forced to strap on skis to get around. On June 4, 1872, the weight of spring snowmelt caved in part of the mine. The mine manager informed Park, who kept it quiet. The news was dismissed in London as another bear rumor, but the directors eventually had to confirm it, although they downplayed the damage. In fact, underground work was completely stopped.[36]

The mining law gave ownership of an ore body to the mining claim that covered the outcrop, or apex, and allowed the owners of the out-crop to follow the vein as it dipped under property owned by others. The law assumed that all ore bodies were neat and tabular veins, without branches. This *law of the apex* was a recipe for legal hell, as proved on the Comstock lode, and now at the Emma, where the ore body was an irregular mass. Following the cave-in, owners of a nearby mining claim now tried to steal the Emma ore body with an apex lawsuit.

The Cincinnati & Illinois Tunnel Company had bored into the Emma ore body in April 1872 and tried to mine Emma ore, but they were blocked by Emma miners. Bridges Willyams had assured the May shareholders meeting that the two companies were in harmony. Harmony was thrown aside when the June cave-in shut the Emma tunnel and gave the Cincinnati & Illinois miners the opportunity to steal Emma ore unmolested. Emma miners finally cleared out the tunnel cave-in and reentered the mine, but they found their works in possession of rival Cincinnati & Illinois miners, who repulsed the Emma miners with firearms.[37]

The Emma was forced to sue in court to regain possession and stop the Cincinnati & Illinois from stealing more ore. The trial brought rival platoons of eminent mining engineers, each testifying to the absolute truth as represented by his employer. Senator William Stewart was back in harness representing the Emma Company in the courtroom in Salt Lake City, but legal decorum wilted when opposing counsel referred to one another in court as "scoundrel," "thief," and "swindler." The court found for the Emma but assessed only nominal damages of $5,000 against the Cincinnati & Illinois Tunnel Company. Having demonstrated how easy it was to remove Emma ore through their tunnel, the owners of the Cincinnati & Illinois then sold their tunnel to the Emma Company.[38]

13

These are the Dupes who insanely tore, to subscribe the Sum the Directors swore, was worth a Million of Pounds and more, on the return of the Swell who was sent to explore, the Mine by a General of the Army Corps, who was backed by the Britishers, one, two, three, four, who shared with the Yanks from the Eastern shore, who joined with the Men who had cleaned out the Ore that lay in the Mine that Lyon struck.

14

This is the way the Water did pour, into the Mine which *somebody* swore, was worth a Million Pounds or more, on the return of the Swell who was sent to explore, the Mine by a General of the Army Corps, who was backed by the Britishers, one, two, three, four, who went in with the Yanks from the Eastern shore, who joined the Men who had cleaned out the Ore, that lay in the Mine that Lyon found.

15

These are the Faces the Subscribers wore, after the Water began to pour, into the Mine that *somebody* swore, was worth a Million Pounds or more, on the report of the Swell who was sent to explore, the Mine by a General of the Army Corps, who was backed by the Britishers, one, two, three, four, who winked at the Yanks from the Eastern shore, who joined with the Men who had cleaned out the Ore, that lay in the Mine that Lyon struck.

16

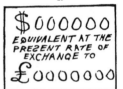

This is the value that Dividends bore, such a bust the Subscribers much deplore, but what could they do when the Water did pour into the Mine that *somebody* swore, was worth a Million Pounds and more, on return of the Swell who was sent to explore the Mine by a General of the Army Corps, who was backed by Britishers, one, two, three, four, who were squared by the Yanks from the Eastern shore, who joined the Men who had cleaned out the Ore, that lay in the Mine that Lyon found.

End of the " *Mine that Lyon found.*"

Titled A New and Instructive Nursery Ballad, *an anonymous pamphlet published in Salt Lake City in 1872 spoofed the Emma mine swindle. Courtesy Bancroft Library, University of California, Berkeley (BANC MSS P-F 317:67(4)).*

Ugly Rumors II

Trenor Park wrote to London that a new vein had been discovered, over twenty feet thick and growing in richness with depth. Emma prices rose on the good news, but new rumors circulated that the mine was exhausted. James Lyon decided that Stewart and Park had swindled him, and determined to get his revenge, and his money, by shorting the stock and telling all who would hear that the Emma was an exhausted mine not worth £5 a share. Others in London echoed Lyon's warnings, but they appeared to be malicious manipulators.[39]

Once more, the conflicting statements prompted the Emma directorate to send one of themselves, this time the company's chairman, George Anderson, whose qualification to judge mining property seems to have been his election to Parliament. In August 1872 Anderson left for Utah, where he was as highly taken with the mine as his predecessors had been. The rumors were attributed to jealous Americans trying to force prices down so they could buy control and to London bear brokers caught short on the stock.[40]

False Dividends

Trenor Park traveled to Salt Lake City and learned that there was no money to pay the usual monthly dividend for November 1872. The grade of ore had fallen off precipitously and could not pay the cost of shipping to a smelter. Since selling his Emma shares in April, Park had repurchased thousands of shares, which had reached £26.50. To support the high price while he sold his shares, Park lent £36,000 to the company and ordered monthly dividends continued, even though paying dividends not taken from profits was against company bylaws.[41]

Another false dividend, the thirteenth consecutive monthly payment, was made on December 2. On December 10, 1872, Chairman George Anderson reported to shareholders that he had seen for himself that the Emma was a great property and that its promoters Park and Stewart were fine and honest gentlemen.[42]

Park telegraphed General Schenck in Paris in December 1872, advising him of the sad state of the mine and urging Schenck to buy a short contract on Emma for their joint account, so they could profit by the coming drop in share prices. "Poker Bob" Schenck was an avid and expert gambler and had quickly adapted to stock speculation. He wired his embassy from Paris to buy a short contract to deliver two thousand Emma shares. The inside dealing would have made Schenck a large profit, but the telegram went to the second secretary of the American legation in London, who refused to execute the order. Instead of a large profit, Schenck was caught holding five hundred shares when Emma prices plunged, and lost money. Trenor Park was nimbler and was left holding only twenty-five shares, the minimum required for a company director.[43]

When the company first organized, Trenor Park had complied with the legal requirement that he sign an affidavit testifying that everything in the prospectus was truthful. Now that he had quit his directorship and was about to quit England, he asked the company to return or destroy his affidavit, but the company refused.[44]

The company announced shortly afterward that there would be no monthly December dividend, as dividends would be paid quarterly. The quarterly dividend to be paid in February 1873, was later also canceled. The false dividend of November 1872 turned out to be the last. The Emma still had ore, but it was low grade and couldn't be mined at a profit without an ore-concentration mill nearby. Share prices plummeted. The London directors had to admit in their 1873 annual report that not only was there no money to resume dividends but the company was in debt to Trenor Park. Professor Silliman had badly overestimated the value of the ore. The Emma dividends had in fact been paid on the basis of ore that had already been mined and piled on the surface at the time the company was floated in London: out of the £193,532 in dividends, ore actually mined during the past year had amounted to less than £30,000 . The shareholders had paid £1 million for a worked-out mine.[45]

The Emma company hired a new mine manager, George Attwood, who examined the mine and confirmed that it was exhausted. Not trusting anyone, the directors had three more experts report on the mine. All three concluded that the great ore body was worked out and that any future profit would depend on finding new ore.

Attwood found some high-grade ore, but the proceeds were absorbed in repairing the mine. He paid a county tax bill long overdue. He had to re-timber much of the mine workings because of sloppy work by his predecessor. He built an ore-concentration plant for the lower-grade ore left in the mine and extended the mine downward looking for new ore bodies but found only barren limestone.[46]

In 1873, Salt Lake City newspapers printed lies of fabulous new discoveries in the Emma that would soon return it to the glory days. The reports were contradicted by the new manager, George Attwood, who wrote to London: "The future prospects of the mine are most gloomy." Attwood was astonished at how his predecessor had gutted the best ore without even providing proper timbering to prevent the mine from collapsing. It had clearly been worked not for long-term production but for a quick profit to boost the stock. He used unusually strong terms for an engineering report when he wrote to the directors "the former management cannot be too highly censured." He also told them that ex-director Trenor Park had acquired the Illinois Tunnel Company and was renewing the old apex litigation against the Emma. Confused shareholders did not know whom to believe, because those trying to boost share prices accused Attwood of bearing the stock.[47]

The Emma company settled the matter by hiring the geologist Clarence King to examine the mine. King was chosen because his recent exposure of the great diamond swindle had given him an international reputation. He reported back that the Emma was "with insignificant exceptions worked out."[48]

Lawsuits Start to Snowball

An unhappy Emma shareholder mailed a pamphlet to other equally unhappy shareholders, in which he accused Trenor Park and the mine manager Silas Williams of having cemented rich silver ore to the walls of the Emma mine before Professor Silliman and the two members of Parliament inspected it. The accusation was disbelieved by mining people, who knew that such fakery would be obvious, even to an academic such as Silliman. When the *Nation* published the accusation in the United States, Silas Williams sued for libel.[49]

The Emma management refused to repay Trenor Park the money he had advanced to pay the false dividend. The new company chairman went to Utah to see the mine for himself. On his way back through New York, he hired an American law firm to file suit for $5 million against the principal promoters, Park, Stewart, and Baxter, charging fraud and conspiracy. Some stockholders suspected that litigation was only another plot to drain the treasury and petitioned unsuccessfully for a compulsory windup of the company.[50]

Minister Schenck's Undignified Exit

Shareholders in Glasgow, Scotland, ever-suspicious of the "spivs" of the London financial district, elected one of their own as the Emma company's chairman. Chairman MacDougall was eager to press charges of criminal fraud against many involved in the Emma sale—including the American minister. However, the Grant administration did not want the spectacle of an American minister pleading diplomatic immunity in order to escape criminal prosecution. The Democratic majority in Congress was eager to investigate Schenck's role in the Emma and started hearing testimony in February 1876. Secretary of State Fish at first told Schenck that President Grant did not want him to resign but then wired Schenck that a resignation would end the embarrassment.[51]

Schenck mailed his resignation and then telegraphed a request for a leave of absence to attend the hearings in Washington, asking that his resignation not be acted upon in the meantime. Schenck, the expert gambler, knew when to fold his hand. He had planned to leave on March 11, 1876, but instead left London on March 3, without fulfilling the diplomatic protocol of taking leave of the queen in person. The reason for his hurry was revealed when solicitors for the Emma Silver Mining Company tried to serve a writ on Schenck as he passed through Euston Station but were prevented by his diplomatic immunity. Schenck lost diplomatic immunity three days later when President Grant accepted his resignation, but by then Schenck was at sea and safely out of British territory.[52]

Schenck's welcome in Washington was less than he had wished. He was an embarrassment to the Grant administration; his former Democratic colleagues in Congress were all too eager to talk to him—once they had him under oath. Ill, Schenck took to bed, and the hearings went on without him. When he finally appeared, he was untruthful and evasive. He stubbornly played dumb when he insisted that his sweetheart contract with Park was not a payoff for serving as director. When the committee confronted him with the evidence, he pleaded lapse of memory.[53]

Professor Silliman refused to retract or apologize for his woeful overestimates of the quality and quantity of ore. He told the committee that he was not responsible because he had not estimated that the mine was worth $5 million—or any number whatever, and so if the share buyers relied on the vague superlatives in his report, it was their own folly.[54]

The committee decided that Schenck was not guilty of fraud but unanimously criticized him for becoming a director of the Emma company. He retired from public life and wrote a classic book on the art of playing draw poker.[55]

An Avalanche of Lawsuits

The Emma company went after former directors and agents it alleged had taken funds improperly from the promoters. In 1876 the company received £3,500 from its claims and continued to press a claim against the former company chairman and member of Parliament George Anderson, whom it accused of taking £1,000 from the Emma promoters. For a company capitalized at £1 million, however, this was small change. The company spent £4,100 during the same period and was left with a scant £1,840 in the treasury.[56]

The lawsuit against Park, Stewart, and Baxter came to trial in New York on November 11, 1876. The defendants portrayed themselves as innocents who were as surprised as the shareholders when the Emma ran out of ore. They called the shareholders sore losers trying to welsh on a bad investment. Park swore that he had nothing to do with the contents of the prospectus and had not even seen it until it was in print—a lie that exploded when the plaintiffs produced a printer's proof sheet of the prospectus with edits in Park's handwriting. The Emma Company tried to substantiate the charge that Park had caused his employees to cement the walls of the mine with silver ore to fool Professor Silliman, but he had no evidence to support the accusation.[57]

Park, Stewart, and Baxter had left out of the prospectus any mention of the £100,000 fee paid to Baron Albert Grant. This was fraud in Great Britain (as Grant was learning to his sorrow in the Lisbon Tramways fiasco), but not in the United States, where the trial was being held. The plaintiffs pointed out untruths large and small in the prospectus, from suppressing the unfavorable Blake report to misrepresenting the distance to the nearest railroad.[58]

The trial dragged on for months, and the jurors complained of the hardship of being away from their occupations. The defense lawyer tried the unique tactic of standing in open court to offer $2,500 to the jurors as compensation for the long trial—strictly as a humanitarian gesture. The prosecution lawyer could not afford to be outdone and immediately offered the same amount. The judge had other ideas, however, and disallowed private payments to the jurors.[59]

Testimony finally ended on April 18, but summing up kept the case from the jury until April 27, 1877. The jury began with seven for acquittal, and by the end of the day there was only one juror holding out for a guilty verdict. The jurors persuaded their reluctant member at 1 p.m. the following day and returned to the courtroom to announce their decision for the defendants.[60]

The Emma company's chairman, MacDougall, commented bitterly: "For the moment, Park and his confederates have got off with both the mine and the million, and a jury of his country, after a trial of unprecedented duration, have shown a patriotic appreciation of his conduct by deciding that he was quite right in doing so."[61]

The Emma Company appealed, but in March 1878 the trial judge denied a motion for a new trial. The bad end of the Emma affair convinced many London investors that American mining men were swindlers and that the American legal system would always rule against Britishers trying to gain redress.[62]

After the Emma Silver Mining Company Ltd. refused to repay the money provided by Trenor Park to pay the false dividend, Park obtained a court judgment against the mine. The court foreclosure halted mining in December 1874, and Park bought the mine in the sheriff's sale in 1876.

Since he was also the principal creditor, almost all of his purchase price was paid to himself.[63]

Park sold whatever could be carted off: surface machinery, mining supplies, and office furniture. He then leased the mine to tributers, who picked the bones to find small amounts of overlooked ore within easy reach. Timbers rotted and allowed the mine to cave in from the surface. Water flooded the lower workings. The usual rumors of new bonanzas contrasted with the small silver production.[64]

James Lyon was still convinced that he had been swindled out of the Emma mine profits, and he filed another suit against Trenor Park and Henry Baxter for $350,000 in June 1878, but the court did not give him satisfaction.[65]

When Chairman MacDougall threatened legal action against Baron Albert Grant, the baron accused him of being a litigious speculator. Grant said that it would be better for the company to spend its funds on further exploration of the Emma property; he offered to lend £10,000 for this purpose—if the company would drop legal action. The directors knew that they had a powerful case against Grant and refused the offer. Grant sent a letter to all shareholders asking them to make him chairman of the Emma Silver Mining Company, but few trusted him.[66]

The law moved slowly, but on June 29, 1880, a London court found that Grant had committed a breach of trust in making an illegal profit from the company. Grant was assessed £120,000—money he no longer had, since he had entered bankruptcy. Grant's Lisbon Tramways and other schemes had also fallen apart amid charges of fraud.[67]

The court appointed a trustee to sell Grant's £1 million home in London's West End, and his Norfolk country estate. Before any of the proceeds were distributed to creditors, however, the untrustworthy trustee absconded with the money. Baron Grant remained impoverished the rest of his life.

The Emma company brought another American lawsuit, this one in equity, against Park, Stewart, and Baxter, in October 1879. The case inched its way inconclusively along legal paths over the next several years.[68]

Trenor Park had possession, but the mine itself was a worthless hole in the ground. There was a chance of finding new silver bonanzas on the property, but that would require capital, and the property was burdened by liens, legal threats, and the bad reputations of both himself and the mine. At the end of 1880 Trenor Park proposed that the Emma Silver Mining Company drop all remaining legal action against himself, his partners, and Albert Grant; in return, he would give them their mine back. Company directors delayed and negotiated but had little choice but to get back in bed with Park. The company reorganized in 1882 as the New Emma Mining Company, with 70,000 shares, of which 50,000 went to shareholders of the old Emma company, 11,000 to Park, and the rest to settle debts and raise new capital.[69]

The mine was in bad disrepair. A snowslide had destroyed the surface equipment, and much of the mine had caved in from neglect. It took several years to repair the damage, but the managers, thoughtfully and frugally this time, began orderly exploration for new bodies of silver ore but found none large enough to return the company to profitability.[70]

The new company entered into the old legal tangles with its neighbors over who owned what. The court issued an injunction against Emma miners working areas claimed by others, but the suit was finally settled and work resumed. The mine also suffered from the excessive snowfall of Cottonwood Canyon, when in March 1884 a snowslide destroyed the mine buildings and killed thirteen people.[71]

James Lyon was still convinced that he had been swindled out of $650,000 by Stewart and Park. Although he had fallen out with Robert Chisholm back in 1870, he and Chisholm joined to sue Stewart and Park for cheating them. Chisholm settled separately with Stewart and Park, so Lyon sued him as well. In 1891 Lyon sued William Stewart for $1 million, which he said he could have made out of the Emma mine sale if Stewart had not betrayed him. Stewart called Lyon a blackmailer and said that he had paid Lyon $50,000 years before for a signed release.[72]

1916: Swindling Returns to the Emma Mine

Most mining engineers concluded that the Emma ore body was exhausted, but J. H. Morton studied the mine and theorized that the ore body had been cut by a fault and that its continuation was still hidden under the mountain. The New Emma Silver Mining Company explored for the lost lead but never found it. Work at the Emma was finally suspended in 1894.[73]

The Emma mine passed through various hands. In 1916 the infamous mine promoter George Graham Rice took control of the Emma and began another round of swindling. But that is another story.[74]

The fortunes of the town of Alta, which had been built to house miners from the Emma and other mines, sagged until in the 1930s someone built a ski tow using old mine rails, and the ghost town grew into a ski resort. The heavy snowfalls, which had always been a plague on mining in Little Cottonwood Canyon, have become its glory. Alta, Utah, is as internationally known as a ski mecca as the Emma mine—now forgotten—was infamous.

A Brief History of Diamond-Mining Frauds in America

T he site of the great diamond swindle is little changed after more than a century. Remote in the northwestern corner of Colorado, Diamond Butte is still an uninhabited flat a few miles across, dotted with sagebrush, clumps of grass, and low outcrops of crumbling sandstone. A wire fence to the north marks the Wyoming border. A dirt road crosses the site: just an unmaintained pair of ruts, though any automobile can follow them across the flat plain. A diamond prospector today would look at these unpromising rocks and keep driving.

In 1872, however, this unremarkable cattle range was the most searched-for place in the United States. Its dull soil was thought to sparkle with an incredible fortune in diamonds. Hundreds of prospectors chased rumors, as newspaper headlines placed it variously in Arizona, Colorado, Kansas, New Mexico, Utah, and Wyoming.

From Diamond Peak you can look down on the butte, still undisturbed by human hubbub. The planned metropolis of Diamond City never saw a human habitation. Many miles from the nearest house, the site is unmarked, seldom visited, and all but forgotten.

School for Swindle: Ralston, New Mexico

Philip Arnold and John Slack were cousins. After they returned home to Kentucky from the Mexican War, Slack served a stint as county sheriff, then they joined the gold rush to California. They worked a placer gold mine in California in the 1860s, before Arnold took his share of the gold back to Kentucky to buy a farm. Arnold found farming less to his liking than he had remembered, and returned west. Slack followed the gold discoveries to Arizona, where he served in the territorial legislature. By 1870 Arnold, age forty, and Slack, age fifty, were working together again, still itinerant miners and prospectors.[1]

San Francisco mine promoters Asbury Harpending and George D. Roberts employed Arnold and Slack at the mines at Ralston, New Mexico, in 1870. At Ralston, Arnold and Slack met James B. Cooper. Cooper had prospected in Nevada, and dealt in real estate in Denver. As with Arnold and Slack, fortune had eluded Cooper.[2]

When Roberts and Harpending realized that the Ralston mines were uneconomic, they tried to sell them in London, with a prospectus that Harpending himself said "put the tales of Baron Munchausen in the shade." But the promotion crashed when Londoners learned the truth. Roberts and Harpending were masters at mine promotion, and their outrageous puffing of the Ralston deposits cannot have escaped Arnold, Slack, and Cooper.[3]

The diamond swindle grew out of disappointment. An Indian at Fort Defiance, Arizona, showed Arnold some garnets. Arnold, Slack, and Cooper prospected the area until they had high hopes and a sack full of stones that they thought were rubies, diamonds, and sapphires. In March 1870 Phil Arnold showed a few stones to the editor of the Arizona *Citizen* in Tucson and told the editor that a Boston jeweler had identified the stones as diamonds. But when Arnold, Slack, and Cooper brought the stones to San Francisco, jewelers dashed their dreams—the stones were worthless.[4]

Where to Hunt for Diamonds

Diamonds form deep in the earth. They are carried in molten magma that pierces upward toward the surface then hardens into the dense dark rock called kimberlite. Diamond prospectors search for the elusive surface exposures of kimberlite "pipes," narrow chimneys of rock that stretch far into the earth.

In 1871, however, what was known of diamond deposits was very little, and that little was mostly wrong. South African diamonds, discovered in 1867, were still being washed from stream gravel and shallow diggings. The only other diamond mines, in Brazil and in Golconda, India, washed diamonds from river gravel and sandstone. In America a few loose diamonds had been found by happenstance, and placer gold miners in California had discovered a handful of diamonds in their sluice boxes.

Diamonds seemed to be associated with sand and gravel, so the prevailing theory was that they formed right there, precipitated from water seeping through the riverbed. It was natural that Arnold and Slack would locate their bogus deposit where the geologists of their day would find most credible—in a sandstone.[5]

Asbury Harpending's role in the great diamond swindle is still controversial. He lived high and low in mining finance, but fortune eluded him in the end. From Asbury Harpending, The Great Diamond Hoax and Other Stirring Incidents in the Life of Asbury Harpending. Courtesy Denver Public Library Western History Collection.

Baiting the Trap

The new South African diamond fields were causing a great stir, and there seemed no reason that America was not similarly favored. Arnold, Slack, and Cooper realized that just a few genuine diamonds could coax a great deal of money from San Francisco capitalists. Arnold, Slack, and Cooper held unblemished reputations—better reputations, in fact, than some of their victims.

James Cooper took a job with a San Francisco manufacturer of diamond-tipped drill bits and gave Arnold and Slack some inexpensive industrial-grade diamonds taken from his employer. Arnold and Slack mixed the diamonds with the Arizona stones and showed the lot to George D. Roberts but refused to divulge the location. After this, Cooper apparently dropped out of the plot. In December 1870 Roberts wired his partner Harpending in London: "ARNOLD HAS RETURNED HE MADE A GREAT DISCOVERY OF DIAMONDS NEAR BURRO ARE KEEPING IT QUIET." Harpending left London in May 1871 and in June he was in San Francisco, meeting with Arnold and Slack.[6]

Harpending and Roberts showed the stones to the mining financiers General George Dodge and William Lent. Harpending and Roberts told the new investors that Arnold and Slack demanded an initial $50,000, which Dodge paid. Arnold and Slack later received another $50,000, but their San Francisco backers wanted to know the location of the diamond deposit before they would part with any more cash.[7]

Arnold and Slack realized that they needed more and better-quality diamonds to really cash in. They left for a London shopping spree with the $50,000 they had received from General Dodge.

In July 1871 an American entered the office of the London diamond dealer Leopold Keller and asked to see uncut diamonds and rubies. When Keller hesitated, for the American was by speech, manners, and appearance a laboring man unlikely to have the money, the American, Mr. "Burcham" (later identified from a photograph as John Burchem Slack), showed a letter of credit from a London bank. Doubts overcome, the dealer showed the gems. Knowledgeable buyers would always carefully inspect and weigh packets of gems. Slack did neither, buying packets of inferior uncut diamonds and rubies.[8]

Slack returned in five days with Mr. "Aundle" (Arnold) and bought more gems. The two bought still more a week later, and Slack returned to buy one last lot before the two left London. Slack and Arnold turned up in San Francisco in September 1871, where together with Harpending they showed the gems to General Dodge. This time there could be no doubt about the richness of the deposit, although they still refused to divulge the location. The conspirators traveled to New York, where Arnold signed an agreement to sell his diamond bonanza.[9]

The Third Man

Soon after the agreement, Harpending traveled to London. Arnold, alias Aundle, checked into a London hotel with a man giving his name as "Lock." Aundle and Lock showed up in Paris, where they bought uncut gems on three occasions before returning to London. Harpending and Arnold were seen in London together, walking arm-in-arm.

In the spring of 1872 a third American entered Keller's London dealership to buy diamonds and said that he had been referred by "Burcham" and "Aundle." In contrast to those two, the new buyer was a "gentleman." He bought two thousand carats of rough diamonds and nine hundred carats of rubies. The dealers were surprised by the eccentric purchases but thought no more of it until the news of diamond discoveries in America.[10]

The American gentleman who bought rough diamonds in spring 1872 was never identified. The *Times* later suspected Asbury Harpending but could not prove it. Back in America, diamond fever heated up. General Dodge wired Harpending in London in May 1872: "Important you come immediately."[11]

Fooling Tiffany

The San Francisco diamond men decided that the enterprise was too gigantic for them, so they went to New York to bring in bigger capitalists. They offered to sell a one-quarter interest to the influential New York lawyer Samuel Barlow, if he would become president. Barlow met the San Franciscans at his home and invited General George McClellan, the publisher Horace Greeley, Congressman Benjamin Butler, the jeweler Charles Tiffany, and a banker named Duncan. These were exactly the sort of powerful allies the diamond men needed.

The San Franciscans rolled the stones out and watched the sparkle in the eyes of the newcomers. Tiffany examined the diamonds, rubies, sapphires, and emeralds, and confirmed that they were genuine but said that he would have to let his employees appraise their value.

The diamond men met a few days later at the home of Charles Tiffany, where the nation's foremost jeweler told them that his employees had valued the sack of stones at $150,000. The diamond men had brought to New York only about a tenth of the stones, making all the gems in their possession worth a whopping $1.5 million. Since Slack and Arnold had only paid $35,000 or so, Tiffany woefully overestimated the value. The overestimate squashed any doubts about the genuineness of the deposits, for Arnold and Slack could never have bought such a fortune in stones. It later turned out that for all their expertise in cut diamonds, Tiffany and his employees had never before examined uncut gems.

Barlow and General McClellan agreed to buy into the company and serve as trustees, if they could first send their own man to inspect the mystery diamond deposit. They picked the mining engineer Henry Janin. The San Franciscans agreed, for a favorable report from Janin would convince investors, as would General McClellan's involvement.

The diamond men had already paid Arnold and Slack $100,000. Now Arnold demanded and got another $100,000 immediately; he would be paid another $150,000 and a significant share of the new diamond company only after he showed them the diamond deposits.

Changing the Law

A great problem was that no one could gain the right to mine diamonds on public lands, because the law allowed only gold, silver, mercury, and copper deposits to be claimed. Diamonds had simply not been contemplated when Congress passed the mining law in 1866.[12]

Fortunately for the diamond syndicate, a rewrite of the mining law was being considered off and on by Congress. A mining bill had passed the Senate in the previous session but died when the House recessed. Like the 1866 law the proposed mining bill restricted the minerals that could be claimed, adding only ores of lead and tin to the list.[13]

The diamond syndicate, however, included men accustomed to working the levers of power. The mining bill had languished for over a year but was now introduced in the House. The diamond men gave a thousand shares of stock to Congressman Benjamin Butler, who saw to it that the law was amended to their satisfaction and passed. The bill was essentially the same as that in the last session, but the key phrase "or other valuable deposits" had been added to the list of locatable minerals. The bill passed in May 1872 and was signed by President Grant. After a year of nervous secrecy, the syndicate could now stake their claims on the diamond fields.[14]

The mining law still retained the democratic forty-niner principle that restricted the acreage that an individual or company could claim. A company could claim no more than 160 acres, but the diamond magnates were clever enough to get around the law by using a holding company. The New York Mining and Commercial Company would reveal the ocation of the diamond fields to the companies doing the actual mining, each mining company being able to claim its own 160 acres. In return the mining company would give half its stock and a portion of its profits to the holding company. The holding-company plan required continued secrecy. The diamond men followed the common practice of the time and created a code for use in telegrams. Diamonds, rubies, and sapphires were called "sugar," "salt," and "pepper." To disguise the location, Wyoming became "Reno," and Cheyenne and Denver became "Lancaster" and "Philadelphia."[15]

The First Expedition

The New York investor Samuel Barlow chose Henry Janin to study the diamond fields, because of Janin's reputation and integrity. Janin was honest but too often assumed that others were as honest as he. In 1871 he had inspected a California gold mine for English capitalists but neglected to take his own samples or make his own assays. Instead, he relied on the mine superintendent. Based on Janin's high recommendation, the property was floated in London as the London & California Mining Company. Janin's carelessness was discovered only after the ore proved much poorer than advertised, and the company lost money.[16]

By this time Slack had sold out his entire interest and was working for his cousin Phil. They had prepared a patch of remote wilderness in northwestern Colorado, where the soil sparkled with London diamonds, and they were now ready to show it.

Among the diamond expeditioners were Asbury Harpending and Alfred Rubery, an Englishman who had been Harpending's friend since they had tried to outfit the sailing ship *Chapman* as a Confederate privateer during the Civil War. They were arrested and imprisoned before they could sail out of San Francisco Bay. Rubery was eventually pardoned, and returned to England. Marital problems prompted Rubery to seek a change of scenery, so he accepted Harpending's invitation to America in 1872.[17]

The diamond men traveled from New York to Saint Louis, where they waited two weeks for Arnold and Slack. Then all went by railroad to Rawlins, Wyoming, where they mounted horses and rode south. Arnold led them by a circuitous route that took three days, but finally in the late afternoon of June 16, Arnold brought them to an outcrop of sandstone about one acre in area and invited them to look for diamonds. Rubery quickly found a diamond, and then others began to find them, prompted by Arnold's suggestions about where to look. Ten men searching for just an hour found 285 diamonds before dark.[18]

Janin ordered batches of the sand and gravel to be hauled to the nearest creek and washed with gold pans. He accepted Arnold and Slack's offer to do the physical labor, and they proved their skill by finding a steady succession of diamonds and rubies in their gold pans. Arnold and Slack worked the gravel for four days and hauled and washed about a ton and a half of dirt. The result was more than 1,600 carats of diamonds (seven-tenths of a pound) and four pounds of rubies. Janin estimated the recovered gems to be worth more than $8,000.[19]

The diamond men spent the next six days setting stakes in the ground to mark claim boundaries. There were eighteen subsidiary companies in the works, and each one of them was to have its own 160-acre claim.

After nearly a week on the diamond butte, Janin wanted to stay for more tests, but Harpending insisted that they were running low on provisions, and he wanted to rejoin his wife in San Francisco. The expedition returned to civilization, convinced that they owned the greatest mineral deposit in the world. Arnold and Slack knew that they had successfully hoodwinked some of the most astute mining men in the nation.[20]

Slack and Rubery stayed behind. Although he camped on a supposed huge diamond deposit, Rubery did not bother to gather any diamonds but relaxed and enjoyed the scenery. Some said that Slack and Rubery had stayed to guard the diamond fields, but they left only a few days later and returned by a longer route. The longer route was meant to confuse Rubery, because some of the diamond men did not trust him.

Janin's Report

Henry Janin's report was an eager and unqualified endorsement. Janin calculated that more than $5,000 of gems could be washed out of each ton of gravel, and he extrapolated many millions of dollars of profit. He recommended that his employer, Barlow, buy shares based on a total value of $4 million, as a "safe and attractive" investment.[21]

Arnold collected the $150,000 promised him, then he sold his share of the diamond company to Lent and Harpending for $400,000 and went back to Kentucky.

After staking their claims, the usual procedure was to secure title by recording the location with the mining district recorder, then start mining. The diamond company, however, kept the secret to themselves, because the San Francisco and New York Mining and Commercial Company still needed to find seventeen other companies, each one of which would pay it $5 million for the secret of the location, putting $85 million in the diamond company treasury.

The Diamond Rush Is On

When the diamond men incorporated the San Francisco and New York Mining and Commercial Company on July 20, the story leaked out and the San Francisco *Chronicle* hoisted the story into headlines the next day. The principals admitted that they knew the secret location of the world's greatest deposit of precious gems. They put Arnold's diamonds and rubies in a glass case in the Bank of California for all San Francisco to gawk. The company offered fifteen thousand shares for sale on August 1 and took in $600,000 within twenty-four hours. Another five thousand shares offered on August 2 collected an additional $250,000. The diamond men had gained $850,000 in cash but still owned eighty percent of the company.[22]

The diamond men gave interviews about the diamond field but kept the location secret—leaving that to rumors and the imaginations of newspapermen. The diamonds were variously said to be in Arizona, Colorado, New Mexico, Utah, Wyoming, and even Kansas. Many diamond seekers went to Ralston, New Mexico, where Arnold and Slack had recently prospected. But by accident or design, most rumors placed the diamond fields in the Navajo country of northeastern Arizona and northwestern New Mexico. Within a few weeks, hundreds of prospectors were searching for the diamond fields—hundreds of miles from the real location.[23]

At least six diamond expeditions wandered the vast desert Four Corners region. Some reported that they had found claim monuments with notices bearing the names of Lent and his associates 150 miles from Fort Defiance, Arizona. Most of them found what they thought were rubies and diamonds but were only garnets and quartz.

Word of the American diamonds panicked the South African diamond diggings in October 1872. The diamond buyers closed shop for fear that the American diamonds would glut the market. Calm returned, especially when London newspapers that pronounced the whole thing a fraud arrived and reassured the diamond buyers.[24]

The Second Expedition

The diamond men sent a second expedition to the diamond fields less than two weeks after the public announcement. The ten expeditioners slipped out of town on various pretexts, and before anyone realized it, they had all disappeared to the mystery mines.[25]

This time, neither Arnold nor Slack went along. The only member of the second expedition who had visited the diamond fields before was Harpending's English friend Rubery, who had traveled to and from the field only by confusing roundabout routes. The party got off the train at

Green River, Wyoming. Despite the secrecy they were shadowed by diamond hunters. To evade pursuers, the group split in two. The followers chose one group to shadow, but only the group not followed continued to the diamond butte. Because of the pursuers and Rubery's unfamiliarity with the country, it took them thirty days to arrive at the diamond fields.[26]

The expeditioners searched for diamonds immediately, and after an hour had gathered 285 of the gems. Completely satisfied, they spent eight days surveying mining claims, water rights, and timber claims, then returned to Green River in October and boarded the train to San Francisco. They told a reporter in Elko, Nevada, that they found no diamonds and that the whole affair was a fraud; to another reporter in Sacramento, they said that they had run into hostile Indians and turned back without visiting the diamond fields. A San Francisco *Call* journalist, however, reported that they had reached the diamond fields and confirmed the rich deposits.[27]

Tall Tales

In the excitement, purveyors of tall tales saw their yarns published in newspapers. A New Yorker named Morehouse related how in a trackless area of northeastern Arizona, he found a maze a mile across of giant anthills that sparkled with precious stones. The Denver *Rocky Mountain News* credited Morehouse with "a lively imagination" but found space for his yarn on page one.[28]

Legends were quickly remembered or fabricated about the pioneers Kit Carson and Jim Bridger having discovered precious stones in the Four Corners area. The diamond excitement even reached the East Coast, where newspapers printed a story about a man in New York State who supposedly died from a peculiar arthritis caused by diamonds spontaneously crystallizing in his joints.[29]

Other Swindlers Join the Parade

The diamond company spawned imitators, many claiming to know the secret location. The previous year, a blowhard named Thomas Miner had said that he had discovered huge gold deposits in Arizona. He was so compelling that the territorial governor and 260 other hopefuls accompanied his wanderings for weeks until they realized that Miner had never even seen that part of the territory. Miner was in San Francisco when the diamond news broke, and he told all who would listen that he was the original discoverer of the diamond fields, which Lent, Harpending, Roberts, and company were trying to steal from him. Arizonans, however, knew Miner from bitter experience. The Prescott *Miner* called him a "Bilk," and the Arizona *Citizen* hinted that a lynching would be appropriate.[30]

Miner was small and unassuming, and his sincerity convinced some San Francisco businessmen to back him. Only five days after the news of the diamond discovery, they incorporated the Original Diamond Discovery and Mining Company, and Miner began planning an expedition. While he was still organizing, however, company officers sent a group of prospectors without his knowledge to his location in Pinal County, Arizona. When he learned of the trick, Miner severed connections with the Original Diamond Discovery and Mining Company.[31]

Another diamond hunter told the newspapers that he had seen the diamond fields in northeastern Arizona, hundreds of miles from Miner's location. He incorporated the Arizona Diamond Mining Company on August 5. In less than a month at least eight other diamond companies formed, and all sold stock. Many claimed to have inside knowledge of the location of the discovery and called the others frauds, and soon many searchers were reporting their own phony diamond discoveries.[32]

The Truth Emerges—And Is Denounced

The London diamond merchant Keller realized that his mysterious customers must be the source of the American diamonds and told the newspapers. The London *Times* and the London *Telegraph* hammered the affair as a fake. The *Times* wondered how American financiers could believe such known sharpers as Harpending and Roberts. However, Keller's exposé was denounced in America as an attempt to keep the diamond trade in the hands of London and Amsterdam dealers. Even the usually careful *Mining & Scientific Press* was so consumed by diamond mania that it dismissed Keller's revelations as the desperation of threatened monopolists.[33]

The London diamond merchants Pittar, Leverson & Co. obtained some of the Arizona gems, and their examination confirmed the fraud. Not only did the diamonds resemble Brazilian diamonds, but the rubies and sapphires looked like those from Burma and Ceylon. In addition, diamonds, rubies, and sapphires had never been found together in nature.[34]

San Francisco papers promoted the excitement uncritically. However, newspapers in Arizona, Colorado, and New Mexico—closer to and more familiar with the supposed diamond fields—saw swindlers at work and urged caution. The *Optic* in Las Vegas, New Mexico, said that the excitement "originated in fraud, is prolonged by fraudulent means and for the sole purpose of committing a fraud by some heartless speculators." Even in San Francisco, the fever cooled enough for the San Francisco *Bulletin* to declare that selling stock in diamond companies was "a put up job."[35]

In Prescott, Arizona, where Slack was well known, most thought the thing was a fake, because they knew Slack to be too canny to sell a genuine diamond bonanza for such a low price. The Prescott, Arizona, *Miner* was skeptical of a boom being pushed by "professional stock gamblers," and recognized as obvious lies the reports by various diamond-hunting companies of gathering diamonds by the quart.[36]

When his report was finally made public, however, Janin's endorsement convinced many skeptics. After learning of Janin's endorsement, the editor of the *Rocky Mountain News*, only two months after denouncing the affair as a fraud, joined in forming another company to locate and mine the diamonds.[37]

The Two Philip Arnolds

Philip Arnold tried to divert attention from the true location of his diamond claims. In August 1872 he told a Laramie, Wyoming, newspaper that the diamond fields were in Colorado's San Luis Valley, which he wrongly described as being several hundred miles nearer the border of Arizona and New Mexico. Arnold left Laramie supposedly en route to southern Colorado but curiously rode off in the wrong direction.[38]

Arnold or someone like him got off the train in Denver in September and denied the statements made by the Philip Arnold in Laramie, Wyoming. He showed rough diamonds to Denver jewelers, who confirmed their genuineness. After a week of going about in ostentatious mystery, the Denver Arnold took the train to Wyoming to confront his Laramie double. Arnold arrived in Laramie and was identified as the same Arnold previously interviewed by the Laramie papers. He left without explaining the discrepancies between his Laramie and Denver selves.[39]

The Gray Expedition—A Diversion

The diamond company feared that outsiders were close to finding the diamond field so decided to misdirect attention from northwestern Colorado. Harpending gathered about twenty adventurers in his San Francisco office and promised to lead them to fabulous wealth. Harpending would pay expenses, and they would spend four months locating and mining deposits of gold and precious stones, for which they would receive a half-interest. Harpending was playing them for suckers, but he knew that if he could divert attention until snowfall, the secret would be safe until spring. News of the secret expedition spread quickly, as Harpending no doubt hoped it would.[40]

The well-publicized secret expedition was headed by Captain Mike Gray, a hotel proprietor and former sheriff. The one man in the group who supposedly knew the secret location was a Mr. Jones, who claimed to be the sole survivor of a band of prospectors wiped out by Indians. Such characters popped up regularly in the old West, telling tales of incredible gold and silver deposits in uncharted wilderness and promising to lead others to the source. Again and again expeditions set out confidently following men of this strange breed, but nearly every time the groups wandered until it became clear that they were following a deluded fool.

Gray and his men left San Francisco by train on August 30, 1872. They arrived in Denver, sold the remainders of their Chicago-bound tickets, and took the southbound train to Pueblo, Colorado, the next morning. Every move of the secret expedition was reported in the newspapers.

The Gray expedition bought provisions in Pueblo and continued on horseback toward the remote Four Corners area where Arizona, Colorado, New Mexico, and Utah meet. Gray was followed at a distance by a small army of Colorado prospectors. The direction taken by the Gray diversion was reinforced by William Gilpin, ex-governor of Colorado, who in a burst of home-state boosterism, asserted positively that the gem fields were in the San Juan Mountains of southwestern Colorado.[41]

No one had revealed the destination to the members of the expedition, but of course they expected to go to the diamond fields. Jones now told them that they would soon arrive at a great gold deposit where his friends had washed a $1,000 a day in gold before being killed by Indians. After a month on the trail, however, they discovered themselves back at a spot they had passed two weeks before and realized that Jones was lost. The Colorado prospectors who had followed them turned homeward in disgust. The Gray expedition wandered through labyrinths of canyons on half rations and began trading horses and mules to the Indians for food. A Navajo offered to guide them to Fort Defiance, Arizona, but led them through a succession of small Navajo settlements until they realized that they were getting no closer to the fort and their guide was running them from one village to another so that his tribesmen could trade food for the expedition's horses and mules.

Jones insisted that his great gold deposit was real, but when the men threatened to kill him, he fled and disappeared from history.

The diamond company declared in October 1872 that field work would be suspended until the following spring and that the diamond site would remain a secret. The editor of the Arizona *Citizen* smelled "the perfect bilk" and declared that no legitimate mining operation would stop for the winter.[42]

Arnold and Slack had gotten away with the swindle. With the onset of winter, travel would become difficult and the swindle would not be discovered until spring. In the meantime the organizers could sell more stock.

Clarence King Tracks Down the Diamond Fields

As the diamond excitement mounted in the fall of 1872, thirty-year-old Clarence King was finishing his five-year project examining the geology and mineral deposits along the transcontinental railroad, from Cheyenne, Wyoming, westward to the California state line. A well-connected Yale graduate, King had conceived the project and talked Washington officials into funding the exploration and putting himself in charge.

Now after years on foot and horseback in unmapped territory, King and his men feared that they had missed the greatest mineral deposit within their area—the new diamond field. They knew that the secret expeditions had embarked from railroad stations in southwestern Wyoming. The Green and Yampa Rivers were impassable at that time of year, and so the diamond fields could not be in Arizona or New Mexico but had to be somewhere in the thousands of square miles of southwestern Wyoming, northwestern Colorado, or northeastern Utah. But where?

As the weather grew colder, King and his geologists halted field work and gathered in San Francisco to compile notes. They discussed what they had heard about the diamond discovery, and by piecing together scraps of information, they realized that the diamond company's "Diamond Peak" was a certain mountain just south of the Colorado–Wyoming border. King knew the peak and the flat treeless butte just north of it, but his geologists had somehow missed the diamonds. King's love of science, adventure, and self-promotion propelled him to return to the diamond butte to see for himself.

Two of King's subordinates, geologist Samuel Emmons and topographer Allen Wilson, set out for Fort Bridger, Wyoming, to buy supplies and hire mule drivers; King left a day later so as not to arouse suspicion. They set out with their retinue—the cowboys driving their pack animals were told nothing of the purpose of the expedition, although with all the talk of diamonds that fall, they must have known. They left Fort Bridger on October 29 hoping for good weather, because an early-season blizzard would stop their expedition and send them struggling back to civilization until spring.

The temperature fell below zero, and the hard Wyoming wind punished them, but the winter snows held off, and on the afternoon of the fifth day, they came to the diamond fields. They were elated to find the prospect pits, wooden claim stakes marking off the claims of the diamond company, and, before long, diamonds in the decomposed sandstone. Hundreds of prospectors had searched fruitlessly for this spot, but King, Emmons, and Wilson alone had found it.[43]

The geologist Clarence King won instant celebrity
for his quick exposure of the diamond swindle.
Courtesy U.S. Geological Survey.

The trio found diamonds in the loose dirt, along with rubies and even a few small emeralds. They crushed and sieved the crumbling sandstone but were unable to find a single diamond in place. They panned the streams draining the diamond mesa, which should have been rich in placer diamonds but found none. Their suspicion was aroused and it became evident to them that the diamonds had been planted. After three days on the mesa, King, Wilson, and Emmons agreed that the diamond deposit was a fraud.

A tale would grow that King's party found diamonds hidden in the crooks of trees, as if Arnold and Slack had treated their swindle like an Easter-egg hunt. In fact, no trees grew, then or now, in the diamond field. Another doubtful story, which Asbury Harpending spread years later while trying to belittle King, was that King's German assistant found a diamond already cut. Like the diamonds-in-trees tale, the precut-diamond story lacks credibility, first, because no first-hand account mentions it, and second, because there was no German in King's group.[44]

While King and his men studied the diamond fields, up rode J. F. Berry, a mine operator from Salt Lake City. Berry and two others had followed their trail to the site and had watched their movements with a telescope. Now he introduced himself, and the geologists told him of the fraud. While Berry satisfied himself that King's suspicions were correct, yet another group of four more diamond hunters arrived. All agreed that the diamonds had been planted.[45]

Stopping the Swindle

King realized that while he was geologizing, people were being swindled in San Francisco. The trio started back. For greater speed Emmons took the pack animals back to Fort Bridger, while King and Wilson rode to Black Butte Station, which was closer. They reached it after a day of hard riding and caught the next train to San Francisco. King didn't know the diamond-company owners, but he knew Henry Janin, so he immediately tracked Janin down late at night.

King detailed his findings to Janin into the small hours of the morning, and the older man recognized that the company owners must be informed at once. He gathered company officials the next morning, November 11, 1872, and King told them his findings. King had returned as the diamond men were forming new companies to work some of the 160-acre parcels they had staked out in the field. Just the previous week, they had formed the First Choice Diamond Company and announced plans for eighteen more.[46]

Roberts and Harpending were adamant that the diamond deposit could not be fake. But the other directors listened to King and Janin and realized that they had probably been fooled. Several of the diamond directors suggested to King that it would be worth his while to keep the matter quiet for a few days, but King refused, saying, "There is not enough money in the bank of California to make me delay the publication for a single hour!" Despite the bribery attempt, he agreed to keep the matter quiet for two weeks, on the condition that company officials agree to halt stock sales.[47]

Secrecy was thrown to the Wyoming wind. Prominent diamond men such as Janin and General Colton, traveling with the government geologist King in a group of twenty, attracted notice. Newspapers reported their departure from Black Butte Station, and the party was followed by diamond seekers braving snow flurries and subzero temperatures. They arrived on the ground one week after King had returned to San Francisco with the bad news. They spent two days confirming the fraud. There were still a few rubies and diamonds lying about waiting to be discovered, but extensive tests of the bedrock and nearby creeks found absolutely no diamonds.[48]

General Colton remembered that the company's second expedition the previous summer had discovered 285 diamonds in just one hour over a small area; he sadly noted that the second party had spent eight days at the site surveying claims and, had they spent just one more hour searching for diamonds over a wider area, they would have come up empty-handed and the swindle would have been revealed.[49]

Newspapers printed the bombshell on November 26, 1872. Henry Janin had been fooled, along with some of the sharpest capitalists in San Francisco. All the other diamond companies that had reported genuine diamond discoveries suddenly dropped from sight. Newspapers that predicted that San Francisco would become the world's leading diamond-dealing center now hunted for explanations.[50]

The Aftermath

The Gray Expedition, Harpending's grand diversion, cold, fatigued, and hungry, straggled in small groups into Fort Defiance, Arizona, during a snowstorm. At the fort they learned that the diamond affair was a swindle and that they had been the worst humbugged of all. Gray and his men were stranded without funds or supplies. Some made their way back to San Francisco as best they could; two died of exposure.

Some principals in the burst bubble tried to repair the harm. Milton Latham and George D. Roberts refunded payments that they had received for now-worthless stock. Henry Janin, who had made $30,000 by selling some of his shares, gave the money to an investor whom he had encouraged to invest in diamond stock.[51]

Clarence King's dramatic exposure won him instant acclaim as an example of the best in scientific men: incorruptible, a scholar not burdened by disdain for practical knowledge, equally at home in eastern universities and in the rough camps of the West. The San Francisco *Chronicle* editorialized, "We have escaped, thanks to God and Clarence King, a great financial calamity."[52]

King's superior, General Humphries, disliked being upstaged by a subordinate. He privately criticized King for examining the diamond fields without permission and denied King's request to publish an account of the adventure in a magazine. But King was too popular to chastise publicly, and after Humphries's retirement King was made head of the United States Geological Survey.

Arnold returned to Kentucky as a local hero who had "been successful out-Yankeeing the Yankees." He insisted that the diamond fields were real, if perhaps too poor to work. William Lent sued Arnold for $350,000, but Arnold insisted that Lent, Roberts, and the others were merely trying to welsh on a bad bargain. After Lent attached his personal belongings, Arnold agreed to return $150,000 in exchange for dropping all charges. He used the remainder to start a bank in Elizabethtown, Kentucky. However, he found banking in the Bluegrass State to be highly competitive: A rival banker wounded him in a gunfight in 1879. He survived the shotgun wound but died soon after of pneumonia.[53]

John Slack quietly took his $30,000 from the swindle and moved to Saint Louis, Missouri, where he bought the Coffin and Manufacturing Company. His coffin-making firm died, and in 1880 Slack moved to White Oaks, New Mexico, where he lived modestly as an undertaker and coffin maker. At his death in 1896 his estate was worth only $1,611.[54]

Soon after the swindle was exposed, James B. Cooper came forward with the supposed inside story of the swindle. However, parts of Cooper's story are doubtful, and it is a mystery why Cooper would have confessed when he was not even a suspect. A San Francisco grand jury quizzed the principals, minus Arnold and Slack, who had left the state.[55]

San Francisco newspapers openly speculated that George D. Roberts and Asbury Harpending had conspired with Arnold and Slack. Harpending hurriedly sold his California assets, and, like Arnold, returned to Kentucky. The London *Times* tried to implicate Harpending when his friend Alfred Rubery sued the newspaper for libel, but no witness could identify Harpending as the American "gentleman" who bought rough diamonds in association with Arnold and Slack. After several years in Kentucky, Harpending returned to San Francisco and reentered mine promotion.[56]

Suspicion of involvement with the diamond hoax dogged Harpending for the rest of his life. He wrote his autobiography, *The Great Diamond Swindle and Other Stirring Incidents in the Life of Asbury Harpending*, to absolve himself of blame. He said that it was "absolutely false" that he had ever employed Arnold, even though Arnold had worked for him in the Ralston mines. Harpending dated his first knowledge of the diamond affair to May 1872, when he claimed he was still profoundly skeptical. In fact Harpending had learned of the supposed diamond discovery in December 1870. Despite the inaccuracies Harpending's autobiography largely succeeded in rewriting history, because most writers on the hoax have uncritically copied Harpending's version.

The location of the false diamond field in Colorado was well documented at the time, but since then people have moved it to New Mexico,

Utah, or Wyoming. The involvement of the diamond swindlers in the Ralston mines in New Mexico led some to confuse the two and to believe that the diamond swindle took place near Ralston City, which is now a ghost town named Shakespeare.[57]

Real Colorado Diamonds—1976

"In my opinion, the Great Diamond Hoax delayed the discovery of diamonds in Colorado. For decades after, nobody in his right mind would have believed diamonds could have been found anywhere in the region. People would have been laughed at for proposing that there were diamonds in Colorado. In 1977 and 1978, corporate geologists told me that what was going on was just another hoax."

—Dan Hausel, geologist, Wyoming Geological Survey,
Westword, October 30, 1997, p. 25

Clarence King worried that the supposed diamond discoveries within the area of his Fortieth Parallel Survey would impugn the accuracy of his work. A hundred years after King and his pioneer geologists rode over the mountains and plains, diamond deposits were indeed discovered within his Fortieth Parallel Survey.[58]

The only American diamond mine worked in recent years is in Larimer County, Colorado, just south of the Wyoming border, though several hundred miles east of the site of the 1872 diamond swindle. The kimberlite outcrops are small and eluded discovery until a hundred years after Clarence King passed through. In 1910 a prospector claimed he had found a diamond in the mountains of Larimer County many years earlier, but he had lost both the diamond and the locality. His revelation seemed far-fetched and attracted little attention, but diamond-bearing kimberlites were finally discovered in Larimer County in the 1970s, and a mine was started. At last report, it had shut down.[59]

Governor Lyon's Idaho Diamonds—1865

The great diamond swindle is the most famous, but it was not the only American diamond fraud, or even the first. That distinction goes to the phony diamond rush started by Idaho's territorial governor Caleb Lyon.

Lyon hailed from Upstate New York. He earned a degree in civil engineering at age eighteen and became a popular lecturer. He was appointed American consul for Shanghai, China, but left his post to join the California gold rush of 1849. He was secretary to the California constitutional convention, then returned east to serve in the New York legislature before serving one term in Congress.[60]

Lyon went to Idaho in 1864 as the new territorial governor and superintendent of Indian Affairs. He stayed only four months before sneaking out of the territory on a "duck-hunting trip" to avoid taking sides in the clash over moving the territorial capital from Lewiston to Boise. He was absent for nearly a year, during which he returned to New York to lecture.

Lyon's absenteeism led President Lincoln to favor another candidate for territorial governor; however, Lincoln's assassination allowed Lyon's political allies to secure his reappointment as territorial governor, and he returned to Idaho in November 1865. He excused his neglect of duty by saying that he had been granted a "leave of absence" to spend time in New York promoting investment in Idaho mines; his charm and forceful personality won many supporters.[61]

This time Lyon had business in Idaho other than politics. Soon after his return he was in the mining town of Ruby City to make a florid speech on the glorious future of the mines and the nobility of the miners. He showed the editor of the Owyhee *Avalanche* what Lyon said was a rough diamond that a prospector named Sam Wilson had found nearby. Lyon's story was that he had met Wilson on a steamship bound for New York the previous May. Wilson, ignorant of the nature of the stones, had shown them to Lyon, who recognized them as diamonds. Later in New York a jeweler supposedly confirmed their value and bought the largest stone for $1,000.[62]

Despite the exciting find, Lyons and Wilson remained in New York for months before returning to Idaho. According to Lyon, the exact location of the discovery was lost when Wilson perished en route when the steamship *Brother Jonathan* foundered off California—Wilson, however, is not on the passenger list.[63]

Lyon and the newspaper editor D. H. Fogus formed a partnership and went together to where Wilson had supposedly found the gems. But word leaked out, and a large throng followed Lyon and Fogus.

The women of Ruby City were particularly affected. They roused husbands from their beds and sent them out in the cold night to find diamonds. Men rushed about in the dark to stake claims. An eyewitness wrote that "each man had a wife or sweetheart who must be the owner of a diamond claim." Unattached men with claims in the diamond fields were suddenly popular with single ladies. The mines and mills at Ruby City shut down for days while their employees hunted for diamonds.[64]

The prospectors organized the Diamond Basin mining district, and the future metropolis of Diamond City was laid out. Lyon returned to Boise to attend to state business, leaving a horde of diamond hunters to deal with his "discovery."[65]

The prospectors knew enough to identify the larger crystals as quartz, but smaller crystals were mistaken for diamonds. Probably none of the prospectors could identify a rough diamond, though self-proclaimed diamond experts showed up to tell the prospectors that their stones were genuine diamonds. But those who proclaimed that the stones were diamonds were selling, not buying. Jewelers spurned the stones.[66]

Even with no buyers, new rumors started a fresh stampede in January. Snowfall hindered the search, but optimists predicted that spring snowmelt would reveal the diamonds. Single ladies still found claim owners irresistible, and several joined future diamond barons at the altar. The editor of the Owyhee *Avalanche*, who owned a diamond claim and some lots in Diamond City, ridiculed the idea that so many diamond "experts" could have been fooled, and concluded that the finders were keeping their discoveries quiet. As late as April the *Avalanche* insisted that genuine small diamonds abounded. Spring melted the snow, but no diamonds were discovered. By the rules of the district, each claim must be worked four days each month to remain valid. But the claim holders no longer considered the claims worth that much of their time, and the district evaporated.[67]

Caleb Lyon tried to have Idaho admitted as a state so that he could become its United States senator, but his ambition was frustrated by his rapidly sinking popularity. The Owyhee *Avalanche*, which had hailed Lyon in November 1865, just four months later called him "an old imbecile." Lyon abruptly abandoned Idaho for the second time in April 1866, after only five months in the territory, taking with him $46,000 in Nez Perce Indian funds. He later reported that the money had been stolen when he left it unguarded on the train en route to Washington, DC. Of Lyon's brief second stay in Idaho, the historian Bancroft commented, "He lent his signature to any and every bill of the most disloyal and vulgar-minded legislature that ever disgraced the legislative office, except the one that followed it."[68]

An Idaho man visiting New York later that year investigated Lyon's diamond story and discovered that it was "false in every particular." He said that New Yorkers dismissed it laughingly as one of "Lyon's big lies." Lyon's motive remains a mystery.[69]

Return of the Idaho Diamond Frenzy

After nearly thirty years had erased the disrepute of Caleb Lyons's scheming and embezzling, people remembered only that a former governor had vouched for discovery of diamonds in Idaho. In 1892 some prospectors claimed to have found diamonds, and people in Nampa and Boise left their jobs to rush to the site of the first diamond delusion. An estimated thousand people staked claims. The town of Diamond City sprang again from nothing, with a restaurant and a saloon.

A previously unknown but self-proclaimed expert, "Professor" Plast Byrne of Melbourne, Australia, arrived to pronounce the Idaho stones "silicon diamonds, about nine and a quarter pure." The professor said that although harder than regular diamonds, the silicon diamonds were only worth half as much. He did not gain a following, because knowledgeable people knew that there was no such thing as a "silicon diamond," and Idaho prospectors thought that Byrne was out to swindle them out of half the value of their diamonds.[70]

The second Idaho diamond excitement petered out like the first one, as prospectors found that no matter how ardently they believed that they had found diamonds, no one would buy their quartz crystals.

The Diamonds of Tonopah, Nevada

E. W. Hews was a thin Louisianan given to spouting erudite nonsense. He had prospected for twenty years without success in Montana and Colorado, and arrived in Tonopah, Nevada, in 1905. Gold eluded him once more, but Hews told the Tonopah *Sun* that he had found diamonds outside Tonopah. "Within a month," he wrote, "Tonopah district will be counted among the diamond producers of the world."[71]

Hopefuls streamed to the site all night and the next day and staked the area around the Hews discovery claim. Hews displayed his "diamonds" at a Tonopah drug store, and although the stones didn't resemble diamonds, few if any in town had ever seen uncut diamonds. Hews explained his peculiar theory by which diamonds transformed into topaz, even though the two have entirely different chemical compositions. He decided that the blue shale at his claim was really the blue earth known to South African diamond miners and said that topaz on the surface was a sure indication of diamonds at depth.[72]

No one cared about Hews's prattle about "corrosive sublimates" but listened when he said that he had sold an option for a one-third interest to San Francisco capitalists and that he valued his claim at $4 million. Tonopah jewelers said that the stones weren't diamonds, but Hews's confidence and scholarly speech gained him an enthusiastic following.[73]

Promoters laid out the rival town sites of Kimberly and Ladysmith and sold lots in the instant tent villages. Armed groups threatened one another for jumping diamond claims. Hews upped the value of the claim in his own mind to $5 million and permitted Tonopah's leading businessmen to invest in his Tonopah Diamond Mining and Exploration Company.[74]

John Hassell and John Shay, who said that they were familiar with the South African diamond mines, arrived in Tonopah to great fanfare. The Tonopah *Sun* revealed that Hassell and Shay had tested the stones and proven them to be diamonds, but the truth was the opposite: they tested one stone and declared it not a diamond.[75]

The unfavorable test killed the diamond excitement a week after it began, although some claimed that the critics were agents of DeBeers trying to suppress rival diamond-mining areas. E. W. Hews left Tonopah and was not heard from again.[76]

Other Diamond Delusions, Other Places

Diamonds, like gold, have a powerful hold on the imagination. Small crystals are often mistaken for diamonds the world around. Mistaken diamond booms have occurred as far afield as Honduras and Bosnia. Mexicans search for the lost diamond mine of Vicente Guerrero.[77]

Rivers or glaciers can carry a loose diamond far from its source, so that the discovery of a single genuine diamond may produce a frenzy of searching but no more diamonds. Such diamond excitements in Eagle, Wisconsin, in 1884 and Dysartville, North Carolina, in 1886, inspired promoters to buy rough African diamonds to support their pretense of owning diamond mines.[78]

False diamond discoveries popped up here and there across America in the nineteenth and early twentieth centuries. The "South African diamond expert" became a familiar feature of diamond frauds, like the "Cornish tin miner" in tin-mine frauds. The DeBeers cartel was accused of squelching American diamond mines. Diamonds were reported in Georgia in 1871, and the silliness there persisted off and on for the next two decades. Baseless diamond excitements flared in South Carolina in 1881, Utah in 1882, Montana in 1891 and 1901, Arizona in 1899, New Mexico in 1901, Colorado in 1909, Texas in 1911, and Pennsylvania in 1925.[79]

More recent mine swindles have involved gold mines, silver mines, platinum mines, and (before the price crashed in the late 1970s) uranium mines. Diamond-mine swindles, however, have not been reported in America for many years.[80]

6

Tin Men

Although America is poor in tin deposits, it has a history rich in tin-mine frauds. Mere lack of tin ore did not inhibit the imaginations of promoters. Again and again, investors spurned the dull truth and chose to believe the beautiful lies of swindlers.

America's First Tin Swindlers: Kansas Indians

In the 1840s, members of the Kansas tribe showed the Baptist missionary Johnson Lykins pieces of tin ore that they said were found in great quantities. Lykins sent a piece of the tin to the scholar of Indian studies Henry Schoolcraft with a map of the location described by the tribe. Lykins said that his duties and the danger of Comanches and Pawnees prevented his personally investigating.[1]

But Schoolcraft knew that the sedimentary rocks of the plains are not likely to host tin ore and suggested that the tribe had lied out of "gross self-interest." Nevertheless, he published Lykins's map in his 1851 book on Native Americans.[2]

Schoolcraft's publication spurred an expedition from Missouri that searched for weeks in the Smoky Hill valley. Although the expedition gave presents to the Indians, they refused to reveal the location of the lode. A Wyandotte named Mudeater showed white men a piece of cassiterite (the most common tin ore mineral), bringing more fruitless searches in 1860. Native Americans harvested blankets and other goods from another group of hopefuls a few years later but then led them to a place near the Smoky Hill River, where there was only fool's gold.[3]

Schoolcraft's suspicions were correct: Native American headmen admitted to the pioneer T. F. Huffaker that they had obtained the ore specimens from white men and were using the tin rocks to obtain trade goods in return for bogus information on the source.[4]

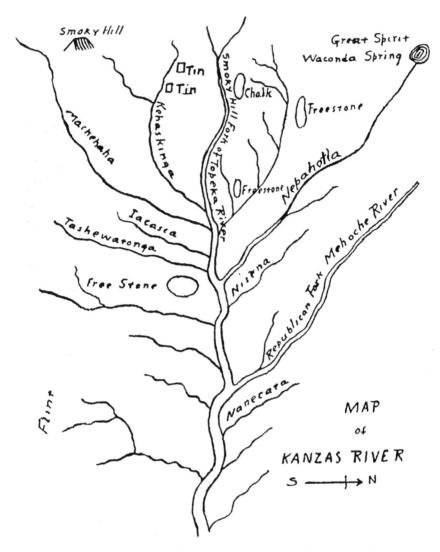

Kansas Indians fooled greedy white men into believing that there were tin mines on the western Kansas plains. From Henry Schoolcraft, Historical and Statistical Information *1, pp. 157-59, 1851.*

The Indians worked their tin game off and on for about twenty years, whenever a white man thought he could buy a bonanza for a few trinkets. It ended only when the Native Americans were displaced by settlers. The swindle took on a life of its own years later, when a prospector using the map from Schoolcraft's 1851 report didn't find tin but started a false gold rush and a new round of mine swindling. But that is another story.

Tin Mountain, Missouri

"The oldest and rankest mining dragon of all is the tin ore fraud. It began with the lie that the United States government had offered a large reward for the discovery of a tin mine. This led to an immediate discovery in Missouri, and the investment of $250,000 in a worthless mill which stands idle and rotted today."

—*Mining Industry and Tradesman*, March 10, 1892

The Missouri tin mines sprang from the fertile imagination of the German immigrant Albert Koch. He ran a Saint Louis, Missouri, museum featuring everything from live alligators to carnival acts. In 1840 he assembled a skeleton—mostly of mastodon bones—that he dug up at various sites and exhibited the behemoth to his thrill-seeking museum visitors. Koch enlarged upon nature by adding ten extra vertebrae and bones from at least five other genera of animals (including a ground sloth) and by padding the skeleton with wooden blocks between the bones. He proclaimed that his skeleton was from a web-footed aquatic monster that walked along the bottom of rivers, rising occasionally for air.[5]

Koch sold his museum in 1841 and toured America with his prehistoric monster. Knowledgeable scientists pointed out his errors, but he knew what the public wanted. The British geologist Charles Lyell met Koch and described him as "a mixture of an enthusiast and an imposter, but more of the former, and amusingly ignorant." Koch took the traveling wonder to Europe. Rather than ship the beast back to America, in 1842 he sold it to the British Museum, which removed the extra bones and assembled the mastodon properly.

Koch dug up his next fossil circus–wonder in Alabama, where he found a nearly complete fossil whale skeleton, and shipped the bones to New York in 1845. Koch knew that people would rather see a sea serpent, so he added extra vertebrae, creating a 114-foot-long sea monster. "Doctor" Albert Koch toured the Northeast with his creation. Paleontologists recognized it as a humbug, but the public was dazzled. Koch escorted his sea serpent to Europe, where he eventually sold it to the King of Prussia, then returned to Alabama for bones to fabricate another sea monster, which he brought back to Europe.[6]

In 1859 Koch told the world that he had discovered a huge vein, forty to sixty feet wide, that carried an astounding $107,252 in gold and platinum per ton—in Madison County, Missouri. He mentioned as an aside that the rock also had a little tin. But despite his efforts and fawning local press, the deposit, by Koch's description surely the richest gold and platinum deposit in the world, remained idle.[7]

Koch was back at it in 1867, except that gold and platinum were forgotten, and the great ore body in Madison County had transformed in his mind into a mountain of tin ore—enough to supply the entire world. Soon that whole world knew of his discovery, on what became known as Tin Mountain. Koch died in late 1867, but his discovery was taken up by other promoters with similar disregard for the truth.[8]

English miners from Cornwall were the world's foremost tin miners, and a Cornish endorsement was needed to attract investors. It is a mystery where American promoters found Cornishmen so ignorant of tin as to report favorably on every fraudulent or uneconomic tin deposit through the years. In the case of Koch's discovery, the inevitable Cornishman was R. W. Dunstan, who examined the prospect and enthused about the "new Cornwall" in Missouri. Prospectors swarmed the hills looking for tin ore, though almost none knew what it looked like.[9]

There are no tin minerals on Tin Mountain. The "ore" was gabbro, a common dark igneous rock. Gabbro contains mineral grains that can form globules of iron in a test furnace. An assayer who, by ignorance or deceit, failed to test the metallic globules could mistake the iron for tin. Such an assayer was "Professor" Hy. Marec de Beauregard, supposedly a graduate of the Paris School of Mines, who produced the metal droplets and pronounced them tin. De Beauregard, whom the *American Journal of Mining* dismissed as "an incompetent witness," spread enthusiastic nonsense about Tin Mountain, saying that he could detect tin all over the mountain by looking at its soil and plants and tasting the water of its streams.[10]

The tin promoters exhibited a bar of tin—falsely said to come from the Tin Mountain ore—at a Saint Louis city fair. The fair judges awarded a prize for the tin bar and gave de Beauregard a diploma for his phony assays. The promoters led tours to Tin Mountain for Saint Louis businessmen, who were dazzled into investing their money. At least three companies formed to exploit the veins. The Missouri Tin Mining Company kept thirty workers driving a tunnel and building an ore furnace under the direction of another wandering Cornishman of questionable competence.[11]

News of the discovery reached London, where the *Times* and the *Mining Journal* discussed the potentially depressing effect of Missouri tin on the Cornish mines. The Lynchburg, Virginia, *News* crowed that the Missouri tin mines would force the shutdown of mining in Cornwall.[12]

A chemist for the Government Land Office in Washington wrote that he had examined ore supposedly from Tin Mountain but concluded that the tin ores had come from elsewhere—probably Cornwall. He tested many genuine Tin Mountain specimens and found no tin at all. He detected a trace of tin in only a single specimen. Professor de Beauregard appealed to American resentment of English commercial power when he hinted that unfavorable assays were a conspiracy by the "gilded specter of perfidious Albion."[13]

To further convince investors, a Saint Louis chemist promoting the tin mines offered the use of his laboratory in assaying the ore. Skeptics could select their own ore from the mine, guard it against salting, supply their own chemical reagents to preclude adulteration, and conduct their own tests. The tests in the promoter's laboratory always found tin. It was not until later that investors discovered the trick: the laboratory was supplied with graphite crucibles, and at the bottom of each crucible was a piece of tin foil hidden under a smear of graphite. Each time a rock sample was melted in one of the graphite crucibles, the tin melted into the sample and produced a favorable test.[14]

After two years of loud promotion and large expenditures, the tin men grew more strident to cover up the obvious lack of tin pouring from the furnaces. They said that they needed new and more costly ore-processing equipment. A singular feature of the Madison County ore was now that the tin was supposedly of a peculiar type undetectable by standard chemical analysis, and one promoter declared that tin was not a chemical element at all but could be synthesized from other more common substances. The *Engineering & Mining Journal* sighed, "The age of alchemy is not yet past" and warned potential investors away, but naïve belief in nonsensical chemistry proved that not all Missourians are "from Missouri."[15]

By 1870 the Missouri Tin Mining Company had driven four tunnels into the mountain but had produced no tin. The promoters lured still more money out of Saint Louis by leading tours to the mine. The excursionists were given food and liquor and watched as tests drew between five and eleven percent tin from rock that the best government assayers said contained no tin. Seeing is believing, and investors put up the money for the expensive new ore-processing plant.[16]

The Missouri state geologist Albert Hager wrote in his 1871 report that there was no tin in the mines. But a state senator denounced Hager for harming the mining interests of the state, and had the state government suppress the report. The politically influential tin promoters had Hager fired.[17]

Promoters finally started their costly new mill, but it recovered no tin because there was no tin to recover. Saint Louis investors lost an estimated $500,000 in the failed ventures. One of the promoters was later reported plying his swindles in Denver. The site of the fraud is still known as Tin Mine Mountain.[18]

A Canadian Sojourn

In late 1871 a swindler claimed to have discovered rich tin deposits near Otter Head, Ontario, on the northern shore of Lake Superior. He sold part interests to some men at Silver Islet, Ontario, but the investors examined the area and found no tin.[19]

The swindlers redoubled their efforts. In fall 1872 a man named Woods inquired in London about buying a few tons of high-grade Cornish tin ore for shipment to America. This was during the height of the great diamond swindle of 1872, which had used diamonds bought in London. London mining men refused to help him, so he went to Cornwall, where he bought all he needed and shipped it back to America. Woods and his confederates took the tin ore by boat to Otter Head. There they built artificial tin veins by filling a pair of natural cracks in the granite shore with the Cornish tin ore, mixed with crushed waste rock from Lake Superior copper mines and sodium silicate. The mixture hardened into a tolerable imitation of tin veins.[20]

The swindlers showed specimens of their English ore back in the States. In November 1872, Detroit investors sent Captain William Harris to investigate. Harris was a foreman in a copper mine at Ontanogan, Michigan, and since all mine captains on the Michigan copper range were Cornishmen, he was thought just the man to examine the veins. It was late in the season; snow was already on the ground, and he feared being stranded on the remote shore all winter. Harris studied the veins for only two hours before returning to Michigan with more samples. Assayers reported that it was as good as the finest tin ore from Cornwall itself, and the tin boom was on.[21]

On the basis of Harris's two-hour examination, his clients bought the phony tin deposits for $200,000 then reportedly went to London and resold it for $1 million. While Lake Superior was icebound, and Otter Head was covered thickly in snow, the number and width of the tin veins multiplied in the fevered imaginations of the investors.[22]

Word of the tin reached miners at Silver Islet, Ontario. A mining boom is a powerful incentive, and fortune seekers trekked three hundred miles to Otter Head and back on snowshoes to stake their own claims in the new Cornwall before the spring thaw allowed the world in. The winter prospectors examined the veins more closely than Captain Harris, recognized they were artificial, and even found a barrel of sodium silicate at the swindlers' camp at the site—one version said that there was even some straw mixed in with the "vein."

Captain Harris revisited the site the next spring and admitted his mistake. He publicly blamed his error on the hasty examination; but privately, by one account, he laid the blame on the brand of champagne served on the trip. When word reached Detroit, the swindlers fled.[23]

Ogden, Utah

Saint Louis, Missouri, promoter Dr. Washington West announced in 1871 that some rock from the mountains just east of Ogden, Utah, averaged a fabulous twenty-five percent tin. Prospective buyers collected rock samples that assayed rich in tin; they produced the necessary Cornishmen, who inspected the phony deposits and pronounced them the new Cornwall, et cetera.[24]

The mountains east of Ogden swarmed with prospectors who imagined that every dark or heavy mineral was tin ore. The Ogden *Junction* marveled, "Every sixth man you meet has a tin mine in his pocket." When government assayers could not find even a trace of tin in the rock, the promoter accused the assayers of incompetence.[25]

The Utah "tin ore" was only dark igneous rock that contained no tin. In the end, there were too many assayers to fool, too many samples to salt. Dr. Washington West, when confronted with the evidence, admitted that he had spiked the assays with powdered solder, and he quickly left town headed south.[26]

Harney Peak, South Dakota

The Harney Peak tin mines were one of the most spectacular failures of nineteenth-century American mining. It is difficult to know where foolishness left off and swindling began, but both were there in abundance.

In 1883 the tin mineral cassiterite was discovered on the Etta property, near Keystone, Dakota Territory, igniting a tin boom in the Black Hills. Investors wanted to see cassiterite, so prospectors whose claims lacked the mineral collected specimens from the Etta and showed them as coming from their own property.[27]

Mining-savvy capitalists sent engineers, geologists, and Cornish foremen to inspect the tin discoveries; they reported back unfavorably. The owners of the Etta and other claims could not sell them to knowledgeable miners, so they offered them to New Yorkers with no mining experience.[28]

The Harney Peak Mining Company

"The idea of there being any tin property there worth $10,000,000, or any thing like that sum, is simply absurd."

—*Engineering & Mining Journal*, June 4, 1887

The owners of the Etta mine pooled twenty mining claims into the Harney Peak Tin Mining Company. The nominal capitalization was $5 million, but since the organizers gave most shares to themselves, the money available for development was much less. They brought in the English tin "expert" Professor G. F. Randall to inspect the deposits and tell investors that they could not afford to miss the great opportunity.

Randall, with no regard for the truth, wrote:

> The rock needs no mining whatever. It is simply throwing down the greisen rock into chutes and ore-bins and carrying the same to the mill for crushing and concentrating. To form any estimate of the extent of this formation is impossible, the magnitude is so great, and it is no exaggeration to state that it requires a fortnight even to survey this mass of rock. These mines will certainly rule the tin markets of the world.[29]

The *Engineering & Mining Journal* advised,

> The mining community of the Black Hills should raise a subscription to pay the expenses of the 'Professor's' trip back to England on condition that he will never come back. He is hurting the prospects of their country as a future tin-producing district. Turn him out!

The people of the Black Hills, however, welcomed and believed Randall's report.[30]

The promoters sold shares with much fanfare. Characteristic of a company operating for show, the Harney Peak's management built a large mill but neglected to develop the mines. The company started the ore mill in 1886 but shut it down a month later. Although it had a 240-ton-per-day capacity, the mill crushed and processed only four hundred tons of rock in its thirty-one days. The mill had consumed the best ore produced by the mines over a two-year period but recovered less than seven tons of tin concentrate, worth about $1,200, or less than four days of the company payroll.[31]

The New York directors realized that their $5 million company was just another unprofitable property and wanted out of a losing deal. The obvious solution was to double the price and sell it to the British.[32]

Their plan became apparent when they issued another $5 million of stock and began a campaign to sell it in London. *The Engineering & Mining Journal* criticized the stock issue and asked, If the American Harney Peak managers really thought that the mines were so valuable, why were they selling? The *Mining & Scientific Press* pointed out the vast difference between the mountain of rich tin ore claimed by the company and the complete lack of results after more than four years.[33]

Champagne-Tour Geology

"Vincent they have kept completely under the influence of champagne all the way out and back."

—Observer of the British inspection party in the
New York Times, December 1, 1890

The tin lodes of Cornwall were being depleted, and London financial men realized that the future of tin lay overseas. They were excited about controlling this new Cornwall in America but first sent a delegation to see if it was as rich as the sellers promised. One was a member of Parliament; another was a stockbroker who knew little about mining but assured his customers that, based on what he saw, the company was worth £2 million. Leading the party was Professor Minos C. Vincent, a legitimate but very optimistic mining engineer. He had reported favorably on the Flagstaff mine in Utah shortly before that property exhausted its ore and cruelly disappointed the English owners. The Harney Peak promoters relied on Vincent's optimism for a favorable report. They also reportedly relied on a man whom they had paid $100 per day to accompany Vincent during his visit to the Black Hills, for the purpose of keeping him drunk.[34]

Two years later another English mining engineer described the familiar method in connection with a different swindle:

> Usually the English expert is taken charge of from the moment he lands. He is kept well charged with champagne and whisky, he is driven out to the property at night, and when he has learned his lesson well, he is given a few hours to make up his mind about a well-salted mine. He then cables home his report, and is not allowed to speak to a single disinterested individual until he has returned to New York.[35]

The promoters gathered a selected load of fine tin ore and placed it prominently on the rock pile in front of the first prospect hole to be visited by the English inspectors. As soon as the visitors had been impressed by the rich ore and had moved on, the promoters "recycled" the ore by gathering it up and carting it to the next prospect ahead of the inspectors, so that it was viewed over and over.[36]

Vincent's 120-page report was a masterpiece of grandiose projections of the great number of miners to be employed and the enormous tonnages of tin ore—but it was based on very little data. The holy grail of nineteenth-century mining was the "true fissure vein," a body of mineral filling a crack in the earth that would continue undiminished as deep as anyone cared to follow it. Vincent found a number of Harney Peak properties to contain true fissure veins of tin. The *Engineering & Mining Journal*, however, derided his inconsistencies and made-up jargon.[37]

The Purloined Ore Sacks

The official British inspectors were easily convinced, but another English visitor was more skeptical. He witnessed Vincent's wine consumption and noticed that the professor took rock samples only in places suggested by their hosts.

To collect representative samples, the Englishman quietly hired a miner to assist him and returned to the mine in the middle of the night to take samples, hoping that he would be undetected. He and his assistant chipped samples every twenty feet along the tunnel and returned to his room at 4 a.m. with two sacks of rock samples. Soon after he went to sleep he was awakened by the sound of his window being raised. Thinking it safer to allow the ruffians to steal his money, he did not stir from his bed. To his surprise, the burglars took only his ore sacks and went back out the window. The visitor—still feigning sleep—soon saw the men return as stealthily as before and replace the ore sacks. When he examined the bags in the morning, the barren contents had been replaced with rich tin ore.[38]

Vincent's report convinced many in London that the Harney Peak Company would yield fabulous profits. The American promoters shipped boulders of ore thick with dark glistening cassiterite to London. The promoters followed the traditional practice of bribing, more politely called "subsidizing," the financial and mining periodicals of London. Newspapers in the Black Hills were eager to promote investment and required no bribes. The historian W. Turrentine Jackson noted: "In all the annals of mining investment, no clearer evidence of corruption and abuse of the newspapers and the mining press to promote the cause of a mining speculation can be found than in the events associated with Harney Peak tin." But despite the bribery, through 1887 and most of 1888, the English investors withheld final agreement to buy Harney Peak.[39]

The British syndicate finally agreed, bought $2.4 million in shares, and formed the Consolidated Tin Mining Company to buy $5 million in bonds issued by the Harney Peak Company. The new company was headed by retired diplomat Lord Thurlow, who thought that the Harney Peak mines were so rich that they would flood the tin market. As did American tin boosters, Thurlow blamed opposition to the Harney Peak Company on jealous English tin interests. The Harney Peak Company issued more shares to increase its capitalization to $15 million. Control still resided in New York, however, where the management ran the company to sell stock rather than to mine tin.[40]

One of those criticizing the deal was William Blake, who said that it was folly to spend large sums when the prospects justified only a small and frugal operation. Blake's report stirred controversy in London, for he was an acknowledged expert on the Black Hills tin deposits, but, after all, what did Americans know of tin?[41]

The company bought new claims, built new mills, but mined little. It owned four hundred unpatented mining claims, each requiring annual work to maintain ownership. In the meantime development on the few really promising claims was neglected. When American tin dealers reported that they had investigated the deposits and declined to invest, Harney Peak president Samuel Untermeyer accused them of trying to sabotage the company because it would soon flood the market with cheap tin.[42]

That the supposedly fabulous profits were fictitious was tacitly admitted by the company when it petitioned Congress for a tariff on tin. Some wondered what had happened to the rich tin ores to make them suddenly too poor to operate without a tariff, while others questioned why the government should tax American consumers to assure profits for a company soon to be owned by the British. Nevertheless, Congress passed a four-cent-per-pound tariff, to take effect in 1893.[43]

The Harney Peak Company took options on tin claims at high prices. Even the incurably optimistic prospectors were surprised at the high prices they were promised. When the options came due, however, company representatives offered only a fraction of the option purchase price. One claim owner who had given the company an option to buy his claim for $15,000 was offered only $4,000 when the option came due. Most claim owners realized that the lower prices were still the best they could get and sold to the Harney Peak Company. The sales documents still bore the original high option price, however, and the Harney Peak employees pocketed the difference.[44]

Punishing Honesty

One critic of the overpriced promotion was Thomas White, a Cornwall native who worked as a surveyor in the Black Hills. When White wrote letters to the London *Financial News* denigrating the Harney Peak deposits, his Dakota neighbors vilified his character. First, 175 Dakotans petitioned the surveyor-general of Dakota to revoke White's commission as deputy United States mineral surveyor, which he did, taking away White's livelihood. White was also charged with perjury in connection with some surveying, was arrested, and was released on bond. A Rapid City grand jury then indicted him for criminal libel.[45]

White refused to be cowed. The perjury charge died when his accuser refused to testify, and the criminal libel charge was dropped. White then turned the tables by suing the New York *Herald* for libel.[46]

Some faculty at the Dakota School of Mines were also not as enthusiastic about Harney Peak as the boomers would have liked, so three of the skeptical professors were encouraged to leave the college. Over faculty protest, the tin-boomer Gilbert Bailey was appointed a professor of metallurgy. Bailey was a geologist but was not trained in metallurgy. He had helped promote the Harney Peak bubble and a pair of other dubious Black Hills promotions, the Lookout and Sullivan mines, but to local boosters, these were exactly the qualifications they wanted in a professor.[47]

The *Engineering & Mining Journal* didn't hesitate to criticize the Harney Peak bubble. For its honesty the journal was accused by the Rapid City *Journal* of being "the blackmailing tool of an unprincipled clique" of tin importers. The Rapid City *Journal* happily predicted, "The Black Hills will be a tin producing business long after the *Engineering & Mining Journal* has ceased to exist."[48]

The Strange About-Face of the London *Mining Journal*

In 1888 the London *Mining Journal* suddenly halted its criticisms of the Harney Peak Company and began giving the properties favorable press. Both the editor and assistant editor resigned from the *Journal* and disavowed responsibility for the about-face. The *Mining Journal* was thenceforth in the pockets of the Harney Peak boomers and gave the company uncritical publicity.[49]

Harry Marks was the publisher of the London *Financial News* and was a notorious financial blackmailer. The American tin men charged that Marks had demanded $25,000 as the price of his support; when they refused, he attacked the Harney Peak Company in print. Marks may well have tried to blackmail the Harney Peak Company, but they clearly had much to be blackmailed about.[50]

Back in the Black Hills, however, the editor of the Deadwood *Pioneer* urged the Harney Peak Company to quit the publicity campaign of sunny articles in the *Sun* and other New York papers and start producing tin. The Deadwood *Pioneer*'s good advice met with only criticism from local tin boosters.[51]

When the Harney Peak Company was on the brink of full-scale operations, the English shareholders belatedly sent a real tin expert to the Black Hills: Captain Josiah Thomas, a man with great experience in the Cornish mines. Considerably more level-headed than the academic inspectors who preceded him, he recognized that the ore was poorer than it appeared and described the properties as unproven. He was noncommittal when interviewed by the Black Hills *Times*, intending his report only for his clients. The reporter, however, put extravagant praise in his mouth for the great tin deposits around Harney Peak. The Harney Peak Company likewise twisted his report around and claimed that Captain Thomas was "confirming beyond question" the great value of the property.[52]

A Bright Future Assured

"We are told that the value of the mines is no longer problematical. With that we agree. To begin with it was nothing and it has succeeded in 'holding its own.'"

—*Mining Industry & Tradesman*, March 10, 1892

Harney Peak went into full production in 1892. When Lord Thurlow presided over the annual meeting of the Harney Peak Company in London in December 1892, he reported that the mines were producing an endless supply of three- to four-percent tin ore. The mill was running, and the railroad lines were in place. He spoke confidently of having two ore mills in operation by the following spring, merrily crushing one thousand tons of ore per day. Thurlow planned a triumphant inspection tour of the mines and mills in the spring. All shareholders had to do was enjoy dividends paid from promised profits of £300,000 per year for many years.[53]

The Harney Peak Company continued to issue optimistic statements right up until February 1893, when the superintendent returned to the Black Hills from a company meeting in New York and abruptly shut down all mining and milling. In a year of full-scale operation, the company had produced 180,000 pounds of tin, worth only $37,800. This was the result of a $15 million investment in stock and another $5 million in bonds. Some speculated that the company was holding back until the tin tariff was to take effect later in the year, but the mines stayed shut.[54]

End of the Harney Peak Company, and Lawsuits

"The main trouble with the company and with the whole enterprise has arisen from the dishonesty and bad faith of the English people."

—Samuel Untermeyer, president, Harney Peak Tin Mining, Milling, and Manufacturing Company

English investors petitioned the courts to appoint a receiver and accused company president Samuel Untermeyer of issuing stock to himself and overcharging the company for legal fees. Untermeyer protested, but the court found a preponderance of evidence showing that the management had misused their positions to profit at the expense of the shareholders. Critics on both sides of the Atlantic hailed Dr. Albert Ledoux's appointment as receiver, as he was a mining engineer and noted tin skeptic.[55]

Untermeyer and the rest of the ousted management tried to regain control of the company, which despite their twelve years of failure, they insisted was extremely valuable. They charged that the receivership was a plot by minority shareholders to gain control of the tin bonanza.[56]

While the lawsuits progressed, Ledoux investigated the properties. Since the extravagantly capitalized company was broke, he sold the mills and machinery to pay taxes. Suits against Untermeyer and the rest of the New York management were eventually dropped, for while their actions seemed unethical, it could not be proven that they had broken any laws.[57]

Lord Thurlow, who assured British investors of the great value of Harney Peak right up until it shut down, went on to promote another dubious American financial scheme. The New York lawyer Samuel Untermeyer, who engineered the overblown London promotion, later made a name as a reformer for his work in the congressional "money trust" investigation of 1912–13. By that time few remembered his questionable role in Harney Peak.[58]

There is tin around Harney Peak but not much. All the talk of "true fissure veins" was so much moonshine. The tin was in pegmatites of limited extent, where cassiterite was mixed with other dark minerals, giving the rock the appearance of being richer than it really was. The receiver Albert Ledoux administered the properties until the company wound up in 1909. He found no mineable tin but made some money mining a small amount of the lithium mineral spodumene. Years later other companies bought the Harney Peak properties and for a few years managed to frugally produce small amounts of tin.[59]

Broad Arrow Mine, Alabama

In 1874 William Gesner reported that he had found rock containing metallic tin, lead, and bismuth in Clay County, Alabama. The Alabama state geologist dismissed the "rock" as man made, since those metals do not occur naturally in their metallic state. Gesner continued to search Clay County and in 1879 told a farmer near Ashland that he had found tin on his farm. His brothers George and John Gesner joined William in Alabama. The Gesners secured mining rights to Callaway's eighty acres and to more than four hundred acres surrounding it.

All was done in secret until November 1882, when the Gesner brothers announced that their property, named the Broad Arrow mine, would soon produce tin. A letter to the *Engineering & Mining Journal* lied that the Gesners had a forty-five-stamp mill in operation and that the Broad Arrow mine was in "one of the finest mineral regions in the world." Visitors were banned, however, so the public depended on the Gesners for information.[60]

In December 1883 the Gesner Mining & Smelting Company announced that the mine and mill had started to produce metal. One of the Gesners traveled to New York with a bar of Alabama tin and a prospectus. The tin bar spurred stock sales, but the mines and smelter did not produce any more tin. The Gesners spoke optimistically of their plans to install more and better machinery; meanwhile they sold more stock.[61]

William Gesner sold shares in another tin concern, the Alabama Tin Mining Company. One of his partners was former Alabama governor William Smith, who sent a sample of metallic ore to William Gesner for analysis. Gesner returned a button of tin he said had been extracted from the rock. However, when Smith sent the tin, along with a specimen of the same rock, to the Alabama state geologist, he was informed that the rock sample contained no tin at all and that the tin bead could not possibly have been extracted from it. The state geologist commented on the piece of tin: "If anyone says he has melted it from the ore sent me, have him repeat the process before your eyes." Ex-governor Smith realized that Gesner had "greatly deceived" him with the phony assay but took no action.[62]

The Gesners promised to reopen the Broad Arrow mine in the summer of 1888; fall came, and they blamed vague "metallurgical problems." Investors wearied of delays and promises, and eventually even the Gesners quit their rosy predictions. After the Gesners left, geologists examined the Broad Arrow property but could find no tin whatsoever.

Santa Ana Tin Mining Company, California

J. A. Comer formed the Santa Ana Tin Mining Company in about 1903 to exploit a great tin lode in Trabuco Canyon, Orange County, California. His company supposedly controlled eleven square miles of tin-bearing ground of astounding richness. When it became plain that the property had no tin, the desperate promoters made promises that were even more unbelievable. Their new secret process could increase recoveries more than tenfold. Among the metals now in the Santa Ana ore were platinum (more than ten ounces per ton), nickel, uranium, and cobalt. The tin mine was bought by the Borden Milk Company, which intended to use the tin for its milk cans; the company built an ore mill at the site, but the mine produced no tin.[63]

The Myth of the Tin Conspiracy

"Our Federal government insists on throwing the tin business and its control to the British and has issued an edict that there shall be no tin mining in the U.S.A."

—J. P. Hall, "Short History of the Struggle to Mine Tin in the West," *California Mining Journal*, August 1952

Tin promoters have long spoken of a conspiracy against tin mining in America. One of the Missouri tin swindlers intimated in 1869 that skeptics were in the pay of English tin interests. In 1876, advocates of a would-be tin mine in Maine likewise charged that tin monopolists had sabotaged development. In 1888 a correspondent blamed English tin interests for buying American tin deposits for the sole purpose of shutting them down. In 1903 the Santa Ana, California, tin promoters revived the anti-English accusation.[64]

After the Second World War, tin promoters in California and New Mexico charged that the US government was deliberately preventing American tin mining by withholding subsidies and loans. Government geologists who found the tin deposits uneconomic were denounced as part of the cover-up.[65]

The author Howard Clark charged in 1967 that the federal government had conspired to prevent tin mining in America. He said that, but for the conspiracy, tin mines would be operating in California, New Mexico, North Carolina, and other states. Little has been published in recent years of the supposed tin conspiracy, but it is no doubt still a topic in barrooms near the suppressed bonanzas.[66]

7

Professor Samuel Aughey:
"A First-Class Charlatan"

Many mining swindlers called themselves "professor," but Samuel Aughey was one of the very few with genuine academic credentials, for he was a professor at the University of Nebraska, and the territorial geologist of Wyoming.

Aughey actually started as a Lutheran minister. In 1871 the University of Nebraska hired him for a professorship of natural sciences. In the nineteenth-century West, where boosterism was valued more highly than scientific method, Aughey's missionary zeal pushed him to the forefront of acclaim. He became the first secretary of the Nebraska State Historical Society and the first president of the Nebraska Academy of Sciences. His lecture classes overflowed, and he even addressed the appreciative Nebraska legislature in 1873.[1]

Aughey preached the "rain follows the plow" theory, popular with Western farmers and land promoters. The theory predicted that agriculture on the great plains would cause rainfall to increase, so that western Nebraska would soon have the same climate as Illinois. Aughey's optimism extended to Nebraska's mineral wealth, and he long insisted that the state was rich with coal deposits, despite their observed absence. He wrote the *Catalogue of the Flora of Nebraska*, which won effusive praise at the time but was later labeled the work of a "first-class charlatan" who simply copied every western plant from standard reference works and claimed that they flourished in Nebraska.

He augmented his university salary with private work, but the board of regents questioned the time Aughey spent away investigating Wyoming coal deposits. Financial difficulties led to four court judgments against him in 1883. When he was accused of forging letters of credit, he blamed the trouble on enemies, resigned from the university, and moved to Wyoming. Aughey was among the first scientists to examine petroleum seeps over the Salt Creek oil field in Wyoming.[2]

Reverend Samuel Aughey was a professor at the University of Nebraska, then the state geologist of Wyoming, before swindling people in Wyoming and Arkansas with false assays. From Western Historical Co., History of the State of Nebraska, *1882.*

In October 1884 the governor of Wyoming appointed Aughey as territorial geologist. Aughey's zealous optimism led him to endorse a few mining frauds, but those eager to attract capital to the territory accepted these as mistakes made for a good cause. Aughey was reappointed in 1886 and was named a founding trustee of the University of Wyoming.[3]

Learning the Art of Fraud

In 1885, prospectors searched the foothills west of Cheyenne, Wyoming, where the Silver Crown mining district had been the site of repeated frauds over the past four years. Most assayers could not detect much of value in the Silver Crown rock, but those less scrupulous said they could find gold aplenty. One of those shamelessly exaggerating ore values was the territorial geologist Samuel Aughey.[4]

When wealthy cattlemen considered purchasing the Carbonate Belle prospect, they gave rock samples to different assayers but found that none but Aughey could detect any gold. Aughey told them that his assaying success was the result of a secret process of his own invention that could recover $200 in gold and copper per ton of ore.[5]

To prove his process, Aughey took a load of Carbonate Belle ore to the Hartsfeld smelter in Newport, Kentucky. The choice of a smelter so far away is puzzling, but its owner, Charles Hartsfeld, was well known in mining circles as a fraud. Aughey pronounced the Newport experiment with his secret process a complete success, but by the time he returned to Cheyenne, his body had been poisoned by smelter fumes. His Cheyenne backers feared that his secret process would die with him, because he refused to reveal it. They sent him to recuperate at Hot Springs, Arkansas, but not before he left a description of the secret process in a Cheyenne safe-deposit box, to be released to the investors only when they paid him or his heirs $100,000.[6]

The spas of Hot Springs gave Aughey back his health, and he returned to Wyoming, where the Carbonate Belle investors insisted on a test of the Aughey process. They chose Aughey's assistant territorial geologist, Wilbur Knight, as a scientist of unquestioned integrity. Knight was suspicious of his own boss and set exacting conditions. Aughey tried to supply the ore, but Knight insisted on collecting samples himself. No one but Knight was permitted in the laboratory during the test, so Aughey had to give directions from an adjacent room. Just before the test, Aughey specified some supplies not on hand, and Knight had to leave the laboratory to buy them.

At the end of the tests, Knight had extracted a gold bead from each of the duplicate tests. The weight of the beads showed that the ore was worth an astronomical $5,000 to $6,000 per ton. The investors congratulated the inventor and one another. Aughey immediately went to a Cheyenne bank to borrow $500, and two of his investors cosigned the note.[7]

But Knight was suspicious. He questioned the witnesses, who told him that while Knight was out buying the missing supplies, Aughey had needed to take some medicine and had entered the laboratory for water to wash it down. Knight closely examined the remnants of the crushed ore samples and found mixed in with them filings from a gold coin. Aughey had brazenly and crudely salted the samples and had nearly gotten away with it.

Knight resigned as Aughey's assistant, later citing the "crookedness of [his] superior, S. Aughey." Aughey was a popular figure, however, and both he and his intended victims resolved the situation quietly. Aughey resigned his position as territorial geologist, and newspapers reported that he had returned to Hot Springs, Arkansas, for his health.[8]

Arkansas Mining Frauds

Samuel Aughey landed in Arkansas amid one of the precious-metals frenzies that plagued that state in the late 1800s. He decided to nurture his finances as well as his health and promoted the most worthless mining schemes.

Reported silver strikes had lured prospectors to western Arkansas in the late 1860s, but no profitable mines resulted. Prospecting again surged in 1878, this time around the optimistically named Silver City, Arkansas. More than a dozen prospects were excavated and three smelters built, but the boom died. Silver City, once the center of Arkansas's would-be Comstock lode, is now just a spot along the road thirty miles west of Hot Springs.[9]

Another and equally false excitement now centered around the new mining boomtown of Bear City, Arkansas, west of Hot Springs. In 1887, encouraged by assayers who seemed to be able to find gold and silver in any rock, many people abandoned their work to prospect for gold and become rich. Mining companies capitalized for a total of $11 million organized and sold stock. At least seven prospects were excavated and three ore-processing mills or smelters built. Company promoters started a mining stock exchange in Hot Springs, Arkansas, to sell shares to one another and to tourists relaxing at the spas.[10]

One of the optimistic assayers of Bear City was "Professor" R. R. Waitz. In the early 1880s Waitz had promoted his own ore-treatment process at Silver Cliff, Colorado, a mining camp particularly victimized by "process men" like Waitz, who promised to extract more gold and silver than could be done using conventional methods. Waitz formed the Waitz Milling Company, took investors' money, and with much fanfare built an ore mill based on his new process at Silver Cliff. While the mill was under construction, Waitz attracted clients by giving them assays unusually high in gold and silver and promising similar recoveries once his smelter was running. The first results from the Waitz mill showed that it produced even more gold and silver than could be found in the assays. But despite years of tinkering, the Waitz mill never performed as promised. Waitz showed up in Arkansas, no doubt to repeat his Colorado performance. Word of his failure either never reached or never bothered the Bear City mine promoters, who were charmed by his ability to assay high gold and silver values from rock that other assayers said contained no precious metal.[11]

Aughey Endorses the Central Continental Fraud

The very profitable investments Saint Louis businessmen made in Colorado mines led to a mining-stock boom in Saint Louis in the mid-1880s. Shady promoters formed the Central Continental Gold and Silver Mining Company, ostensibly to mine properties in Colorado and Arkansas but really to fleece Saint Louis investors. For a measure of legal protection, the promoters located the company across the state line in East Saint Louis, Illinois.[12]

The Central Continental promoters hired Professor Samuel Aughey to promote its property in Montgomery County, Arkansas. Aughey obligingly found that the property was worth an astounding $50 million—a high price, but it would have been more than justified had the property contained the ore body Aughey described: three-quarters of a mile wide, five hundred feet deep, and miles long, all averaging $44.30 per ton in metals, a size and richness to make it one of the world's greatest precious-metals deposits. A. B. Webster, another "professor," joined in Aughey's fraudulent endorsement.[13]

Recognized and reliable assayers reported that they could detect little or no gold or silver in the Arkansas rock samples. The promoters' excuse was that the standard fire assay vaporized the gold and silver telluride minerals and so did not detect the metal. Modern-day prospectors sometimes give the same excuse. There is a grain of truth to this, for gold and silver tellurides do suffer some vaporization loss in the fire assay unless the method is altered slightly, but the loss is generally small. At any rate, the Arkansas ores did not contain enough tellurium to affect the results.

The Lost Louisiana Mine

Halfway between Silver City and Hot Springs, Arkansas, is a depression where a hot spring once flowed to the surface and deposited sooty black manganese oxide. Extinct hot springs are common in that part of Arkansas, and their manganese deposits have no precious metals. Prospectors dug a shaft down the throat of the old hot spring and drove tunnels branching off from the shaft but found no ore and abandoned the hole.

The owners of the property announced in 1886 that the old workings were the Spaniards' fabulous "Lost Louisiana" mine. That Spanish records mention no such mine did not limit the promoters' imaginations. They sent five tons of rock to the Hartsfeld smelter at Newport, Kentucky, where the swindler Charles Hartsfeld reported, as always, that the barren rock was rich in gold and silver. The Lost Louisiana promoters announced plans to build a mill large enough to process one thousand tons of ore per day.[14]

The only trouble was that most assayers could not find any silver or gold in the Lost Louisiana. Not to worry, however, for the assayers for the Lost Louisiana mine were Professor Samuel Aughey and Aron Beam, who reported an average of two and a half ounces of gold per ton in the Lost Louisiana ore. When the Saint Louis Assay and Testing Works failed to find even a trace of gold in the Lost Louisiana ore, Aughey and Beam went up to Saint Louis to demonstrate their superior methods to assayers in the big city.

On May 1 and 2, 1888, Aughey and Beam assayed thirteen samples of the ore under the close supervision of Saint Louis assayers but could not show more than a trace of gold. On May 3, however, they seemed to get their bearings and in the second sample managed to find eight ounces of gold per ton. The next day Aughey and Beam could find no gold in three assays, but in the fourth they showed over eleven ounces of gold per ton. All the tests over the four days were performed on the same ore. Aughey and Beam had shown a respectable average gold content, though very unevenly distributed.

Aughey and Beam's seeming success was punctured, however, when one of the observers, H. A. Wheeler, disclosed that in mixing the chemicals for his second assay of May 3—in which Aughey had found eight ounces per ton—Aughey had, incredibly, forgotten to put in any ore. Wheeler further revealed that on May 4 he had surreptitiously replaced the Lost Louisiana ore sample with iron ore, but Aughey and Beam had supposedly produced gold even from the ordinary iron ore. During the first two days, Aughey and Beam were watched too closely for them to perform their usual tricks. The pair had evidently introduced gold into two of the later samples by some sleight of hand, but only when the observers' vigilance was relaxed by the knowledge that there was no Lost Louisiana ore in the samples. The results were nationally publicized in mining circles.[15]

Since numerous Arkansas mining corporations were being floated in Saint Louis and New York, the Arkansas legislature appropriated money for a state geologist. Governor Hughes asked John Wesley Powell, director of the United States Geological Survey, to propose a candidate. Powell recommended Professor John C. Branner of Indiana State University, and the governor persuaded Branner to take the job. Branner studied Arkansas deposits of coal, oil, aluminum, lead, and zinc but refused to endorse the imaginary gold and silver.[16]

Aughey, Beam, and other boomers were causing enough excitement and controversy that Governor Hughes asked State Geologist Branner to investigate. Branner had already studied the Lost Louisiana and other mines and replied in writing on the same day that the gold and silver "mines" were worthless frauds. The Little Rock *Arkansas Gazette* printed Branner's letter.[17]

Branner's exposure of the fraud aroused the indignation of Arkansawyers with an interest in promoting the boom. They wanted a shameless promoter in the office of the state geologist (as Samuel Aughey had been in Wyoming) and urged the governor to fire Branner. The threat was no joke, as not many years earlier the state geologist in neighboring Missouri had tried to prevent a fraudulent tin-mining scheme and as a reward for his vigilance had been fired. The boosters in Bear City hung Branner in effigy.[18]

On September 13, 1888, Governor Hughes forwarded to Branner letters and petitions from detractors charging Branner with libeling their properties. Branner named eleven nationally respected mining experts and challenged the promoters to hire any of the experts to examine their properties. He promised that the state survey would publish the resulting reports verbatim, but no schemer dared take up Branner's challenge to allow honest engineers to evaluate his property. Governor Hughes kept Branner as state geologist, and the mining boom faded.[19]

Although exposed publicly as a cheat, Aughey stayed in Arkansas for a few years trying to keep his swindles alive. He was reportedly later caught trying to salt a mine in Missouri. He returned to preaching for a time in Alabama, then joined the faculty of a small Alabama college in 1898. Aughey moved to Spokane, Washington, in 1902, and reestablished his practice as an over-optimistic mining consultant, which he continued until his death in 1912.[20]

8

Swindle in South Park

Mining swindles sometimes discredit specific localities. Wickenburg, Arizona, for example, was so plagued by exaggerating mining promoters that the joke grew that whoever drank from the Hassayampa River was thenceforth unable to speak the truth.[1]

The Cripple Creek district of Colorado is one of the greatest gold producers in the United States. Yet the surface outcroppings of ore at Cripple Creek, neither remote nor inaccessible, were not discovered until 1891, years after the other great Colorado mining districts. The discovery of Cripple Creek was delayed partly because the area was associated with the infamous Mount Pisgah hoax.

S. J. Bradley—Drug Clerk, Prospector, and Swindler

In 1880 S. J. Bradley moved to Rosita, Colorado, where he was known for his whiskey drinking. He and a partner panned for gold in the summer of 1883 around the high mountain valley of South Park. His partner tired of prospecting, stole some horses, and fled to Montana. Unwelcome in both Rosita and South Park because of his partner's crime, Bradley moved to Leadville, Colorado.[2]

Bradley believed that he had found a rich gold placer in South Park, where Little Cottonwood Creek wound through ranch land more than eight thousand feet above sea level. In March 1884 he convinced some Leadville men to back him for a half-interest. Bradley and one of his backers traveled to his South Park prospects and carried samples back for assay. But the samples contained not even a trace of gold.

However, the following night one of his backers, George Nichols, invited Bradley to join in a swindle, and he immediately agreed. They told the other partners, most of whom were not in on the swindle, that the placer contained $900 per ton in precious metals.

The group organized the Hidden Treasure Mining Company, intending to sell stock and divide the proceeds. It would be Bradley's job to salt the samples. The evening before they returned to the site, Nichols and Bradley walked into a drugstore and bought nine bottles of gold dissolved in aqua regia solution.

How to Salt a Sample

"Salting," secretly adding a valuable substance to a sample, is a practice whose origin is lost in history. There are many ways to do it, most of them no doubt dating back at least to the alchemists of the Middle Ages. No mining country has been free of salting, although in the late 1800s, an English mining engineer proclaimed American swindlers unsurpassed in the dark art.[3]

In early Colorado salting was not even a crime. It was not until 1873 that a Colorado judge declared that salting was a crime under common law. The territorial legislature clarified the situation the following year by making salting a crime punishable by up to fourteen years in prison.[4]

Only a few solutions will dissolve gold; the most common in use at the time was aqua regia, a mixture of acids. Dissolved in aqua regia, gold can easily be added to a sample, and with a hypodermic needle can even be introduced into samples sealed inside bags. Gold dissolved in aqua regia was the means chosen by Bradley and his fellow swindlers to remedy nature's oversight.

S. J. Bradley, L. E. Parker, and George Nichols traveled to the prospect and took pairs of samples to be assayed by both Parker and Nichols using different assayers. Since Nichols was in on the swindle, Bradley salted only the samples to be tested by Parker. Parker nearly derailed the plan when he took all the samples as the three left the site. If Parker chose the unsalted samples for assay, the scheme would die. The next day, Parker and Nichols met to divide the samples. While Nichols distracted Parker in conversation, Bradley opened Parker's valise and divided up the samples, giving Parker the salted ones.

Parker gave his salted samples to a legitimate assayer. Nichols left his samples behind the bar in a saloon. Parker's assays, of course, found large values. When the group met to compare assays, Nichols made excuses as to why his assays were not yet completed, while another conspirator wrote down Parker's results. Nichols and Bradley later filled in blank certificates signed by a cooperative Leadville assayer, falsifying results to nearly match those of Parker's salted samples.

The fake assays were exciting: the gravel on Cottonwood Creek ran between 20 and 107 ounces of gold per ton. The elated group arranged for Bradley and Butters, another partner, to return the next day to stake more claims.

The celebration of the Hidden Treasure Mining Company alerted others to the new bonanza. In their greed Bradley and company had added too much gold to the samples. First from Leadville, then from Colorado Springs, Denver, Pueblo, and other towns, the footloose would hurry to stake claims along Cottonwood Creek. The Hidden Treasure Mining Company had hoped to entice faraway investors but did not anticipate crowds at the discovery site.

Bradley and Butters were to set out the next morning to stake mining claims over the prospect, but Butters learned that people intended to follow him, and he disappeared. Bradley went alone by train to Canon City, where he bought a bottle of aqua regia, in which he dissolved a $5 gold piece, to prepare for more salting.

Two men watched for Butters at the Leadville train station all day, intending to follow him, but Butters eluded them by renting a team of horses to an outlying station, where he boarded the train alone. The livery stable owner, however, chanced to remark on Butters's exit to one of the pursuers in a saloon that evening. Butters arrived at Canon City at midnight, with the surveyor David Miller. Also at the station were five men determined to follow Butters to the new field. Bradley, Butters, and Miller left Canon City in a wagon at 1 a.m., closely followed. Six miles outside Canon City, however, the pursuers' horses gave out, and they returned to Canon City.

The Hidden Treasure group arrived at Cottonwood Creek the next morning and staked claims. The pursuers finally arrived and found the Hidden Treasure people, staked three claims of their own, and took samples from Bradley's prospect pit.

Both parties of prospectors slept in a single ranch cabin. While they slept their samples were unguarded, usually in the pockets of coats left hanging on the wall, making it easy for Bradley to salt the samples in the night.

The Rush to Golden Valley

The Fairplay *Flume* skeptically noted on April 10 that Leadville promoters were attempting to generate a mining excitement. Gold hunters began leaving Leadville for the diggings on April 15. Many were jobless newcomers—the Leadville *Chronicle* called them "hayseeds." The exodus from Leadville increased over the next few days. Newspapers across Colorado picked up the story, and hundreds of searchers traveled to the camp via Canon City, the nearest railroad point.[5]

In August 1874 the geologist H. T. Wood had led an unsuccessful prospecting party to the eastern flank of Mount Pisgah. Colorado newspapers now remembered the Wood expedition and named the new rush the "Mount Pisgah excitement," even though Mount Pisgah was eight miles east and barely visible from the new camp.

Many, but not all, assays from the new field showed excellent gold values, but prospectors were puzzled that the ore looked like common gravel, without free gold or any of the usual signs of mineralization. A Leadville assayer found tellurium in one of the samples and concluded that the ore was in the relatively unfamiliar form of gold tellurides.

Canon City merchants sold provisions at inflated prices, to the envy of storekeepers in Pueblo and Colorado Springs. With profits at stake, promotion of the new gold fields became the civic duty of merchants and public men. Newspapers of all three towns ridiculed suggestions that the new district might not be a bonanza.

The Colorado Springs *Gazette* inaccurately claimed the better route to the gold fields and bitterly complained that Canon City merchants had bribed the venal Denver press to support Canon City's claim to the easiest route. The Pueblo *Chieftain* also insisted falsely that its town was very nearly as close to the new gold mines as Canon City.[6]

Villainy Detected

The assayers Robert Bunsen and W. B. Page became suspicious seeing the owners of the Hidden Treasure Mining Company celebrating in the streets of Leadville rather than developing their supposedly valuable claims. They put one of the South Park samples in distilled water and found that the water turned acidic. They added drops of silver nitrate solution to a tube of the water and watched a white precipitate of silver chloride form in the clear liquid. They notified another assayer, who tested the solution and found dissolved gold. Clearly the sample had been salted with gold in aqua regia.

At 9 p.m. Friday night the editor of the Leadville *Herald* was writing up the latest developments in the gold rush to Bradley's claim for the next morning's paper, when Page and Bunsen strode into the office carrying a test tube with the telltale white precipitate of silver chloride. The editor looked at the tube and realized that something was up. "Well, what new developments?" he asked the pair. "Here is an evidence of chloride in the new gold ores," answered one, recounting how their suspicions led them to test for signs of salting. "I tell you, it smacks of the syringe."[7]

News of salting in Leadville newspapers prompted recollections that Bradley and Nichols had bought gold chloride solution from a druggist. But the evidence was dismissed by a Canon City assayer, who noted that chloride was naturally present in parts of South Park. He accused Leadville of suppressing a rival mining district, while Leadville people pointed out that Canon City merchants and assayers had an interest in promoting a false gold rush.[8]

The name of the place changed from "Cottonwood Creek" to "Golden Valley." Tent saloons poured liquor for the thirsty. Newspaper headlines attracted people unfamiliar with prospecting or mining. A Leadville correspondent noted an "appalling" number of Denverites in the camp.[9]

Bradley returned to Canon City for more aqua regia. He took the opportunity to celebrate in Canon City saloons and returned to the camp that night, the bottle of aqua regia in his pocket. In his drunken state he fell out of the wagon, breaking the acid bottle and burning himself and his pants. The acid burns aroused suspicion, and the assayer Arthur Chanute tested a piece of Bradley's ruined pants for traces of gold. Luckily for Bradley, he had not yet dissolved gold into the acid, so the test found no gold, but suspicion increased.[10]

On Saturday a storm covered the ground with snow. Three to four hundred prospectors camped in the inclement mountain weather, impatient to know if they had been duped. The crowd pressured Butters and Bradley to allow sampling of the discovery shaft. Bradley and Butters objected that their shaft was full of snow and, besides, the windlass at the top of the shaft had disappeared. An old miner suggested they descend the eighteen-foot shaft on a rope and dig out the snow, and the crowd forced Bradley and Butters to relent.

The prospectors quickly cleared the shaft of snow and bullied Bradley into descending the hole with the miner to show him the supposed ore gravel. The old miner dug three feet into the wall to collect a sample that could not have been salted, to Bradley and Butters's visible consternation. The assayer Harris took some of the samples and left for Canon City to assay them, promising to return as soon as he had finished testing the samples.

That evening Harris had not returned. Arthur Chanute grew impatient and left for Canon City with the remainder of the samples, promising to return immediately if the samples showed gold. A prospector volunteered to accompany Chanute, and if the samples showed no gold, he promised to return with a bucket of tar to apply to Bradley and Butters. Although it was late night when he reached Canon City, he found Harris, who had just finished testing his samples without finding gold. They immediately began to assay Chanute's samples.

By 5 a.m. Sunday, Chanute and Harris had assayed all samples and found not even a trace of gold. The negative results immediately made Harris and Chanute unpopular with Canon City storekeepers. A Canon City assayer admitted that he had faked results because the boom was good for the town. Chanute sent a letter to the prospectors at the site, slept until Sunday afternoon, and caught the train back to Leadville.[11]

Hopeful fortune seekers continued to stream in from all points, and by Sunday night the population had swelled. Eyewitnesses estimated the crowd at no more than five hundred at any one time, although many came and went.[12]

The Hidden Treasure Company men met quietly Sunday night for an organizational meeting of the new mining district. They passed regulations providing for claims of twenty acres each, allowing unlimited time to prove up the claims and promising to shoot anyone trespassing on the original discovery claim. The rules favored the Hidden Treasure group by giving them large claims, allowing them to hold the claims without working on them, and preventing further sampling of the discovery shaft.[13]

Goodbye Golden Valley, Hello Suckertown

On Monday, April 21, the gold seekers learned of the previous night's mining district meeting and immediately acted to nullify it. More than two hundred men met in a mining district meeting that morning and voted to limit claim sizes to ten acres and to require immediate work on each claim. During the meeting the assayer Chanute's letter arrived from Canon City and was read to the assembly, informing them that none of his samples showed any trace of gold.

The crowd repaired to the discovery claim to sink another shaft thirty feet from the discovery shaft. Butters refused, but the angry men were in no mood to be stopped. With many men joining in, the new pit soon reached the ore horizon, and a select committee of six went down to sample the gravel. The committee was composed of two men each from Lake County (representing Leadville interests), Fremont County (the Canon City interests), and Park County, where Golden Valley was located. The six brought the samples to a tent where they carefully watched another committee of three assayers jointly test each sample. After each sample had been assayed twice, the samplers and assayers announced to the crowd outside that not a single one of the twelve tests had found any gold.

The crowd immediately searched for Bradley and Butters, who had disappeared. Bradley had received a note from Leadville advising him to leave because word of the fraud was spreading and a wagonload of rum was on its way from Leadville. Bradley fled, afraid that a liquor-filled crowd might lynch him. He lost the trail in fog and nearly died from cold before finding shelter.

Disappointed gold seekers consoled themselves with a drunken spree. The prospectors changed the name of the camp from "Golden Valley" to

"Suckertown," but on their way out they met a like number coming in, most of whom refused to be dissuaded, so by Tuesday there were probably as many searchers as there had been a day or two earlier.[14]

The gold rush was over on Cottonwood Creek but continued in city newspapers. The Pueblo *Chieftain* said that the Leadville *Herald*'s cry of fraud was only jealousy. Reassured by the *Chieftain*, another one hundred Pueblo men left Tuesday for the gold fields.[15]

The Denver *Republican* correspondent refused to believe that it was a fraud, saying that someone would have had to douse gold chloride over an area of more than ten square miles. The Denver *Tribune* writer had been in the camp over the weekend, but he somehow missed the main events of the discovery of the fraud. His wired story from Canon City on Tuesday called the fraud charges "pure idiocy." However, by Wednesday, April 23, Bradley's flight and discouraging reports from camp convinced newspapers in Denver and Pueblo—but not in Colorado Springs—that there was no gold in Golden Valley. According to the Colorado Springs *Gazette*, the stream of people leaving for the camp continued to swell. At the diggings, however, the story was different. Returning prospectors convinced latecomers to turn back, and by Wednesday night, only about one hundred people remained.[16]

Bradley's Surprise Dividend

"I was an equal partner, so here is their share of the first dividend declared by the Hidden Treasure Mining Company."

—S. J. Bradley's confession, *Denver Tribune*, May 12, 1884, p. 2

Bradley resented being a fugitive while the businessmen who had him do their dirty work walked the streets of Leadville. Several weeks later, the Denver *Tribune* found Bradley, who signed and notarized a written confession in Pueblo on May 9.[17]

Bradley's confession named Bradford, Henry, and Nichols as coconspirators . To substantiate his account Bradley gave the *Tribune* a warning note that Nichols had sent him. Bradley went out of his way to say that Butters and Parker were innocent of the fraud. Bradley remarked: "As for Parker—well he is prey for anyone." A Red Cliff, Colorado, correspondent agreed: "His simple and trusting disposition is too fragile for a thin atmosphere. All here pronounce him the most perfect tenderfoot that ever ventured west of Denver. A babe could bamboozle him in salting samples of dirt."[18]

However, the statement of a confessed salter must itself be taken with a grain of salt. The level of detail of Bradley's account convinces one that others joined him, but Bradley may have changed the particulars. He tried to clear Butters, but eyewitnesses described Butters's actions as highly suspicious. In addition Butters had a reputation as an "active schemer and rustler." It is possible that Bradley cleared Butters because Butters had stood by him when the prospectors were becoming hostile.

Bradley blamed Henry, Bradford, and Nichols for setting him up as the scapegoat. S. J. Bradley disappeared and was not heard from again. Nichols, Henry, and Butters also left town. Edward Bradford alone steadfastly denied complicity and remained in Leadville.[19]

The Shadow of Fraud

"There is a new prospecting boom in the neighborhood of Florissant. It is to be hoped that it will turn out better than the late one, which came to so sorry an end."

— Early notice of the Cripple Creek discovery in
Mining Industry and Tradesman, May 7, 1891

The Hidden Treasure Company's swindle was exposed less than a week after the public learned of the supposed discovery. The fury came and went as quickly as a spring snowstorm. Prospectors left behind only their curses, and hustling Golden Valley went back to being peaceful Cottonwood Creek, home only to cattle. No physical trace remained to mark Bradley's claim, but the public memory lasted for years. The swindle was inaccurately publicized as the "Mount Pisgah fraud," and Mount Pisgah became synonymous with chicanery and false hopes. On the eastern side of Mount Pisgah, however, great stores of gold ore remained undisturbed, now protected from discovery by Mount Pisgah's unsavory reputation.

Years later, the cowboy Bob Womack, not knowing any better, stubbornly prospected the high cattle land east of Mount Pisgah and discovered the gold deposits of Cripple Creek. Womack had difficulty convincing mining men that he had discovered gold tellurides near Mount Pisgah, because they had heard it all years before from the Hidden Treasure Company swindlers. Bradley's fraud probably delayed for years the discovery and development of one of the greatest gold-mining districts in the United States.

9

Leadville: A Comstock in
the Colorado Rockies

*"Capitalists came in and worthless mines were palmed off on them,
and as a natural consequence their fingers were burnt clean to the
knuckles. It was then they formed the opinion that all mines were
frauds, and every miner a blackleg and a swindler."*

—George Daly, manager of the Little Chief mine, Leadville

S ilver discovered in the late 1870s in the Colorado Rockies caused
Leadville, over ten thousand feet above sea level and the highest city
in the United States, to spring up near the Arkansas River, there only
a snow-fed freshet. Promoters sold companies at prices beyond the capacity of even the best properties to justify. After the inevitable downfall, the
frauds caused the frenzied purchase of Leadville shares to be replaced by
irrational avoidance of all mining stocks.

San Francisco had been the center of American mining finance since
the California gold rush and still controlled most western mines. San
Francisco mismanagement of the Comstock and other mining districts
had long appalled New Yorkers. They had heard of and disdained "California methods" of abusing mines for stock manipulation, even though
the railroad stock manipulations of Drew, Gould, and Vanderbilt in New
York exceeded anything in San Francisco. Now in the midst of an economic boom, New York investors were determined to save fair Leadville
from the Comstock's shameful fate by bringing it under honest eastern
control—and of course gain enormous profits for themselves. So Leadville, Colorado, became the first great mining district controlled in New
York, whose shareholders were in for an education in the ways of mine
swindlers.

Big Swindle in the Little Pittsburg

Horace Tabor had knocked about Colorado for years, chasing each rumor of gold. He and his wife ran a general store that moved with each gold rush and grubstaked prospectors by giving them food and supplies in exchange for a portion of any discoveries. He grubstaked George Hook and August Rische, and when they found the Little Pittsburg deposit near Leadville in May 1878, they and Tabor were suddenly rich. Tabor bought out his partners and became the sole owner.

But the rush of prospecting had created a welter of conflicting mining claims. Little Pittsburg miners exchanged pistol shots with miners from the overlapping Winnemuc claim. When US Senator Jerome Chaffee bought an interest for himself and the banker David Moffat in the New Discovery claim, which also overlapped the Little Pittsburg, Tabor knew that he faced well-connected foes who could steal the Little Pittsburg in the courts. Tabor and Chaffee engaged in a spirited duel for control in which they soon bought out the other owners in the New Discovery, Winnemuc, and Dives claims.

They faced one another with a ruinous legal contest but wisely agreed to consolidate their interests. Tabor, Chaffee, and Moffat agreed to jointly buy out the remaining interests. They combined their claims with three adjacent properties to form the Little Pittsburg Consolidated Mining Company. To pay for the buyout, they needed to sell nearly half the shares to outside investors.

David Moffat and Jerome Chaffee were experienced mine investors—some said manipulators. They had been accused of duping Dutch investors into paying an astronomical price for the Caribou mine, the subsequent failure of which gave American mining a black eye among the Dutch. They were also key figures in an apex lawsuit against the Terrible mine, owned by English investors. The lawsuit effectively forced the owners to give Chaffee and another plaintiff a share in the profitable mine near Georgetown, Colorado. The English saw the lawsuit as nothing more than legalized theft, leading one to comment bitterly, "Any English capitalist is a downright fool to buy a mine in this district."[1]

A Pleasure Trip to New York City

In early 1879 the Little Pittsburg mine was profiting $300,000 per month, and Moffat and Chaffee rode a private rail car to New York City to impress wealthy New Yorkers. Potential shareholders were awed by the display of conspicuous consumption by the newly minted millionaires from the hinterland.[2]

A Pleasure Trip to Leadville

Moffat and Chaffee's pleasure trip had the desired effect on New York: it excited desire to see the mines that paid for such luxury and to emulate the success of the Colorado millionaires. An excursion was organized in the opposite direction, and Moffat and Chaffee led influential easterners through the tunnels of the Little Pittsburg. Moffat hoisted $100,000 of ore from the mine in a single day to impress the visitors, convincing them that this mine was a fail-safe investment.[3]

The mining experts Rossiter Raymond and Winfield Keyes examined the mines. Raymond found that the ore in sight would yield a profit of $10 per share, with another $5 per share probable. Moffat and Chaffee had no trouble finding prominent New Yorkers to serve on the board of directors, and the first issue of shares was quickly subscribed at $20 per share. Horace Tabor took the opportunity to sell his interest for a round million dollars to Moffat and Chaffee. The second issue, in October 1879, sold well at $25, after which the stock was accepted on the New York Stock Exchange. The big board was usually leery of mining stocks, relegating them either to the less fastidious Mining and Petroleum Exchange (the little board) or to the curb market.[4]

From the start, Little Pittsburg shares paid $0.50 per month in dividends, and the company had a growing treasury surplus. Despite the high rate of ore production, the company claimed that the miners were finding new reserves at twice the rate that they were mining and that known ore reserves guaranteed steady dividends stretching five years into the future. Share prices rose to a high of $34.50 in December 1879.[5]

The Inevitable Crash

The ore reserves were a figment of publicity. The mine superintendent was under strict orders to supply $100,000 per month in dividends, whatever the effect on reserves. The high production far outstripped discovery of new reserves, and the ore was nearing exhaustion.[6]

Moffat and Chaffee led the management in issuing happy press releases while selling Little Pittsburg shares as rapidly as possible. The corporation secretary contracted to sell even more shares than he owned and ended up with a very profitable short position. The other directors booted him out for bearing the stock, but his offense was really to have sold his shares before his fellow directors did likewise. The great blocks of shares thrown on the market woke up investors, and word leaked out in March 1880 that the ore was reaching its end. The company barred visitors from inspecting the mine, further inflaming rumors.[7]

When the truth finally emerged in March, everyone denied blame, most loudly Moffat and Chaffee. They proclaimed that they didn't release reports that the mine was nearly exhausted because they believed that further exploration would find still more bonanzas, but their massive sales of shares belied their words. By mid-March, Moffat, Chaffee, and the other directors of the Little Pittsburg—all of whom expressed great surprise at the depleted ore reserves—held only token amounts of stock.[8]

Rossiter Raymond, who was blamed by some for his earlier favorable report, correctly pointed out that he had estimated that the known ore was worth only $2 million to $3 million, equal to $10 to $15 per share, less than half the high price reached by the market. The mine had followed Raymond's prediction by producing $2 million before running out of ore. Raymond went back down into the Little Pittsburg but found that the only immediate way produce ore was to rob the pillars, although the Leadville *Democrat* twisted his comments around to headline "Plenty of Ore Still Left in the Mines."[9]

Having deftly used their inside information to sell ahead of the crash, Chaffee and Moffat pocketed their profits and resigned from the board. The disgraceful management of such a famous mine as the Little Pittsburg brought a general disrepute to mining stocks in New York, and mining share prices suffered generally.[10]

Moffat and Chaffee on Trial

One of the excursionists of 1879 sued Moffat and Chaffee for $11,000, alleging that they had made false statements in 1879 to induce him to invest in the Little Pittsburg. The case was duly heard in the New York courts. The jury deliberated all night but at 8:30 the next morning declared Moffat and Chaffee not guilty.[11]

The Little Pittsburg issued its last dividend in March 1880 and share prices declined to less than $1, despite occasional bonanza rumors. Moffat and Chaffee are remembered today not for their sharp practice in the Little Pittsburg but for their business and political successes. Each has a county in Colorado named after him.[12]

The Chrysolite Mine: Horace Tabor Is Swindled into a Fortune

The prospector "Chicken Bill" Lovell had earned his nickname by being caught in a blizzard while hauling a wagon load of chickens over the mountains. By the time he joined the rush to Leadville in 1879, Lovell had reached fifty-two years of age in a young man's occupation. He was known as a talented prospector, but after years of searching, wealth had eluded him.[13]

Horace Tabor was famously fooled by "Chicken Bill" Lovell
when the latter "salted" his Chrysolite claim and sold it to
Tabor. Tabor had the last laugh when he dug a few feet farther
and found a large ore body. Courtesy Library of Congress.

Chicken Bill and his brother Joe had located two mining claims adjacent to the Little Pittsburg claim, but after digging a shaft through thirty feet of rock and not finding ore, the two concluded that the Little Pittsburg ore body did not extend onto their property and decided to sell it. The task was given to brother Bill. August Rische, superintendent of the Little Pittsburg mine, owed Chicken Bill a favor, so when Lovell asked him for a wheelbarrow of rich silver ore from the Little Pittsburg, Rische agreed.[14]

Lovell took the ore from Tabor's mine, dumped some of it down his own shaft, and piled the rest at the mouth of the pit. He then offered the claims to Tabor himself, who was shopping for mining properties. Tabor examined the claims and personally took samples from the bottom of the shaft. Seeing that the ore in Lovell's shaft was of course every bit as good as that in the nearby Little Pittsburg, Tabor bought the claims from the Lovell brothers for $2,700.

Chicken Bill bragged around Leadville about how he had salted the shaft and swindled Tabor. Tabor realized that Lovell's shaft was not in ore, and confronted Lovell on the street in Leadville. "I thought that you said that there was good ore on that property?" Tabor challenged him. "I know that there was," Lovell countered, "because I put it there myself."15

Having bought a prospect salted with ore from his own mine, Tabor deepened the shaft, and his miners soon found the rich ore body that would become the Chrysolite mine. Chicken Bill had swindled himself and his brother out of a fortune.16

Chicken Bill Lovell continued prospecting. He is said to have swindled an English visitor to Leadville, again using borrowed ore. Although he was credited with finding ore deposits in the Mosquito and Aspen districts of Colorado, the cheater never prospered, and died destitute in 1886, aged fifty-nine. Brother Joe Lovell continued working in the Leadville mines as superintendent. No taint of swindle attached to him, and he eventually retired to California.17

The Chrysolite Mine Again: The Public Is Swindled out of a Fortune

Tabor combined the Lovell brothers' claim with six others and floated them as the Chrysolite Mining Company in November 1879. On the strength of dividends of $200,000 paid each month, shares rose from less than $13 to more than $40 before the end of 1879.18

George D. Roberts Brings Comstock Methods to Leadville

The late 1870s saw a boom in mining shares in New York. Prominent New Yorkers interested in mining formed the Bullion Club to promote mining investment. A cynic might have examined the membership roster and discovered a who's who of notorious San Francisco stock manipulators.

California operators migrated to New York in the late 1870s and brought Comstock-style swindling with them. Years of swindling and a change in the California constitution discouraging stock speculation had depressed San Francisco stockbroking, and many brokers moved to New York. James Keene, always a step ahead of the market, was the first to migrate east. Asbury Harpending and William Lent, both of diamond-swindle fame, also set up shop in New York to take advantage of the new interest in mining shares, and both continued the manipulations they had learned in San Francisco. The American Mining Stock Exchange opened in New York in June 1880: its chairman was formerly chairman of the San Francisco Stock and Exchange Board; the secretary of the new exchange was likewise from California.[19]

After James Keene, the most successful of the transplants was George D. Roberts, a San Francisco stock operator who relocated to New York and began one ill-fated mining promotion after another. He arrived in Leadville in February 1879 with the mining engineer Winfield Keyes, looking for mines to purchase. According to one report, despite Roberts's high stature as a financier, when he arrived in Leadville he was broke and operating on pure bluff.[20]

Roberts bought a number of properties. He floated the Iron Silver Mining Company, the Little Chief Mining Company in Leadville, and the Robinson Consolidated Mining Company a few miles north across Fremont Pass. He bought the closely held Chrysolite Mining Company, reorganized it, and sold the new shares in New York. Since Roberts was bringing in capital and running the mines at full force, the Leadville press hailed him as a mining genius. He appointed his cronies to operate the mines.[21]

Roberts's San Francisco stock manipulations were well known, but the Leadville *Democrat* welcomed what it called the "'gambling' system of mine development" and wished the Californians success in establishing the same in the startup New York mining exchange. The recently amended California constitution, said the *Democrat*, had brought depression to the California and Nevada mining industries by driving out stock gambling. In his Leadville flotations—in contrast to his promotion of the State Line companies in Nevada—Roberts dealt only in first-class properties; in energetically developing the mines, he employed too many of their newspaper readers and advertisers for any Leadville newspaper to oppose him. In fact the Leadville papers had a reputation for uncritically puffing any mine in the district. A prospector who failed to convince the Aspen, Colorado, *Sun* to print praises of his worthless claim scolded the editor that the Leadville papers were more accommodating to promoters.[22]

The new Chrysolite Company began paying monthly dividends of $1 per share, which pushed prices up to $40. The mine manager Winfield Keyes calculated mine reserves of $3 million in January 1880 but revised his estimate upward to $7 million in March, more than enough ore to keep miners busy for the next three years.[23]

School Ties

In March 1880 the Chrysolite company hired Rossiter Raymond to examine the mine. Raymond and mine manager Winfield Keyes had been friends since their student days at the mining school at Freiburg, Germany. Since he knew his friend Keyes to be capable, Raymond accepted the mine manager's information on ore reserves without as careful an underground investigation as he would normally perform. After two days in the Chrysolite, Raymond wrote a highly favorable report that evening and received $5,000 for the examination. As editor of the *Engineering & Mining Journal*, the industry's most reliable and trusted periodical, Raymond was in a unique position to promote the Chrysolite. On April 17, 1880, the *Journal* asserted, "That this stock is selling below its actual value, there can be no question." Raymond did not discover until too late that his friend Keyes had deceived him.[24]

The very week after the *Journal* told readers that $19.75 per share was too low, the Chrysolite Company cut its dividend in half, and the price declined further. The *Journal* was still confident that the Chrysolite had ore in sight to pay dividends of at least $26 per share over the next three years. A representative of a large New York investment syndicate went to Leadville and wired back: "The Chrysolite is all right. Raymond's report fully confirmed." Despite the assurances of the leading investors and mining periodicals, the stock declined further to $13.25.[25]

In mid-May company officials suggested that the May dividend would be omitted but insisted that the long-term prospects were as bright as ever. Share prices sagged further amid rumors that Raymond had overestimated the Chrysolite's ore reserves. Raymond hurried west to Leadville once more and reexamined the Chrysolite. Once again he left confident of the Chrysolite's richness and blamed the stock slump on stock manipulators and "panic-stricken amateur investors." In reality it was the insiders who were unloading their shares as fast as they could.[26]

When ore shipments lagged and the June dividend was omitted, the shares slumped to $6, but the *Engineering & Mining Journal* insisted that the true value of the mine would soon send them back up.[27]

Workers walked out of the Chrysolite mine one morning in May 1880, angry that the manager Winfield Keyes had forbidden the miners to smoke or to speak to one another in the mine. The Chrysolite miners demanded a wage increase and the right to name shift bosses, and miners from the other mines soon joined them and shut down all the Leadville mines. Some concluded later that Keyes deliberately provoked the strike to delay the revelation that he was running out of ore. Certainly his prohibition on workers' speaking to one another was odd, but the ban on smoking was no more than a commonsense safety precaution. At any rate, the speed at which the strike spread to all the Leadville mines shows that the issues went beyond Keyes. The Chrysolite reopened just a week later, but Keyes blamed the strike for interrupting ore shipments, forcing still another dividend to be skipped.

As it became apparent that the mine output had previously been pushed to unsustainably high levels, a number of directors resigned, including George Roberts. Now that he had unloaded his stock at inflated prices, Roberts played the opposite game and tried to drive prices down. After a brief visit to the Chrysolite in July, he pronounced the mine to be without ore and valueless. The new directors discovered that each previous dividend been paid partly out of borrowed money at high interest rates, so that while paying $1.1 million in dividends, the mine had piled up a hidden debt of nearly $500,000.[28]

The Hidden Tunnel

Chrysolite shareholders weathered another misfortune when a shaft house caught fire and the fire spread down the timber-lined shaft into the mine. The repairs sunk the mine still further in debt. Despite the mine's problems and obvious mismanagement, the *Engineering & Mining Journal* persisted in its deluded faith in the Chrysolite and advised shareholders not to sell their shares. At the annual meeting in November 1880, the *Journal* editor Rossiter Raymond was elected president of the company.[29]

In the mine the new management discovered a hidden tunnel that led into an unmined block that they had counted as ore. The walls of the secret tunnel showed only barren limestone and proved that the mine had much less ore than they had thought. The mine manager Winfield Keyes had driven the tunnel to prove up the ore in that section, but when it found only barren rock, he covered up the entrance to the crosscut and lied to his friend Raymond about the ore reserves.[30]

The Chrysolite mine continued to ship silver ore, but the proceeds went to rebuild from the fire and to pay off the debts incurred to pay past dividends. It was not until August 1881 that the company could resume the dividends suspended in May 1880. Two dollars and fifty cents in dividends were issued in 1881, after which they stopped. After Raymond left, new management picked up the old practice of false press releases, stating that the mine "never looked better." The miners continued to explore underground, but the mine was practically exhausted. A final dividend of $0.25 was paid in 1884. Out of the $26 per share in future dividends that Raymond had calculated as the "minimum" in March 1880, only $2.75 had materialized.

Gutting the Robinson Mine

"There has been a great deal of pretty tall lying going on some-where."

—Brayton Ives, president, Robinson Consolidated Mining Company

The Robinson Mine was in the Tenmile District, just over Fremont Pass from Leadville. George D. Roberts bought a major interest in the property in March 1880, incorporated it, and put it on the market as a dividend-paying silver and lead mine.[31]

The former owner, George Robinson, stayed on as a director and major stockholder in his namesake mine, but after his death in November 1880, the Robinson heirs sold their interest to a group of investors headed by the New Yorker Wilson Waddingham and the San Francisco mining man Thomas Ewing. Roberts appointed Ewing as the new mine superintendent then stepped down as company president.

Roberts's man Ewing sent sensational telegrams from Colorado, hailing new ore discoveries in the Robinson that established it as the "greatest mine in the West." The *Engineering & Mining Journal*, having been burned by mistaken confidence in George Roberts's Chrysolite boom of the previous year, was skeptical. The magazine wrote that it had reliable information that the supposed new discoveries were merely the same well-known vein, without any increase in ore grade. Despite the *Journal's* warnings that $10 was too high, share prices climbed to $14.50, and the monthly dividends and increased ore shipments seemed to vindicate the boomers.[32]

Roberts and Ewing were playing the familiar game of overproducing the mine for a splash. The monthly dividend doubled, and share prices boomed. The miners had nearly exhausted the ore body, but the mine manager Ewing claimed that there was still ore aplenty. Silver output plummeted, but Ewing covered this up by secretly borrowing $90,000 from a Denver smelter, money he used to pay the November 1881 dividend.

The Robinson allowed occasional visitors underground, but Ewing would not permit outsiders to measure or sample the veins. Company president Brayton Ives began to suspect his own mine manager and instructed him to allow the mining engineer William Ashburn to study the Robinson mine. Ewing used what little ore was left to "dress" the mine, skillfully exposing rich but thin veins in such a way as to appear as solid masses of ore. Ewing also bribed miners on the lower level to report that they were finding large bodies of high-grade ore. Ashburn directed the sampling and assaying but felt too ill at the eleven-thousand-foot altitude to do the actual work. Ewing selected the miner who took the samples;

the assayer was likewise Ewing's employee. Ewing had every opportunity to substitute higher-grade ore samples and to salt the assays. Ashburn reported back to the New York management and investors that the Robinson mine contained ore in sight worth $2 million.[33]

Robinson shares had been riding high on the stock exchanges on the strength of Ewing's double dividends. Ashburn's report should have pushed shares even higher; instead share prices dropped drastically, because Ewing's partner, Wilson Waddingham, was selling huge quantities of stock.[34]

As soon as he had arranged the November dividend and Ashburn's mistaken report, Thomas Ewing discovered that he had a heart condition that required removal to a lower altitude, and he returned home to San Francisco with the universal thanks of the company directors and investors.

The new mine manager quickly discovered that the mine was exhausted and that the company had borrowed the money used to pay the last dividend. On November 30, 1881, he wired the findings to his superiors and began laying off miners to save money. For weeks afterward the New York headquarters received a telegram every day with the latest sad revelation. The directors appeared as confused as lost children and could only mouth hollow words of confidence in the future of the mine. The New York *Tribune*'s stock market correspondent clung to the belief that the bad news were falsifications to start an "unwarranted 'bear' attack." Some correctly suspected Roberts of dishonest manipulations but wrongly deduced that he was orchestrating the bad news to buy back his shares at a lower price.[35]

The directors immediately ordered the mine examined by more experts. Since the directors themselves were now widely mistrusted, some shareholders sent their own mining engineers to Colorado. Soon five experts and a director were roaming the tunnels of the Robinson mine. The company president traveled to Colorado to investigate Ewing's secret loans, which the company now had to repay. William Ashburn hurried back to Colorado and agreed with the other experts that his previous report was completely wrong. The $2 million of ore had evaporated, and the mine was nearly worthless. Most miners were laid off at the end of December 1881.[36]

Ewing was denounced in New York yet had his defenders. The Leadville *Herald* said that in lying and taking advantage of inside information, Ewing had simply done what any other shareholder would have done if given the chance. Ewing retained the trust of George D. Roberts, who hired him to inspect his latest property at Lake Valley, New Mexico.

The Robinson Company lost one journalistic supporter when the new management refused to pay a $60 advertising bill from the New York

Stock Report. The "advertisement" was actually an editorial supporting the company. The *Stock Report* thereafter had little good to say about the Robinson mine.[37]

The company explored deeper hoping to find more silver, while manipulators spread rumors of new ore bodies and spread money for favorable publicity. In February 1882 the Boston *Advertiser* predicted that the share price would soon double, for which the *Advertiser* was taken to task by the Leadville *Herald*. The *Herald* evidently saw irresponsible Robinson rumormongering as its own exclusive prerogative, because a short time later it also began publishing false reports that the Robinson mine was about to return to its glory days of rich ore and large dividends. The original schemers Thomas Ewing and Wilson Waddingham even had the daring to return to inspect the Robinson mine in April, enthusing about the future of the worked-out property. The rumors and pretense continued until June 1882, when the owners laid off the remaining miners and turned the mine over to tributers.

Wash sales and other manipulations kept the shares alive on the New York market for years afterward, as stock sharpers jiggled the dead bait hoping to entice some sucker fish to bite. The Robinson Mining Company was finally thrown off the exchange in a general housecleaning of worthless stocks in 1896.[38]

The Olathe Silver Mining Company

London investors were not neglected in the Leadville boom. Promoters bought an unproven claim and floated it in England as the Olathe Company. The claim was inconspicuous and practically unknown in Leadville, but its not highly regarded promoter trumpeted it in the London press as one of Leadville's great mines. When inquiries reached Colorado, Leadville newspapers objected that they had never heard of the Olathe Company. After they determined that there was such a claim, the papers noted that it had escaped notice because no more than minimal development work had been done on the claim and it was no better than a hundred other undeveloped tracts. The Leadville papers were quick to extravagantly praise any working mining company, but since the Olathe Company was selling shares in London without even employing Leadville miners, the local papers heaped abuse upon it.

Advertisements contained telegrams that had supposedly been received from Leadville detailing fabulous discoveries in the Olathe workings. Some ads included a letter from a Leadville woman describing the great promise of the mine, but locals recognized the lady's return address as a Leadville bordello.

More Manipulation in the Little Chief Mine

The Little Chief was another mine run by George Roberts and his "California crowd." Roberts's man Keyes was general manager, but daily operations were in the hands of the superintendent, George Daly.

Short and pugnacious, Australian-born Daly had arrived in California as a young man and become a mine manager, first at Aurora, Nevada, and then at Bodie, California. In 1879, when another Bodie company tried to jump part of the ground owned by his Jupiter mine, Daly's men fought back, killed one of the claim jumpers and captured the rest in a nighttime gunfight. Although Daly was not present at the battle, he was arrested for murder along with his men. Daly and his men were acquitted by the court, but he left Bodie after the Miners' Union unanimously voted to lynch him and nearly succeeded.[39]

Daly moved to Leadville, where he soon had the Little Chief producing a steady stream of silver and dividends. Roberts and associates, however, made their money not on dividends but on wild fluctuations of share prices. Over Daly's protests, the New York management ordered that he extract ore as rapidly as possible, leaving insufficient manpower for the underground exploration necessary to replace reserves.

Little Chief shares rode high on the wave of dividends, enabling Roberts to sell out at a great profit. When the ore inevitably ran out, Daly reported that the known reserves were exhausted, ore shipments dropped to a fraction of their former tonnage, and share prices crashed. For his candor Daly was accused of sending false reports to drive down share prices. He dutifully offered to resign, but before his resignation letter arrived in New York, the directors fired him. His replacement found that Daly had reported truthfully and that the Little Chief would have to invest time and money in developing new ore reserves before ore shipments could approach their former level.

Daly left for Lake Valley, New Mexico, to direct more silver mines for George Roberts and another East Coast promoter of Leadville mines, Whitaker Wright. Daly died soon afterward when he and other miners rode into an Apache ambush.

Later Days of Leadville

The swindles robbed Leadville of the confidence of investors and made it very difficult for miners to raise new capital. In the depths of its post-boom hangover, the district realized too late the damage done by abusing investors. According to the Leadville *Chronicle*:

Too many men have gone East and told lies about their prospects. Too many mine managers have promised dividends, well knowing that the promises could not be fulfilled. Too many men are walking around Leadville proclaiming the enormous value of prospects which might be dear at $1, and all this wholesale lying has produced its natural fruit.[40]

After 1881 Leadville shook off its frauds and settled into the business of silver mining. Although swindling continued in Comstock shares into the twentieth century, Leadville had no more major scandals. One reason was that the Leadville shares traded principally on the New York market rather than on the smaller and more easily manipulated San Francisco exchanges. Also, in contrast to the Nevada mines, the leading Leadville shares were non-assessable, which removed an aspect of swindling that continued to plague the Comstock.[41]

IO

Tichenor's Gold: Calistoga Natural Springwater

Anson C. Tichenor was born in Upstate New York and moved with his family to Wisconsin as a young man. Acquaintances described him as a good talker with a cultivated vocabulary, although he could barely read and write. He married in 1855 and bought 80 acres of land in 1857. The other Tichenors established themselves as a leaders in the community, but Anson had other ideas. He enhanced the value of his land by discovering petroleum on it then sold it. Tichenor left Wisconsin, and the new owners not quite as quickly discovered that the only oil on the property had been placed by the hand of man.

In 1865 Tichenor announced discovery of an oil spring on a farm seven miles east of Des Moines and displayed bottles of crude oil in the window of a downtown business. When Tichenor led an excursion of suckers to the oil spring, however, one potential investor stayed behind in the farmhouse and talked to the farmer's wife; she was angry about the oil boom and told of a mysterious two-gallon jug that Tichenor had taken full with him to examine the oil spring and brought back apparently empty. Tichenor organized oil companies in Des Moines and Chicago, but one day Tichenor left town to buy drilling equipment and never returned.[1]

Tichenor turned up near Prairie du Chien, Wisconsin, and found yet another oil spring on property that he fortuitously owned. Investors again watched the oil spring dry up, but not before their money had left town in Tichenor's pocket.[2]

A Crooked Lottery in Nebraska

Tichenor moved to Lincoln, Nebraska, in 1868 and started a string of shady business deals. When the state legislature impeached and removed Governor David Butler for numerous misdeeds, the charges included two of his official dealings with Anson Tichenor: leasing Tichenor a state-owned salt-producing property (another would-be lessor had refused to pay the governor a $5,000 bribe) and extending a sweetheart loan to Tichenor from the state school fund. Tichenor resold the salt lease and used the profit to build the Tichenor House Hotel. The hotel lost money, however, so he raffled it off. Lottery winners found out too late that the hotel had been mortgaged for its entire value and that the other promised prizes did not belong to Tichenor. By that time, however, Tichenor was gone.[3]

The World's Only Cement Mine Swindle

"Colonel" Anson Tichenor—the sudden military rank was not explained—was in Salt Lake City in 1872, when he learned that a local man was looking for limestone suitable for manufacturing Portland cement. Tichenor assured him that he owned a deposit of Portland-quality limestone and led them to nearby Parleys Canyon. Tichenor brought back a quantity of the rock, and a few days later presented the potential buyers with samples of excellent-quality cement.

Negotiations for Tichenor's limestone property were well along when one of the investors thought of trying to make Portland cement from a sample of the limestone that he had dropped in his pocket during the visit. Tests showed that it was not suitable for cement-making, and investigation revealed that a sack of Portland cement had been shipped to Tichenor shortly before he had produced the cement supposedly manufactured by himself.[4]

Salting the Stafford Mine, Utah

Tichenor bought an option on the Stafford mine in Utah and promoted it in grand style. A sales trip to Indiana was unsuccessful, but in December 1872 Colonel Anson Tichenor and his partner "Colonel" W. J. Jones moved into elegant suites in the Saint Charles Hotel in New Orleans. They made acquaintance with New Orleans businessmen and showed two silver bricks from the Stafford mine. They also showed a report by the Salt Lake City mining expert William Gardner that confirmed that the mine had hundreds of thousands of dollars of ore already in sight.[5]

A group of New Orleanians traveled to the Stafford mine in February 1873. The mine was sampled and assayed by William Gardner, with excellent results. The businessmen bought the property for $165,000 and organized a stock corporation to operate the mine.

The businessmen appointed one of their own, General I. F. Harrison, to run the mine. He arrived at the mine in April but after two weeks became suspicious of the ore and sent samples to three different assayers.

Two assayers reported no silver at all, while the third assayer, associated with Gardner, reported 171 ounces of silver per ton. Two laborers gave Harrison written statements that the property had been salted with ore from another mine. To make sure, Harrison hired five mining engineers to examine the property, and they each told him that the property was worthless.

Harrison found that Anson Tichenor had earlier bought two silver bricks from a smelter in Chicago, presumably the same bricks that he had shown in New Orleans. In the Salt Lake *Tribune*, Harrison accused Tichenor, Jones, the former owner of the mine, and the mining engineer Gardner of perpetrating a swindle.[6]

Major J. D. Wooley, the former owner of the Stafford mine, sued the Salt Lake *Tribune* for libel. He claimed that by printing General Harrison's charges, the *Tribune* had spoiled an impending sale of another of Wooley's mining properties for $125,000—a mine for which Wooley had paid only $500.[7]

The Arkansas Swindler

In July 1877 Tichenor installed himself at the Palmer House in Chicago and announced that he had a silver mine in Arkansas for sale. The mine was the Kellog lead mine, which Tichenor claimed contained a great wealth in silver. Such was Tichenor's persuasiveness that a Chicago businessman, without further investigation, paid $10,000 cash to Tichenor, with $20,000 to be paid later. The buyer arrived at the mine optimistically bringing ore sacks to haul out the booty, but Tichenor then told him that he could not extract any ore until he paid an additional $10,000. The investor balked at the breach in his agreement and demanded his money back.

Rather than refund the down payment, Tichenor went to Cleveland in the spring of 1878 to look for more suckers. He returned to Chicago and sold another Arkansas silver mine for $15,000. He even leased a city block from the City of Little Rock to prospect for silver.

He confided to a few of his Chicago acquaintances that he had a fabulously rich gold mine in Arkansas—on Granite Mountain just south of Little Rock—that would be one of the world's most important gold mines. Tichenor sank two shafts on the property and extracted rock that assayed very rich in gold—by Tichenor's assayer. Tichenor said that the gold mine was not for sale and started to shop for mine machinery. Some Chicagoans persuaded Tichenor to sell them interests in his mine—but only as a favor.[8]

A trio traveled to Little Rock and returned with glowing reports and rich ore—ore that a Chicago chemist assayed at more than $1,000 per ton in gold. A Chicago mining expert also returned to Chicago and confirmed the richness of Tichenor's mine. Through the summer and fall of 1878, Tichenor never shipped any ore from his fabulous mine—only selling small interests to selected friends.

Tichenor brought a few of his partners to Arkansas to see the great gold mine they had purchased. Two additional prospective investors in the group collected samples of the ore. Back in their hotel room, they mixed the samples and divided them into two piles. One pile they quietly shipped back to Chicago, the other they left in their hotel room while they visited Tichenor's Kellog mine the next day. Tichenor meant to accompany the party to the Kellog, but at the last minute he decided to remain in town while his superintendent led the tour.

The ore samples that spent the day in the Little Rock hotel room later tested very high in gold, but the samples that had been immediately sent back to Chicago contained not even a trace of gold.[9]

Tichenor's Secret Gold Process, California

Tichenor moved with his family to Santa Cruz, California, where he tried to promote mining properties. The residents of Santa Cruz were careful with their money, however, and he exhausted his funds and applied to the county indigent-relief fund. The board of supervisors paid his fare out of the county.[10]

In August 1880 Tichenor and his family took rooms in the Hot Springs Hotel in the resort town of Calistoga, California. He soon arranged financial backing to buy the hotel for $9,000, and the fun began.

First, Tichenor announced that he could harness the steam from the hot springs on his property to power all the machinery in Napa County. He ran nothing but a steam whistle. Then he ignited a small flame of gas escaping from a hot spring and told everyone that the springs had enough natural gas to light the entire town—in fact, not only could he harness natural gas but he claimed to be able to capture valuable metallic mercury from the vapors. Nothing practical came from Tichenor's boasts, but an observer marveled, "Mr. Tichenor is a natural born genius."

More skeptical of Tichenor's ability was the superintendent of the San Francisco mint. In January 1881 Tichenor told the superintendent that he had a secret process to extract dissolved gold and wanted to test it on the wastewater from the mint. The superintendent had humored such inventors before and gave Tichenor a few barrels of his wastewater for experiments. Tichenor returned a few weeks later with a bar of gold he said he had extracted from the water. However, the mint's superintendent dismissed Tichenor and his process as fakes.[11]

Tichenor announced to the press in September 1881 that he had perfected his secret process for sucking gold from water. He told reporters that he had tested his gold-recovery method on Pacific Ocean water but thought the profit of $1.30 per gallon too meager. He found the hot springs of Calistoga richer. The "Chicken Soup" spring was the richest, and he claimed that he could profit $5 from each gallon of "Chicken Soup" water. The soil around the spring also contained $60 in gold per ton—if treated by Tichenor's process.[12]

Tichenor said that his experimental apparatus could treat springwater at a profit of $7,750 per day and promised that as soon as his process received a patent, he would erect a great gold-producing factory. Tichenor was generous enough, however, to sell his process—and said that he had already sold a half-interest to unnamed investors for a modest $1.5 million. Tichenor demonstrated his process to newspaper reporters and Calistoga residents and under their observation drew a quarter-ounce of gold from a five-and-a-half-gallon bucket of springwater. He spoke glibly of creating gold using electricity and magnetism and invited skeptics to come to his hotel to examine his process under the strictest scrutiny. Newspapers across the country printed the news of Colonel Tichenor's gold-bearing spring.[13]

Mark Twain and the Gold-Bearing Wind of Catgut Cañon

Samuel Clemens's years in California and Nevada mining camps had given him a keen eye for chicanery. When he read of the gold-bearing springs in the New York *Post*, he recognized the fraud and dashed off a vintage Mark Twain satire:

To the Editors of the *Evening Post:*

I have just seen your dispatch from San Francisco, in Saturday's Evening Post, about "Gold in Solution" in the Calistoga Springs, and about the proprietor's having "extracted $1,060 in gold of the utmost fineness from ten barrels of the water" during the past fortnight, by a process known only to himself. This will surprise many of your readers, but it does not surprise me; for I once owned those springs myself. What does surprise me, however, is the falling off in the richness of the water. In my time, the yield was a dollar a dipperful. I am not saying this to injure the property, in case a sale is contemplated; I am only saying it in the interest of history. It may be that this hotel proprietor's process is an inferior one—yes, that may be the fault.

Mine was to take my uncle—I had an extra uncle at the time, on account of his parents dying and leaving him on my hands—and fill him up, and let him stand fifteen minutes, to give the water a chance to settle well, then insert him into an exhausted receiver, which had the effect of sucking the gold out through his pores. I have taken more than $11,000 out of that man in a day and a half. I should have held on to those springs but for the badness of the roads and the difficulty of getting the gold to market.

I consider that gold-yielding water in many respects remarkable, and yet not more remarkable than the gold-bearing air of Catgut Cañon, up there toward the head of the auriferous range. This air—or this wind, for it is a kind of trade wind which blows steadily down through 600 miles of rich quartz croppings during an hour and a quarter every day, except Sundays—is heavily charged with exquisitely fine and impalpable gold. Nothing precipitates and solidifies this gold as contact with human flesh heated by passion. The time that William Abrahams was disappointed in love he used to step out doors when that wind was blowing, and come in again and begin to sigh, and his brother Andover J. would extract over a dollar and a half out of every sigh he sighed right along. And the time John Harbison and Aleck Norton quarreled about Harbison's dog, they stood there swearing at each other all they knew how—and what they didn't know about swearing they couldn't learn from you and me, not by a good deal—and at the end of every three or four minutes they had to stop and make a dividend: if they didn't their jaws would clog up so that they couldn't get the big nine-syllabled words out at all; and when the wind was done blowing they cleaned up just a little over $1,600 apiece. I know these facts to be absolutely true because I got them from a man whose mother I knew personally. I do not suppose a person could buy a water privilege at Calistoga now at any price; but several good locations along the course of Catgut Cañon Gold-bearing Trade-winds are for sale. They are going to be stocked for the New York market. They will sell, too; the people will swarm for them as thick as Hancock veterans—in the South.

Mark Twain
Hartford, Conn., Sept. 14, 1880[14]

Clemens's instincts were correct. By the time his satire came out in the New York *Evening Post*, the swindle had already been exposed.

California State Mineralogist Henry Hanks ignored the Calistoga gold excitement until he received inquiries from potential eastern investors who had read about the amazing gold-bearing springs. Hanks went to Calistoga, and in the company of a newspaper reporter and an inquisitive crowd, took up Tichenor's challenge of close scrutiny. Tichenor at first said that he was too tired that day then said that he would demonstrate the process later, to Hanks alone, lest someone in the crowd steal his process before the patent was approved. Hanks said that he had traveled all the way to Calistoga to see the process and refused to be put off.

Tichenor relented and demonstrated the process to Hanks and the curious crowd. Tichenor filled a container with springwater then took a bottle of liquid from a locked cabinet and added a couple of ounces of his secret compound to the springwater. Hanks asked to examine the bottle of secret solution, but Tichenor quickly locked it away. Tichenor then brought gold out of the solution by plating it onto a lead sheet, but Hanks recognized it as the classic reaction of gold dissolved in aqua regia. Hanks was convinced that the affair was a sham. To confirm his suspicions, he brought back large bottles of springwater from Calistoga. In his own laboratory Hanks found that the water contained no gold.[15]

Twain's satire and Hanks's chemical tests finished the brief fame of Calistoga springwater as a source of gold. The unfavorable publicity broadcast the name Anson Tichenor nationwide, forcing him to drop out of the gold-from-hot-springs business. His wife of twenty-five years left him. Ten years later Anson Tichenor was living in Washington, DC, billing himself as an inventor.[16]

Richard Flower: Master Swindler

R ichard Flower is unsurpassed among swindlers in the audacity of his schemes and his agility in eluding justice. He mesmerized investors with the charm and intensity of a faith-healing preacher, which he once was.

Richard C. Flower was born in 1849 in Albion, Illinois, a town that his grandfather had founded in the Illinois wilderness. He graduated from Northwestern University in 1868, after simultaneously studying law, medicine, and theology. He was already such a persuasive orator that in the year of his graduation he was chosen to defend his Campbellite religious beliefs in a highly publicized debate. He practiced law for two years, then followed his father into the ministry. He preached at churches in Illinois, Indiana, and Kentucky before securing his own congregation in Alliance, Ohio, in 1875.[1]

Listeners crowded his church for spellbinding oratory, but Flower's sermons wandered from the teachings of his sect. He wrote to a friend: "I have long given up the sham of shadow and form. To me, religion is to do as you would be done by, to enjoy yourself, and to give to others all possible pleasure." Others charged him with heresy, but his charisma defeated attempts to defrock him. He quit the Christian Church to found his own Independent Church, and most of his congregation followed the young preacher to his new institution.[2]

Doctor Richard Flower of Boston

"'Sickness is a toy in his hands; in a most phenomenal way he sheds light into the darkened eye, life into the dying form, and robs the sick-room of its sufferer and coffin of its prey.'"

—Louisville *Courier-Journal*, March 3, 1886

Flower left the ministry to hang out a shingle as a doctor and soon had crowds of patients. He also traveled to sufferers in other cities. Flower sold

*Doctor Richard C. Flower: lawyer, preacher, doctor, faith healer,
and mining swindler. From* Combined History of Edwards, Lawrence
and Wabash Counties, Illinois, *1883.*

his Alliance, Ohio, clinic in 1880 and moved to New York City, where he
bought a house on Fifth Avenue.

Flower moved to Boston in 1892 and established the Flower Medical
Institute. He practiced a combination of quack medicine and faith healing and referred to his assistants as "electricians and magnetizers." His
methods were dubious, but his public relations machine was formidable.
Newspapers in Boston, Chicago, Cincinnati, and other cities printed the
most ridiculous praise of his abilities—enough to embarrass all but the
most ardent egotist. He supposedly had only to hold a patient's hand to
give an unerring diagnosis and to prescribe the cure. When still in his mid-

thirties, Flower claimed to have cured 4,122 cancer cases, a cure rate of more than ninety-four percent.[3]

Flower also claimed that he could predict future stock prices. A sycophantic biographer wrote, "His intuitional powers in detecting the actual conditions of the markets seem almost infallible."[4]

Flower ventured into mining in the early 1880s by buying shares in the Julianna Mining Company of Rosita, Colorado. The silver mines of Leadville, Colorado, were booming, and the Wet Mountain Valley around Rosita was touted as the next Leadville. The Julianna, however, uncovered no bonanzas. Flower had invested in a loser but would not accept his losses. After years of swindling people one at a time, Flower realized the possibilities of swindling people the efficient modern way: through stock corporations. Mining stocks were perfectly suited to his purposes.

Flower took control of the bankrupt Julianna Company in late 1883 and installed his father-in-law, C. C. Manfull, as manager. Manfull optimistically reported on developments, while Flower sold Julianna shares back east. Manfull reported rich ore finds for over a year, but the Julianna mine kept losing money.[5]

The Security Mining and Milling Company

"An utter Humbug."

—*Engineering & Mining Journal*, February 19, 1887

By 1885 the Julianna Mining Company was going nowhere, so Flower enlarged the scale of his swindle. The Wet Mountain Valley boom had dried up, and Flower bought nearby marginal-to-valueless mining properties and combined them with the Julianna to form the Security Mining and Milling Company with $10 million capitalization. Flower sold stock by promoting the odd collection of prospect pits and failed explorations as a great mining conglomerate. Compliant eastern newspapers such as the Boston *Traveler* printed long elegies to the Security Company, full of lies that the sorry properties were making money, and silly predictions that it would become the "richest company in the world."[6]

The area was no stranger to mining manipulations. San Francisco stock manipulators James Keene and William Lent had both relocated to New York, from where they were mismanaging their respective Wet Mountain Valley mines—Keene the Silver Cliff, and Lent the Bull-Domingo.[7]

The Boston Stock Exchange refused to list Flower's new company, so Flower took his corporation down to Manhattan, where the less-than-discriminating Consolidated Stock and Petroleum Exchange accepted it onto the board.[8]

War in the Newspapers

Out-of-town newspapers reported that the Security Company bought the town weekly *Sierra Journal* to ensure good publicity, but the editor denied it. The sagging fortunes of Rosita needed employment for its miners, and few newspapers would criticize a company bringing jobs to town. Newspapers in towns not benefiting from the Security Company payroll, or from Flower's largesse, doubted that the company could be profitable. Journalists engaged in learned discussion of one another's opinions of the Security Mining and Milling Company:

> "The senseless being and unmitigated liar who issues that weekly eructation of flatulent flatness yclept the *Wet Mountain Tribune*, condemns the corporation known as the Security Mining Company, and warns its friends (?) in the east (has it any friends in the east, or any subscribers east of the east line of Custer County?) 'to investigate before purchasing.'"
>
> —*Sierra Journal*, September 24, 1885

> "The *Denver Times* has taken a contract from the blackmailers and backcappers."
>
> —Pueblo *Chieftain*, February 1886

> "He is simply running his paper in the interests of a blackmailing schemer."
>
> — *The Solid Muldoon,* June 10, 1887, on the editor of the Silver Cliff *Rustler*

> "Mark Atkins, the yahoo who imagines himself what no one else does—an editor—simply because he possesses the ability to impose profanity and blackguardism upon the readers of the *Muldoon,* still keeps up a stream of abuse at the mining interests of this County, thinking, no doubt, in his blissfully ignorant way, to kill the camp in which he, a drunken vagabond, was not wanted to stay."
>
> — Silver Cliff *Rustler,* June 30, 1887, on the editor of *The Solid Muldoon*

Adding a Real Mine

When a curious stockholder went west to inspect the Security mines, he saw only partly explored prospects and not the wealthy mines described in the prospectus. Flower could spin long-distance fantasies, but they would not bear inspection, even from a stockholder unsophisticated in mining. Flower quickly bought the Silver Cliff mine and sold it to the Security Company at an inflated price. Now the company owned a substantial property to show shareholders.

The Silver Cliff mine had already starred in an ill-fated promotion by James R. Keene in 1879. Keene had won a fortune in San Francisco speculating in Comstock shares before moving to Wall Street, where he lost his entire stake. One of the ways Keene recouped his finances was by promoting the Silver Cliff into a company with a $1.5 million capitalization. Out of the original $1.5 million, $500,000 in cash and stock went to the previous owner and $855,000 to Keene and his fellow promoters, leaving only $145,000 for mine development and surface facilities. Despite the large capitalization and optimistic press releases, the company ran out of the money and was bankrupt and idle by the time Flower bought it. The Silver Cliff—or rather its promoters—had already absorbed a fortune without yielding any profit, but it had a mill and low-grade silver ore, perfect for Flower's purpose.[9]

The addition of the Silver Cliff mine changed the attitude of the *Wet Mountain Tribune* in the nearby town of Silver Cliff. Now that its own readers were on the Security Company payroll, it turned from critic to defender of the company. Will Orange arrived in town as owner and editor of the Silver Cliff *Rustler*. Critics said that Orange was Flower's nephew, but Orange denied it. Nephew or not, he published the most servile flattery of Flower—the great genius, et cetera—and extravagant publicity about the fabulous bonanzas owned by the Security Mining Company.

In 1887 the Security boom was on. C. C. Manfull reported unfailingly good developments, but the company never made a profit. Flower blamed the delay in dividends on the need to completely revamp the mill. The modernized mill finally started in June 1887 and Flower brought a trainload of shareholders to Silver Cliff. Of course, Flower made sure that none of the group knew anything about mining. Judging by the glowing reports in the Boston *Globe*, Flower completely hoodwinked the visitors. The *Engineering & Mining Journal* noted that the junketeers were "as competent to pass an opinion on that or any other mining property as is an Egyptian mummy." Nevertheless, the price of Security shares surged to $9 on obviously wash sales.[10]

Flower had now sold all his stock and had no way to profit further—unless he could levy assessments. The problem was that the company bylaws prohibited assessments. Flower tried to change the bylaws, but shareholders were wary of his unfulfilled promises and refused to give him more money. Share prices collapsed.[11]

The Security mill shut down for lack of ore in September, after a run of only three months. The Silver Cliff *Rustler* continued to report immense bodies of rich ore discovered in the Silver Cliff mine, but the mine never produced enough to keep the mill busy. Workers had to wait a few days before the Security Company could make good on payday in April

1888. The company blamed slow mail service but also announced a special stockholder meeting. The meeting cryptically announced that the management was given full power to deal with the "present indebtedness."[12]

Just what indebtedness the company had was revealed when the company failed to make a scheduled $27,000 payment on May 15, 1888, to Flower himself. Flower filed a lien to seize the assets of the Security Mining and Milling Company. Having spent all its funds on rebuilding the mill and enlarging the mine workings, the company was broke and operations stopped. Share prices dropped to $0.04.

Shareholders still believed that they owned a bonanza and that with a little more investment they could rescue their lost money. The Security Company bought back their assets at a sheriff's auction and renamed the concern the Geyser Mining and Milling Company.

The new directors emphasized that Richard Flower was no longer associated with the company, but President A. A. Rowe and Treasurer James Cartwright of the Geyser Mining Co. were Flower protégés: Rowe had been associated with Flower's schemes since the Julianna Company, and Cartwright was the former treasurer of Flower's medical company in Boston. Rowe and Cartwright had been among the original incorporators of the Security Mining Company, and neither had objected to Flower's swindling. With such management, Geyser shareholders were in for another ride.[13]

The Geyser Company continued exploring for ore in the Silver Cliff mine, now renamed the Geyser. The mine had already lost two fortunes, one under Keene and one under Flower, and it was to absorb a third before investors lost heart. The company deepened the shaft, until by 1900 it was the deepest mine in Colorado, at 2,650 feet. It never found the hoped-for great ore bodies, and the cost of extending the underground workings kept the company losing money. The mine was sold in 1901 to satisfy a small debt.[14]

Dr. Flower went back to his faith healing and marketed his patent medicines. The easy money of mining–stock promotions, however, was still a powerful lure to one with his persuasive talents.[15]

A Land Promotion Scheme Gone Sour

In 1893 Flower promoted an irrigation and land development scheme in New Mexico to Midwestern investors, but the following year an investor charged him with fraud. He was arrested but quickly released in Terre Haute, Indiana, in October 1894. He returned to Boston but left before extradition papers arrived. He was rearrested on a train at Conroe, Texas, and brought to Chicago, where he and his secretary were indicted for fraud.[16]

Flower posted bond in Chicago, but when he emerged from the jail, detectives arrested him for a third time on yet another fraud charge, accusing Flower of having swindled an Illinois doctor out of a $350,000 investment in a coal company. Flower again posted bond and returned to Boston, where he declared bankruptcy and managed to avoid further punishment. In 1898 he relocated to New York and started his mining swindles anew.[17]

Spenazuma, Arizona

A group of Kansas men organized the Spenazuma Company in 1898 to mine near Deming, New Mexico. Flower bought the new company, replaced most of the directors, and moved the fictitious "Spenazuma mine" across the border to Arizona.[18]

Flower paid some discouraged prospectors between $500 and $800 for options on a few unproven pits—options being cheaper than buying them outright—and built a phony mining camp. That the rock was barren was a minor problem for Flower. He plotted a town site on a map, drew in the location of a dam straddling a dry wash, with a nine-mile-long artificial lake that he said would provide electricity and irrigate large expanses of desert. By July 1898 Flower's mining farce was in full bloom. The company hired a couple dozen miners to begin driving tunnels. The camp boasted a store and stagecoach service and talked of a future rail line. Press releases told of mining equipment shipped to the mines and of the fabulous ore bodies being uncovered.[19]

Flower supplied the mine with a legend. Professor T. A. Halchu of Longhorn, Montana, supposedly befriended a dying hundred-year-old Mexican (tall tales in the Southwest are thickly populated with dying Mexicans, each one of whom seems to have the map to a lost mine in his pocket). This particular dying old Mexican repaid the professor's kindness by telling him the story of the lost Spenazuma mine, source of the wealth of the ancient Aztecs. The fabulous mine had lain idle in the centuries since the Spanish conquest, protected by the silence of the Indians, who passed the secret down through the generations. The old mine was marked by a great rock outcrop with the facial profile of the Aztec prince Spenazuma. Halchu searched southeastern Arizona until he saw the stony face of Spenazuma outlined against the sky above Black Rock Canyon and found the long-lost greatest ore deposit in the world. However, those who searched for Professor Halchu found that, despite his supposed international reputation, there was no record of the existence of either the professor or of his hometown of Longhorn, Montana—other than in the Spenazuma prospectus.[20]

*Not remarkable to most eyes, but to the fertile imagination of Richard Flower
this rock formation in Arizona bore the profile of the Aztec prince Spenazuma
and marked the long-lost source of golden wealth for the Aztec empire.
Photo by Dan Plazak.*

Flower never lied by half-measure. The Spenazuma Company claimed
a vein "one-half mile in width and two miles in length, every foot of it rich
in gold, silver, and copper."[21] By the end of 1898 Flower was promoting
his new company furiously with one million shares of stock at $1 each.
Flower stayed off the board of directors, but careful observers might have
noticed Flower's in-law John Manfull in the position of company secretary.

Flower brought twelve investors to see the mines. He let them choose
their own rock samples, and the camp assayer tested their samples for
gold and silver while they watched. The investors were agog when every
sample they chose tested high in precious metals, and they put their signatures to an Arabian Nights' description of "literal mountains of ore,
with ore cropping out at almost every turn, visible to the untrained and
uneducated eye, running anywhere from a few dollars to thousands of
dollars." Two months later, Flower brought another batch of suckers to
parade through Spenazuma, with similar success.[22]

The ore "visible to the untrained and uneducated eye" was much less
impressive to a trained and educated eye, however, and the company assayer puzzled over the results from what appeared to him to be barren
rock. Once he discovered the reason, he raced about the camp with a
six-gun demanding to see Doctor Flower. The assayer told the doctor that

someone had "doctored" the assay chemicals to give false results and that the practice had better stop. The assayer then destroyed every bottle of reagents in his laboratory. By this time, however, the investors were headed east to praise Flower's imaginary mines to their neighbors.

Flower's agents hawked shares in small towns in New York, Connecticut, and Ohio. The salesmen earned commissions of $2.50 to $7.50 per share, and their representations strayed from the truth. Flower hired the stockbroker Henry Clifford. In 1883, as a young man of twenty-four, Clifford had authored *Years of Dishonor, or the Cause of the Depression in Mining Stocks*, a pamphlet calling for honesty in the management of mining companies. Perhaps he subscribed to high ethical standards when he wrote the tract, or perhaps he wrote it only to lure suckers. At any rate, Clifford was reportedly bankrupt and ready for dishonest cash by 1899 when he traveled to eastern towns pretending to be the "commissioner of Arizona" and promoting the Spenazuma mine under the guise of an informational program. Flower paid a few dividends and raised the share price steadily from $1 to $15, creating an astounding market capitalization of $15 million and an illusion of quick wealth that attracted more investors.[23]

In March 1899 Flower arranged a special train to take two more carloads of investors to the new bonanza. The gullibles came from eleven different states, from New Hampshire to Texas, from cities as large as Boston, and small towns such as Cowgill, Missouri. Richard Flower and his wife were the charming hosts.[24]

The Spenazuma Company bought rich ore from neighboring mining districts, carted it to Spenazuma, and dumped the glittering ore beside the phony mine openings. The frame of a supposed mill building was erected, although a mining man who visited Spenazuma laughed that the two-by-four lumber was far too flimsy for a mill and that no real ore mill would be built on an earthen foundation. In fact, the total value of goods and equipment at the mine site, as filed with the Graham County tax assessor, totaled only $1,390.[25]

The trainload of eastern investors disembarked at Geronimo, Arizona, and after an "Indian entertainment," climbed into twelve stagecoaches bound for Spenazuma.

Faking a Stagecoach Holdup

Flower, the great showman, wanted to give his guests a taste of the Old West, so he hired a local ne'er-do-well named Alkali Tom to arrange a phony holdup attempt of the Spenazuma-bound stagecoaches. Some cowboys hanging around the saloons in Geronimo considered playing a scene from a Wild West show to be a quicker and easier way to make money than ranching, and Alkali Tom recruited all he needed. When Alkali Tom and his companions rode off toward Spenazuma, however, Sheriff Bill Duncan suspected that something was up and raced after them.

By the time the stagecoaches reached the prearranged "ambush" site, Sheriff Duncan had caught up with the phony bandits and sent them back to Geronimo before they could act out the holdup drama. Flower was disappointed, but the real playacting was about to begin. When the stagecoaches arrived, Spenazuma was a beehive of building, blasting, and hauling. The site that a miner previously described as "the laziest camp I ever struck" had every man swinging picks and shovels. The visitors were impressed with what they saw and what Flower told them. They contracted gold fever, and in that infectious state they returned home convinced that the Spenazuma was about to make them wealthy. Flower capitalized on the word-of-mouth with a new round of newspaper advertising and publicity.[26]

Bad Day at Black Rock

Flower knew that the investors he had brought to Spenazuma were too ignorant of mining to spot the fakery. Eastern journals were openly skeptical of the phony prospectus but wouldn't send anyone all the way to Spenazuma. The residents of Spenazuma were loyal to their paychecks, and Flower knew from his Colorado experience that local papers, such as the Arizona *Bulletin* in nearby Solomonville, would not look too deeply into a scheme that benefited local laborers and merchants. Larger-town Arizona newspapers, however, such as the Arizona *Citizen* in Tucson and the Arizona *Republican* in Phoenix, had only to read the Spenazuma promotional material to know that the company was a fraud, and they let their readers know it.[27]

After the trainload of happy visitors left Spenazuma, the Arizona *Republican* reporter George Smalley was traveling the area gathering mining news. Smalley had already ridiculed the Spenazuma Company as an obvious fake and now wanted to see the supposed bonanza for himself. At Geronimo, Sheriff Bill Duncan volunteered to go out to the camp with him. Smalley and Duncan found Spenazuma hostile to inquisitive visitors—at least to those who knew something about mining. The frenzied activity of the week before was replaced by desultory work. Visitors were not allowed in the mine workings, and the company office offered only minimal information.

When Smalley tried to buy miners candles at the camp store, the storekeeper refused to sell to him. Smalley distracted the storekeeper while Sheriff Duncan himself shoplifted several candles, and they then surreptitiously inspected the mine workings until Smalley satisfied himself that the "mines" contained no ore whatsoever.

The reporter had seen enough mines to know an obvious sham. He spoke to nearby mine owners who had sold their ore to the Spenazuma Company, and he realized how the game was played. Flower's employee Alkali Tom knew that Smalley was hostile to the Spenazuma Company, and for the ride back to town lent Smalley a horse that the cowboy didn't own—knowing that the real owner would mistake Smalley for a horse thief and likely shoot him on sight. Alkali Tom's plan nearly worked, but Smalley's horse managed to outrace its owner into town, and the reporter jumped aboard the train unharmed, just as it pulled out. The Arizona *Republican* of May 17, 1899, printed Smalley's story detailing the fraud.[28]

Trying to Bribe the Press

Other Arizona papers picked up the *Republican*'s exposé, and the bad news traveled east. Flower's lawyer arrived in Phoenix to threaten the *Republican* with a libel suit, but when the paper's business manager dared him to sue, the lawyer instead offered Smalley $5,000 to retract his story, which Smalley refused.[29]

The Spenazuma Company officers contested the exposé in advertisements and press releases in more compliant periodicals. The company said that the bad press came from disgruntled ex-employees. Company lawyers sent letters threatening to sue those who printed articles about the worthlessness of the property. Graham County businessmen wanted the Spenazuma Company to continue to spend money and publicly supported Flower. In June 1899, however, the Arizona territorial governor issued a proclamation against mining swindles because they were giving legitimate Arizona mining a black eye, and cited the Spenazuma Company by name. Some Arizona newspapers criticized the governor for calling attention to Arizona swindles, but most supported his attempt to stop fraud. Even the Arizona *Bulletin*, which had unblushingly printed every exaggerated press release from the Spenazuma Company, belatedly criticized the company for "a mass of exaggerations and misstatements." The district attorney of Graham County, Arizona, offered to furnish evidence that Spenazuma was a "miserable, damnable fake" to eastern victims wanting to prosecute the culprits. The denunciation by the governor decided eastern opinion against the Spenazuma Company.[30]

The exposure brought panic in some eastern towns. Several hundred people in Tivoli, New York, had confided an estimated $50,000—said to be nearly all the savings of those of modest means in town—to Flower. The treasurer of one of the town churches had traveled twice to Spenazuma and was so enthralled by Flower's mining fantasy that he invested heavily himself, urged friends to invest, and became the local agent to sell more shares in his community. By July the people of Tivoli found their Spenazuma shares worthless and their savings gone. By then Flower had reduced the workforce at the great mine to only three. Flower increased the capitalization in August 1899 to sell more stock, but by then Spenazuma was infamous, and he moved on to other swindles.[31]

Secondhand Swindles

Flower passed the Spenazuma property on to smaller-fry mine promoters. The new management emphasized that Flower was no longer in charge, but they too issued obviously exaggerated statements, although their lies were a bit subdued compared to Flower's. Just before his exit Flower had optioned mining properties in the Clark mining district, and the new management touted these as the next Cripple Creek. The Spenazuma mines were now relegated to minor status compared to the great gold mines and town of Aura being built twenty miles south.[32]

The Arizona *Bulletin* pointed out that since the new general manager had visited the year before and attested to the false richness of Spenazuma, he was either ignorant or dishonest. However, a visit to the *Bulletin* office by the new management squared things sufficiently that the publication once again cooperated in fraud by printing the fantasies dreamed up by the new Spenazuma schemers: "an immense body of high grade sulphrets," "property of untold value," "the richest copper areas in the Territory," and so on. The comedy was the same, only the actors were different.[33]

When the Arizona *Republican* reporter George Smalley learned that the Spenazuma Company had shifted operations, he rode out to take a look. In recognition of his honest journalism, a Spenazuma employee ran him off with rifle bullets, but not before he saw that the new property at Aura was just as much a swindle as the one at Black Rock.[34]

The new management had drained the Spenazuma treasury by the end of 1900 and squeezed more money out of investors by reorganizing under other names. Spenazuma general manager Schuyler Moore stayed on the boards of the new companies to show the new directors how to do it. One expert testifying to the worth of the worthless properties was Professor George Treadwell. Treadwell had been a mining engineer of some repute, but in his later years he allied himself with a string of worthless mining companies in the United States and Mexico, often working with the female mine-company promoter Myra Martin. He and his partners used the old Spenazuma properties—now renamed Black Rock—to promote first the Graham County Mining Company, then the Advance Mining Company, both of which quickly became defunct. Neither company ever made any money—at least not for the stockholders.[35]

Practicing Medicine without a License

Flower's medical practice ran into trouble when the state began requiring medical licenses—which Flower lacked. He continued his medical practice in New York City with the Flower–Thompson Medical Company, despite attempts to put him in jail. Caroline Westphalen of the County Medical Society went to the clinic and requested to be treated by Dr. Flower. After she was examined and sold a bottle of medicine, she had Flower arrested and charged with practicing medicine without a license. In court, however, Flower revealed that it was really his associate who had attended Miss Westphalen. Flower sued the medical society.[36]

Arizona Eastern Mining and Milling Company

Thwarted from further swindling with the Spenazuma, Flower jumped to other Arizona mining properties. By the time Spenazuma was exposed, Flower had formed the Arizona Eastern & Montana Company, which bought a number of copper properties from Henry Clifford and his wife, Maude Clifford. Henry Clifford was the swindler who promoted the Spenazuma Company under the fictitious title "commissioner of Arizona." Now Flower made him manager of the Arizona Eastern's collection of marginal and undeveloped copper prospects in Yavapai and Cochise Counties, Arizona. The company bought a smelter in Spokane, Washington, that it promised would be dismantled and moved to Arizona. In typical fashion, Flower minimized his outlay by taking inexpensive leases, options, or partial payments on the properties, allowing him to pocket more of the stock sales.[37]

The most highly touted property in the group was the Lone Pine mine in the Big Bug mining district southeast of Prescott, Arizona. Maude Clifford bought an option on the mine then re-optioned the mine at a sizable profit to the Arizona Eastern & Montana Company, which put it in the charge of Henry Clifford. When critics claimed that Henry Clifford, and not the Arizona Eastern Company, really owned the mine, Clifford semi-honestly stated that he did not own the mine—but didn't mention that it was his wife's. Clifford, as expected, was soon reporting huge discoveries of ore in the Lone Pine.[38]

The thieves fell out over the spoils. Henry and Maude Clifford formed the Henry B. Clifford Mining Company and began acquiring properties in competition with Arizona Eastern. The New York headquarters dispatched an agent to investigate. The representative found that the Cliffords were undercutting Arizona Eastern, and he fired Henry Clifford on the spot—until Clifford pointed out that the mine was still held in his wife's name, forcing the Arizona Eastern Company to part on good terms and to issue a statement absolving Clifford of any double-dealing. Clifford began mine swindling on his own and issued a prospectus full of lies for the nearby Great Belcher mine, which he had optioned for himself.[39]

The Arizona & Eastern Montana Company paid two-percent monthly dividends in early 1900, encouraging investors to buy more shares. In fact, Arizona & Eastern was operating at a loss and paying dividends out of share sales. In addition to his promotion fees and stock, Flower paid $100,000 out of the treasury to himself and associates, for which there was no record that the company received anything in return.[40]

The high finances lasted less than a year. The Arizona Eastern & Montana had run out of money by the end of 1900, and the county sheriff seized the assets to cover a debt of $1,305, which the multimillion-dollar company could not pay. Maude Clifford sued for return of the Lone Pine mine, claiming that the company still owed her $60,000. The crown jewel of the Arizona Eastern & Montana Company, the Lone Pine mine and its equipment, brought $5,200 at auction. Flower vacated his luxury suite at the Waldorf-Astoria and left New York to avoid shareholders.[41]

The Strange Death of Theodore Hagaman

The stockbroker Theodore Hagaman had profited greatly by selling shares in Richard Flower's worthless mining stocks. When Hagaman fell ill, he entrusted his health to the treatment of his partner, Dr. Richard Flower. While he ministered to Hagaman's illness, Flower also ministered to the credulity of Hagaman's wife. Fanny Hagaman had first met Theodore Hagaman when she was Mrs. Smith, the wife of his coachman. Fanny Smith became very friendly with her husband's employer and divorced her husband of thirteen years to become Mrs. Hagaman. Fanny Hagaman now convinced her dying husband to give her power of attorney, which she used to invest $600,000 in Flower's fraudulent schemes.[42]

Flower's unusual medical methods were ineffective in halting his patient's decline. Hagaman died in his suite at the Waldorf-Astoria Hotel on September 11, 1900. The death certificate listed cirrhosis of the liver, but when Hagaman's family learned that there was nothing left to inherit, they suspected poisoning. Hagaman's body was exhumed in 1903 for tests but by then had deteriorated too much to determine if Hagaman had been poisoned. Fanny Lindsey Smith Hagaman had in the meantime briefly married then separated from her third husband and was now Fanny Delabarre.[43]

Lone Pine Mining Company

The Arizona Eastern & Montana went belly-up, but not before directors robbed it of everything of value. Shareholders reorganized the company as the Lone Pine Mining Company, named after Arizona Eastern's most promising property.

Flower tried to take over the Lone Pine Company by staging a corporate meeting dominated by himself and his allies. He apologized for his Spenazuma and Arizona Eastern fiascos and said that it was his life's duty to restore his good name by making a success of Lone Pine. The old swindler's speech seemed so heartfelt that many of his victims were now unsure which side to support. Flower ignored a shareholder's question about how the Arizona Eastern & Montana Company had paid dividends while going broke. Flower made an impassioned speech charging the reform directors with larceny before the meeting replaced them with Flower's own handpicked men. Despite the apparent rout of the reform directors, Flower's meeting had been illegally called, and so its acts were without effect. Within three weeks the reform forces had compromised with Flower by allowing him to name five of the twelve directors.[44]

The Lone Pine company had no more success than its predecessor, Arizona Eastern. Unwilling to let go of a good thing, Flower repackaged the properties yet again, this time as the Pan-American Mining & Smelting Company and sold more worthless stock.[45]

After he assisted Flower, Henry Clifford stayed in Arizona for a few years to run his own swindles before returning to New York, where he again posed as a reformer while shamelessly exaggerating the value of his own mining enterprises. He wrote *Rocks in the Road to Fortune, or the Unsound Side of Mining*, yet another book hypocritically demanding high ethical standards in mine promotions.[46]

Trying to Bribe the Police

Disgruntled Arizona Eastern shareholders hired Andrew Meloy to represent their interest against Richard Flower. Meloy hired the young attorney George Mills to assist him, but within a year Mills was working for the man he was hired to oppose, Richard Flower. Meloy collected evidence about Flower's swindles, and a grand jury charged Flower on five counts. Flower was arrested for grand larceny in selling Lone Pine stock with false representations.[47]

In view of Flower's habit of fleeing, the district attorney asked for $50,000 bail, but the judge reduced the amount to $2,000, which was paid by one of Flower's wealthy female admirers, Mrs. Cornelia Storrs. Mrs. Storrs was convinced that Flower had saved her life with his unorthodox medicine. She had since lost $300,000 in Flower's Arizona swindles and lost another fortune in his Blue Ridge Mining Company, but she still trusted him.[48]

Flower rashly determined to exact revenge on Andrew Meloy. In March 1903 Flower and his lawyer strode into the Tombs criminal courts to swear out a complaint against Meloy for larceny. According to affidavits by Dr. Flower, his son Jewel Flower, and other witnesses, Meloy had told Flower the previous October that a captain of detectives was demanding $2,500, or the police would prevent Flower and Meloy from doing business in the city and that Flower himself would be arrested on a warrant then supposedly in the hands of the police. According to Flower and his confederates, Flower gave Meloy a check for the bribe.

Meloy denied the entire incident, as did the police captain. The magistrate read Flower's sworn affidavit then arraigned Meloy for larceny and set bail at $1,000. Flower enjoyed his revenge only briefly, however, while the magistrate conferred with the assistant district attorney. A detective tapped Flower on the shoulder and told him that he was under arrest for attempted bribery—on the basis of his own affidavit. The court set Flower's bail, which was promptly furnished. The bribery charge was dismissed the following year, but by that time Flower had bigger problems.[49]

Flower knew how cash could salve the consciences of otherwise honest citizens. With the Arizona Eastern legal case going against him, as well as the ongoing investigation into the Hagaman death, he decided on a dramatic move.

At a meeting of the Lone Pine directors in March, George Mills, now a law partner to Flower's son Jewel Flower, told other directors that Deputy Assistant District Attorney Garavan could be "reached." Mills's former employer Andrew Meloy offered to approach Garavan and, oddly, in view of the ill will between Meloy and Flower, Mills accepted. Meloy told Garavan about Mills's wish to give a bribe. Garavan consulted his boss, then asked Meloy to pass along word to Mills that they would halt all legal proceedings against Flower—if the money were sufficient. Meloy told Mills, "They are dead easy."[50]

A few days later Mills entered the office of Deputy Assistant D. A. Garavan and asked what it would take to quash the indictments. Garavan took Mills to District Attorney Jerome, whom Mills offered $1,500 to surrender the indictment papers on Flower and to end the Hagaman investigation with a declaration that Hagaman had died of natural causes. Garavan and Jerome agreed.

On March 3, 1903, the detective Sergeant Brindley was lunching at a Park Row restaurant when the attorney Mills arrived and sat down. Brindley asked, "Have you any money with you, Mr. Mills?" Mills replied that he did and asked Brindley if he had the indictment papers. Mills produced $2,000, subtracted $250 as his fee, and laid out on the table the remaining $1,750: $250 for Brindley, and $1,500 for Garavan and Jerome. Mills, not to miss a chance, suggested that Brindley invest his share of the booty in Mills's financial dealings. Brindley counted the money, gave the papers on the five indictments to Mills, then called over the waiters who had witnessed the transaction, as well as police detectives seated at nearby tables, and told Mills that he was under arrest for attempted bribery. Unable to post the $5,000 bond, Mills remained in a cell in the Tombs.[51]

Mills, of course, claimed that it was all an innocent mistake, because he had thought that Detective Brindley was representing a plaintiff in a civil claim against Flower and that the $1,750 was meant to settle the claim. The person to blame, insisted Mills, was Andrew Meloy, who had entrapped him. The jury heeded neither his alibi nor testimonials to George Mills's good character and judged him guilty. He appealed on the basis of entrapment, but the verdict was sustained and George Mills, who was thirty-four years old and a former professor at Dickenson Law School, was sentenced to fourteen to eighteen months in Sing Sing prison.[52]

Life as a Fugitive

Flower foresaw prison bars for himself if he were to remain in New York. He left the city, jumping out on $23,000 in bail. Flower never again lived openly under his own name. He eluded justice with aliases, disguises, and abrupt changes of residence. One thing he did not change, however, was his habit of swindling people, often with mining schemes.[53]

In October 1903 Flower crossed from Eagle Pass, Texas, into Mexico. The former Protestant minister hurried through the city of Torreón disguised as a Catholic priest and took up residence in the Mexican state of Chihuahua. Soon he was busy with new mining promotions. At Flower's invitation, Mrs. Delabarre's lawyer met him in Juárez, across the Rio Grande from El Paso, Texas. Flower still owned the Sunset mine in Chihuahua, part of his swindling Arizona mining schemes. Flower told Mrs. Delabarre's attorney that he would repay her by giving her the Sunset mine, which she could operate until its profits had repaid her and other victims.[54]

When word spread of his presence in Mexico City selling worthless mining stock to American tourists, Flower fled to Brazil. He was later reported here and there in Central and South America and in Europe. He returned to the United States to sell a fraudulent brick-making process in Passaic, New Jersey. Before the police could close in, Flower fled to Montreal, Canada, then shifted to Buffalo, New York. Police intercepted a letter written to Flower's son Jewel Flower that revealed the elder Flower to be in Philadelphia and up to his old methods in promoting a secret process to manufacture diamonds and rubies.[55]

The New York police lieutenant Barney McConville, who had arrested Flower four years earlier and had pursued the fugitive in Mexico, traveled to Philadelphia in January 1907. With two Philadelphia detectives, McConville dropped in on "Professor Oxford" and listened over the transom as the learned man was talking two Philadelphia suckers out of their savings in exchange for shares of his latest secret process. Flower had grown a flowing white beard, but McConville recognized the remarkable voice able to charm the money out of people's pockets. Flower protested to the Philadelphia policemen that it was a case of mistaken identity but later admitted that he was the long-sought fugitive. Flower fought extradition, but in March a Pennsylvania court ordered him returned to New York.[56]

Despite Flower's history of jumping bail, the Pennsylvania court let the good doctor out on bond pending his extradition. Flower promptly and predictably skipped bail once more and dropped from sight.

Still using the alias "Professor Oxford," Flower convinced businessmen of Boyertown and Pottstown, Pennsylvania, to invest $400,000 in the Virginia Clay and Material Company that supposedly made telephone poles out of clay. The businessmen grew suspicious when the "professor" could not demonstrate the existence of his telephone-pole factory, and Flower hurriedly left the area. A Reading, Pennsylvania, constable tracked him down, however, and arrested him on a Philadelphia street in November 1907. His jailers soon discovered that they held the fugitive swindler Richard Flower.[57]

Once more, a friendly Pennsylvania court released Flower on bail pending his extradition hearing. When the Pennsylvania courts ordered his extradition, he jumped bail for the fifth time and disappeared.

"Horace Courtland" and the Appalachian Mining and Smelting Company

In September 1908 in Richmond, Virginia, police arrested a gang suspected of fraud, stealing diamonds, and practicing medicine without a license. The ringleader, Horace Courtland, evaded arrest, but Mrs. Courtland stayed behind. Police learned that an associate of the gang, "Mrs. Lindsey," was in fact Fanny Delabarre, formerly Fanny Hagaman, now going by her maiden name. She told police that "Horace Courtland" was Richard Flower. Flower had organized the Appalachian Mining and Smelting Company, installed several local men as directors, and been selling stock in the worthless concern.[58]

The case grew more confused as Richmond police searched for Flower. Flower's son and daughter-in-law arrived in Richmond, as did Mrs. Delabarre's cousin from Ohio, who bore a suspicious resemblance to Dr. Flower. Mrs. Delabarre was followed but slipped away from her police tails. Flower was reportedly sighted in nearby Fisherville, Virginia, but avoided arrest.[59]

"Mrs. Horace Courtland" said that she had met and married Flower in Philadelphia, and portrayed herself as Flower's victim. Richmond police suspected her of crimes in Cincinnati, but Cincinnati police declined to travel to Richmond to make an identification. The police let her go, and she left the city.[60]

A New York millinery company seized the belongings of the once-wealthy Fanny Delabarre over a debt of more than $4,300 worth of hats for which Mrs. Delabarre had neglected to pay.[61]

Brought to Justice

After a six-year search Pinkerton detectives tracked Flower to Toronto, Canada, and in October 1914, New York detectives watched their Toronto counterparts arrest Flower as he returned home. Flower had been up to his old profession in Toronto, promoting a home-heating invention.[62]

Flower had become a drug addict. A Toronto police doctor found Flower in such ill health that he prescribed drugs to supply the addiction. Authorities back in New York were not as accommodating. Jailed and desperate, Flower asked his wife to supply him with drugs. Lillian Flower was caught, arrested, and charged with smuggling cocaine to Flower in jail, to which she pled guilty. Shortly afterward, Richard Flower pleaded guilty to grand larceny and received two years in prison.[63]

Out of prison in 1916, Flower faced more charges but posted bond and for the first time in his life did not flee and forfeit bail. He was still out on bond awaiting trial when he was stricken in a Hoboken, New Jersey, theater. Dr. Richard Flower, the inspiring healer of whom it was said "sickness is a toy in his hands," died suddenly at age sixty-six.[64]

12

Thomas Lawson:
Champion of the People

"I have caused my following—the public—to lose millions of dirty dollars that in the end they may be prepared for My Remedy."

—Thomas W. Lawson, *New York Times,* August 23, 1908, part 5, p.3

Thomas Lawson was born in 1857, the son of a carpenter who had emigrated from Nova Scotia to Cambridge, Massachusetts. His father returned from the Civil War in broken health and died soon after. At twelve Lawson ran away across the river to Boston and talked his way into a job as an office boy in a State Street bank. At age sixteen he convinced his fellow clerks to gamble their savings on a railroad stock and cleared $40,000 when shares jumped from $3 to $22.[1]

Lawson plunged his winnings back into the stock market and spun the wheel again, but this time he lost all but $159, with which he treated his friends to a lavish dinner in the restaurant of a Boston hotel. Lawson left the last of his money as a tip and walked out well fed but broke. He later made that hotel his headquarters.

Lawson grew into an impressive man: tall and broad-shouldered, handsome and dark-haired, with a bushy barbershop-quartet moustache. Piercing eyes beneath thick eyebrows conveyed the certainty of his opinions. His speech was hypnotic. He was impulsive, generous, and always enthusiastic. His forceful and fluid speech gave him an aura of brilliance. The mercurial Lawson would contradict himself and reverse his stock market predictions with breathtaking speed yet remain absolutely certain of his opinion of the moment.

Lawson entered business, making sales slips, order forms, and printing machinery. He took control of a Boston publishing house and shook up the lethargic concern by publishing the works of an infamous anti-Semite. He welcomed the publicity from resulting libel suits. When he followed with publication of the anti-Catholic *Why Priests Should Wed,* a protester smashed a store window displaying the book, but Lawson paid the protester's bail and fine. Lawson himself found time to write a history of the Republican Party, a collection of poems and short stories, and *The Secrets of Congress*.[2]

Lawson competed with the Lamson Store Services Company, a supplier of retail-store equipment. He started a publicity campaign portraying Lamson stock as nearly worthless and sent the shares crashing down, earning $700,000 for Lawson, who had shorted Lamson stock. Lawson had discovered his genius for publicity by which he could send share prices up or down as easily as a child does a yo-yo. He was indicted in New York for stock fraud in the Lamson affair but was never extradited from Massachusetts. In 1890 he bet his Lamson winnings on sugar futures and lost them all.

Grand Rivers, Kentucky: The Greatest Iron Mines in the World

"Grand Rivers Furnace Trustee stock is as near absolute security as is possible in the stock of any corporation—it will pay large dividends, thirty to thirty-five per cent—the prospective profits are enormous."

—Thomas Lawson, 1891

"Well, it turned out that there was no iron in the mines—at least not enough to pay for the extraction, and the investment simply disappeared."

—Thomas Lawson, 1905, *Everybody's Magazine*, March 1905, p. 75

Lawson was already a multimillionaire when he entered his first mining fiasco in 1890, with promoters who meant to establish a steelmaking colossus at the small settlement of Nickellville, Kentucky. Lawson rechristened Nickellville as Grand Rivers City and incorporated the Grand Rivers Furnaces Company to make the iron, Grand Rivers Brick and Tile and Grand Rivers Lumber and Manufacturing to make building materials, and the Grand Rivers Company to profit in real estate. Thomas Lawson served as an officer—president, vice president, or general manager—in each of the companies, and was even elected mayor of Grand Rivers, Kentucky.[3]

The Lawson genius for advertising brought investor money pouring in. It was like any number of other risky and ill-considered mining ventures except for Lawson's phony hype. According to his publicity, the area around Grand Rivers held great reserves of rich iron ore that would make it the most profitable steelmaker in the country. Shareholders didn't learn until too late that the iron ore was too low-grade. The company built a

Few could resist the charm and forceful personality of Thomas Lawson.
From Everybody's Magazine, June 1905.

few steel furnaces, but they were mostly idle for lack of ore. The Grand
Rivers companies collapsed in 1893 and their great initial capital disappeared. Lawson himself was appointed receiver for the company. "The
Kentucky experience," he said later, "is one of the pleasantest experiences
of my life." Shareholders did not remember it with such fondness.[4]

In 1894 the Westinghouse Corporation was locked in a fight with the General Electric Company. Lawson saw that General Electric shares were vulnerable, so he took a short position then hammered the company in newspaper advertisements. His gamble gained him $250,000 when GE share prices tumbled from $118 down to $56 in a single day. He used his GE profits to gamble once more in sugar futures. This time he bought a large short position in sugar and watched his investment melt away when the price of sugar rose.[5]

Bay State Gas

"Not one evil of corporate mismanagement, fraudulent promotion, gross overcapitalization, dishonest administration, corruption of politics, secrecy and crime is absent in the amazing record of its history."

— *Wall Street Journal*, March 11, 1904, on Bay State Gas

J. Edward O'Sullivan Addicks, a corporate manipulator and senatorial candidate from Delaware, was maintaining control of the Boston utility Bay State Gas despite a desperate grudge struggle with the Standard Oil director Henry Rogers. Every time Rogers garnered enough Bay State shares to threaten Addicks's position, Addicks countered by issuing more shares to himself. He increased the capital stock of the company, which was initially par value $3 million, to $250 million, causing Bay State stock to be banned from every organized stock exchange. In 1895 Thomas Lawson threatened to "bear" Bay State stock if Addicks did not give him $250,000 cash, a large block of shares, and half the profits. Addicks had no recourse but to accede to the blackmail, and Lawson was soon championing Bay State. Once on the inside, Lawson used the Bay State Gas treasury to fund his stock speculations. Eventually, Henry Rogers made him a better offer, however, and Lawson turned on his former partner Addicks.[6]

Lawson, Rogers, Addicks, and others tangled the finances of Bay State Gas beyond comprehension and plundered what was left of the poor company and its shareholders. When the financial powers contemplated ending the gas war with a merger of the competing Massachusetts gas companies, they knew that the master publicist Lawson could wreck their plans and asked Henry Rogers to bring his attack dog to heel. Lawson agreed—for $1 million. By the time Bay State Gas went into receivership in 1903, $6.1 million had disappeared from the company treasury. By Lawson's own admission, he made $10 million from Bay State Gas.[7]

The Bay State Gas fight introduced Lawson to the "Standard Oil crowd," notably the Standard Oil directors Henry Rogers and William Rockefeller. Though their financial schemes scandalized Standard Oil's founder, John D. Rockefeller (who himself was not thin-skinned when it came to business ethics), Henry Rogers and William Rockefeller flaunted the implied backing of Standard Oil. Both borrowed Standard's surplus funds to bankroll their ventures and made a point to pay for their speculations with checks drawn on Standard Oil.[8]

The Boston Copper King

Boston had been the financial center of American copper mining since the rise of the northern Michigan copper district in the 1850s, and Lawson gravitated toward the "coppers" that dominated the Boston stock market. He publicized his stock tips in newspaper advertisements, assuring his readers that he could predict the value of copper-mining shares with mathematical certainty. Copper-mining stocks boomed in 1899, and numerous new issues were floated to meet demand. Lawson was among the greatest beneficiaries of the boom, and his fortune swelled to $50 million.[9]

Lawson threw money about in conspicuous consumption. He commissioned a 148-foot luxury yacht, *Dreamer*, finished in mahogany and teak, velvet carpets, and Belgian tapestries, with seven staterooms in addition to a library, dining room, and crews' quarters. The builders of the world's first (and last) seven-masted sailing ship named the freighter the *Thomas W. Lawson*.

Lawson spent $205,000 building the yacht *Independence* to compete in the America's Cup. The New York Yacht Club, however, decided that Lawson was socially unacceptable and, after searching for an excuse to bar him, ruled that the Hull Yacht Club, to which Lawson belonged, was not the proper sort of yacht club to be allowed to compete. They need not have feared Lawson's ship, which proved a clumsy racer. After the *Independence* was damaged in its third non-cup race, Lawson ordered it dismantled, and he handed out pieces as souvenirs. He revenged himself by writing the *History of the America's Cup*.

Lawson was a sentimental and devoted family man. At twenty-one he married his childhood sweetheart, and when he made his fortune, he spent $6 million building Dreamworld, a fabulous estate in Scituate, Massachusetts, where they lived with their six children. Dreamworld had its own post office and a water tower built to resemble a castle. On impulse Lawson paid $30,000 to name a new variety of carnation after his wife and always afterward wore the carnation in his lapel.[10]

After losing several fortunes in sugar speculations, Lawson gained sweet revenge in 1899 when he cleaned up over $1 million in a quick gamble on Sugar Trust stock. He swore off further speculation in sugar.[11]

Butte & Boston Consolidated Mining Company

Butte & Boston was a Montana copper-mining concern that had absorbed its initial capital and $3 million in assessments without ever paying a dividend. Lawson learned that insiders were selling, so he took a short interest, then hammered Butte & Boston in his newspaper advertisements. He was so successful that the $25 shares sank to $2, he reaped a windfall, and the company went into receivership. Lawson bought a controlling interest in the company at the bargain price, and now laden with Butte & Boston shares, became its champion. Share prices rose once more.[12]

The manipulators just ousted from Butte & Boston management retaliated with a bear attack. Lawson had financed the venture by borrowing heavily using his Butte & Boston shares themselves as collateral; if share prices sagged, the banks would call in his loans. Fortunately for Lawson he could dip into the Bay State Gas treasury, and the extra million dollars of ready cash defeated the bear raid and left Lawson with undisputed control of Butte & Boston.[13]

Arcadian Copper Company

The most profitable mining company—by far—in nineteenth-century America was not a gold mine on California's Mother Lode, or a Comstock silver mine, but the Calumet & Hecla, a copper mine in far northern Michigan. Other mines of that era boomed and busted with frightening unpredictability as their veins pinched and swelled with depth, but the Calumet & Hecla's lodes held rich and steady, and the company seemed more like a copper factory than a mine. Conservative Boston directors countenanced no stock manipulations, and blue-blooded Boston investors valued their Calumet & Hecla shares not for speculation but as a source of steady dividends. Year after year, under quiet and efficient management, Calumet & Hecla hoisted ore from shafts, poured copper ingots from furnaces, and mailed dividends to shareholders.

The dream of every copper man was to get in on the ground floor of another Calumet & Hecla. Henry Rogers and William Rockefeller decided that Arcadian Copper, which owned land near the Calumet & Hecla, was just the thing to "boom" as the next copper wonder. They bought the company and reincorporated it with a larger capitalization. To sell the new stock offering, they chose Lawson, who promoted it with his characteristic overblown publicity in 1899, sending shares up to $75.[14]

The new company spent money rapidly on a first-class surface plant to impress stockholders. The Arcadian mine and mill began full operation in July 1899. Lawson's advertisements advised investors to buy Arcadian shares, which he exultantly predicted would climb past $250, perhaps to $450.

Instead of rising, the share price dropped by half to $40 over the next few months. Clarence Barron of the *Boston News Bureau* studied production figures and told his readers that, contrary to the promises, the Arcadian mine was losing money rapidly. Arcadian management angrily insisted that Barron was lying and that the Arcadian was a rich mine with a glorious future. Who was lying became apparent just a few months later, when the Arcadian company abruptly laid off half its work force.[15]

Arcadian managers had built grandiose surface facilities before finding sufficient ore—the same bad policy Lawson had followed with the Grand Rivers iron mines. Lawson showed that either he had learned nothing from his Grand Rivers fiasco, or—more likely—that Grand Rivers had taught him that the presence of ore, however essential to the company, was irrelevant to his own promoter's profit. Rogers and Rockefeller also should have known better but were more intent on selling shares than mining copper. More prudent and honest management took over and proceeded with the underground exploration that should have been done first, but since the Arcadian had already squandered the large initial capital, the new management ran out of money and halted work in 1903. Share prices sank to $0.25 by 1904. The company sold assets to finance further exploration but went out of business in 1909, succeeded by the New Arcadian Copper Company, which although better-managed was no more successful in finding copper.[16]

Amalgamated Copper

"Amalgamated is, in my opinion, the best opportunity ever offered the public for safe and profitable investment."

—Thomas Lawson, 1899, quoted in *Success Magazine*, February 1908, p. 107

"The men who control Amalgamated told me it was not worth half the price it was floated at."

—Thomas Lawson, Everybody's Magazine, February 1905, pp. 173–181

In 1899 Lawson convinced Henry Rogers and William Rockefeller to organize a trust to control copper prices and production much as Standard Oil controlled petroleum. It fell to Lawson to promote share sales for the new company: Amalgamated Copper. Controllers of Amalgamated and its corporate descendants would later try to write Lawson out of their history in the style of Soviet purge victims. News accounts of the time, however, support Lawson as the intellectual father of Amalgamated, however much Amalgamated Copper management later came to dislike him.[17]

Lawson used his connection with Amalgamated to play his own stock market games. As rumors flew through the market, copper shares boomed in anticipation of being bought by Amalgamated. Lawson owned a big stake in Butte & Boston, so he advertised that the company would be the crown jewel of the Amalgamated consolidation. It seemed an odd choice given the poor earnings record, but share prices soared on Lawson's announcement.[18]

After two years of rumors and preparation, the Amalgamated Copper Company formed in April 1899. Rogers and Rockefeller had bought control of the Anaconda and other valuable copper mines in Butte, Montana, for $39 million, and sold their interest to the public in the form of Amalgamated shares for $75 million—a nice profit of $36 million. Lawson's overpriced Butte & Boston was left out, despite his public assurances. The Amalgamated prospectus contained only sixty-seven words describing the company, but Lawson's advertisements promised that shares offered at $100 would quickly advance to $150 or $175. Mobs formed outside the National City Bank in New York to buy shares, and the initial offering was five times oversubscribed. Amalgamated, like Standard Oil, chose not to comply with the minimum financial disclosure then required for listing on the New York Stock Exchange, but it was still traded as an "unlisted stock" on that exchange.[19]

The public was enchanted by the prospect of becoming partners with the Rockefellers, despite the fact that Amalgamated sold for nearly double the market value of its assets. More sophisticated financial commentators cautioned their readers to avoid Amalgamated. London's *Financial Times* was "aghast at the sublime impudence of the proposition." Misgivings about Standard Oil secrecy and inside manipulations were confirmed when the Anaconda announced that it would no longer make its financial reports public.[20]

Lawson expected to get $9 million—a quarter of the insider profits. Rogers and William Rockefeller, who had put up the capital, thought his efforts worth only $2.5 million. Lawson threatened to turn his publicity machine against Amalgamated, until Rogers and Rockefeller agreed to cut him in for a larger share, just under $5 million.

Amalgamated did not rise to $150 or $175 as Lawson had promised but paid regular dividends, and after slumping to $75 rose to $130 with some help (Amalgamated's promoters had kept enough shares to manipulate the price). Amalgamated made its second stock offering in 1901, at $120 per share. This time Butte & Boston was finally absorbed into Amalgamated in a one-to-one stock swap. Financial analysts estimated that Lawson and Rogers had profited $20 million on an initial investment of $4 million in Butte & Boston.[21]

Investors still believed that Amalgamated was a monopoly-in-the-making, a foolproof investment managed by the wizards of Standard Oil. The insiders knew better. Amalgamated was maintaining an artificially high copper price only by holding a large amount of copper off the market, and it could not afford to continue paying high dividends. Now that they had floated the second stock offering and no longer needed to support the price, Rogers and Rockefeller liquidated their holdings as rapidly as the market could absorb the shares. This accomplished, they reduced the dividend in October 1901, and allowed the copper price to float in December. Amalgamated shares plunged from $128 to $66, despite Lawson's assurances that Amalgamated would surge to $180. The *Wall Street Journal* criticized Amalgamated for following the long-standing Standard Oil policy of withholding all financial information from the shareholders and urged that the stock be ejected from the unlisted department of the New York Stock Exchange. Amalgamated shares declined to $33 before recovering.[22]

The *Wall Street Journal* called Amalgamated Copper's 1904 annual meeting "an impudent farce" when, as usual, management hurried through the legally required ceremony, while refusing to release any information to mere shareholders. Amalgamated Copper finally ended its secretive ways when Rogers and Rockefeller passed control to legitimate mining men, and Amalgamated issued its first-ever financial statement at the 1905 annual meeting. The company never succeeded in controlling the copper industry, but it was instrumental in combining the fragmented properties of Butte, Montana, into an efficient, consolidated operation under the Anaconda Copper Company.[23]

Booming Trimountain Copper

Lawson scored a coup by manipulating Trimountain Copper, a mining company organized in 1899, with valuable lands in northern Michigan. After the company collected $500,00 in assessments in the first three years, Lawson and the Amalgamated Copper director Albert C. Burrage acquired a majority of Trimountain shares, and boomed share prices. They worked the mine for a quick splash, taking out only the best ore and neglecting development work. In 1903 Lawson and Burrage had their company pay $300,000 in dividends—mostly to themselves—even while running the company $840,000 into debt. The share price soared further, and Lawson and Burrage sold their interest to the Copper Range Company in a one-to-one stock swap, although they had to agree to pay the Trimountain indebtedness.[24]

Copper Range management were no fools. They knew they were buying a gutted mine, but they also knew it was good ground and were confident that they could find more copper. But when Trimountain miners began to uncover new ore bodies under Copper Range management, Lawson and Burrage tried to renege on the terms of the deal and refused to pay the Trimountain debt, alleging that Copper Range had lied about the condition of the mine at the time of sale. Copper Range had to sue to force Lawson and Burrage to pay the debt.[25]

Now heavy with his Copper Range shares, Lawson promoted that company in newspaper advertisements. Part of the stock-swap agreement, however, was that Lawson and Burrage could not sell their Copper Range shares for three years. They violated the agreement when they sold seventy thousand shares a year early, selling Copper Range even as Lawson was advising the public to buy. They disguised the sales so cleverly that by the time the Boston stock market realized that their frozen shares were being traded, they had been swapped so extensively that the market could not unravel the transactions. Lawson and Burrage pocketed the money, but Copper Range sued for damages.[26]

Trimountain was a genuinely rich property, but it took honest management five years to develop the mines for long-term operation and begin to pay dividends out of real earnings.

Trinity Copper Company

"The Trinity is Lawson's scientific little game for parting fools from their money."

—Horace Stevens, *The Copper Handbook*, 1908 vol. 8, pp. 1337-40

Lawson's most bald-faced swindle was the Trinity Copper Company, which owned mining land in northern California. The owners of some undeveloped California copper prospects went to Boston in 1899 to try to sell them for $60,000, but the Boston copper people declined. Tom Lawson bought the properties for $165,000 and capitalized them in 1900 as a $6 million corporation, Trinity Copper. He sold shares through glowing newspaper advertisements and promised a mountain of copper that within less than a year would generate profits of thirty to fifty percent annually on the initial investment. Mining and financial publications warned against it, but the public listened to Lawson's siren song, and oversubscribed the stock offering.[27]

The Boston Stock Exchange at first refused to list the stock but eventually put it on the board. Initially $25, shares advanced to $42 under Lawson's promotion, until late 1901 when investors realized that the promised dividends were not forthcoming, and share prices crashed.[28]

Lawson made himself president of the company, made his son Arnold vice president, and continued to sell worthless Trinity stock and milk the company treasury. He planted favorable puff pieces in the press to keep up flagging hopes, but the facts were dismal. A smelter and a railroad connection promised in 1901 never materialized. The company built a small office building on the site but did no further development with its great capital funds; the money went instead to Tom Lawson. By 1906, the only three employees were an engineer paid $10,000 per year to write encouraging reports, and two watchmen paid to prevent the investing public from inspecting the property. Poor shareholders trying to determine the real financial state of the company were stonewalled by Lawson and his coconspirators in management.[29]

Lawson pumped life into sagging Trinity shares with newspaper advertisements in 1907. The tone was always the same: the straight-talking Tom Lawson promising spectacular profits and breathlessly warning against the cowardly machinations of Wall Street, which was trying to ruin him as the only honest voice in high finance. He dismissed the lack of copper production at the mine by claiming that it was part of his master plan to wait until the copper price rose. He touted a new agreement to treat Trinity ore at a nearby smelter as if it were a great coup and recognition of the richness of the Trinity ore body, which it was clearly not. In fact, the Trinity could not fulfill its part of the contract. He bragged that famous mining engineer John Hays Hammond had examined the property in April 1906 for a powerful mining syndicate. But since Hammond's employers did not buy the Trinity, and the report was never made public, this was hardly an endorsement. Lawson promised that prices would rise to more than $100. Shares that had been selling for $11.00 rose to $38.50, at which point Lawson sold nearly all his Trinity shares, no doubt more than recouping his advertising costs. Those who held on to their shares waiting to sell at $100 watched helplessly as the price sank once again, an occurrence for which Lawson blamed his enemies.[30]

Trinity issued its first public balance sheet in February 1907, but even the company treasurer could not explain some of the cryptic notations, nor what had happened to the large cash surplus with which the company began. Another balance sheet in September 1908 was an exercise in questionable accounting. The mine began producing ore with great fanfare in September 1907 but soon stopped with no reason given. Ore production started again the following year but again stopped. Every time Lawson needed an excuse for another publicity campaign, he reported that the mine was restarting. Lawson himself admitted that Trinity was a high-cost producer that required high copper prices to operate. The company never paid a dividend.[31]

Lawson invested so much of his reputation in Trinity that he could never admit the truth. In 1913, after twelve years of steady unprofitability, in a burst of twisted Lawson logic, he hailed it "one of the great copper successes." The Trinity Copper Company sold its properties in 1920, after years of inactivity. The land was offered for sale again in 1928 "at a very low price and on favorable terms."[32]

First National Copper Company

In 1907 Lawson bought control of the bankrupt Balaklala Consolidated Copper Company, which owned the mine of the same name in northern California. He formed and sold stock in First National Copper with himself as president, using it as a holding company to control Balaklala. The operations of the Balaklala mine seemed irrelevant to the gyrations of First National shares. First National issued unusual and highly contradictory balance sheets, from which cash and assets appeared and mostly disappeared with no clue as to their fate. To those who knew the Lawson shell game, there could be no mystery as to which shell held the pea: all cash and assets would soon be lodged out of sight in Lawson's bank account.

One First National financial statement listed a $1.5 million asset consisting only of the right to levy assessments on shares—which may have been an asset in Lawson's peculiar view, but the assessment-paying share owners may not have considered it so. In 1910 First National boasted a modest profit of $20,000 in one month, then immediately turned and levied a $1 million assessment. Lawson boomed the shares up to $8 at one time, but the company came to a sorry end, like so many of Lawson's playthings, and never paid a dividend.[33]

Other Lawson Promotions

At least one Lawson promotion turned out to be genuinely good. In 1899 Lawson and his Standard Oil friends bought the Santa Rita copper mines in New Mexico, but the company stopped mining in 1904. Lawson said that his partners, with whom he was now feuding, shut the mines to deprive him of cash, but the truth was that there were not enough rich copper veins for the company to mine profitably. The properties were turned over to tributers, and for several years the Santa Rita Mining Company made most of its money—which was very little—by running the company store.

A bright geologist convinced the Santa Rita partners to mine the property as a large open pit. Lawson tried to use the Santa Rita property to float a new company in May 1909 but sold so few shares that he was forced to return what little money he had collected. More reliable financial backers took control and floated the property as Chino Copper later in 1909. Lawson, who owned less than ten percent of the new shares and was no longer in control, published a three-hundred-page book on the Santa Rita mines, promising that the ore would be treated by a secret process. Lawson bought newspaper advertising space in 1910 to tell the public to buy Chino, calling it "The Marvel of the Age," "An Opportunity of a Lifetime," and so forth. Chino Copper management wisely ignored Lawson's mystery process and dealt with their ore by conventional means. Chino could never live up to Lawson's ridiculous hype but proved itself a steady and profitable producer for many years.[34]

In other cases Lawson was spectacularly wrong. The Lawson Mexican Company was formed with $5 million in capital in 1908 to work gold and mercury mines in Mexico. Lawson spent $665,000 buying and developing the properties before the company ran out of money the following year and sold the property to pay creditors.[35]

Win or lose, Lawson could charm money out of people's pockets. A New York man gave Lawson $500,000 to invest for him. After Lawson lost it all in mining speculations, the investor sued to get it back, even though he admitted in court that he had trusted the funds to Lawson with no idea how Lawson would invest them.[36]

Shareholders knew that wherever Lawson went, excitement was just ahead. Lesser promoters trumpeted the mere rumor of Lawson's interest—for example at Greenwater, California, in 1906, and at Silver Plume, Colorado, in 1917—to sell to those hoping to buy ahead of the Lawson carnival.[37]

Seizing Control of Greene Copper

Colonel William Greene, a former New York City clerk turned Arizona cowboy and prospector, had built from nothing a large copper concern in northern Mexico—but he was perpetually short of cash. Greene arranged through Lawson a $1 million line of credit from the Amalgamated Copper managers, on which he borrowed $135,000. Greene soon discovered what it was to make a deal with the devil, however, when an additional draw of $30,000 was refused without explanation. Lawson and partners demanded immediate repayment, giving the stretched Greene only three days to repay the $135,000 plus interest or surrender control of his company. Amalgamated had set the trap to capture Greene Copper at a bargain price. However, Greene raised the money from John "Bet-a-Million" Gates and escaped an Amalgamated takeover.[38]

In December 1904 Colonel Greene blamed Lawson for a bear attack on Greene Consolidated and promised to travel to Boston to confront Lawson in person. Greene was reputed to carry a six-shooter with a handle notched for various western desperados he had killed, and the sensationalist press predicted bloodshed. A crowd gathered in front of Lawson's office to await the gunplay, and swapped rumors that Greene had arrived in Boston armed and dangerous. Neither Greene nor Lawson appeared, and the crowd went home disappointed at day's end. Greene and Lawson met shortly afterward in a Boston hotel room, but the meeting was cordial, and to the disappointment of the newspapers, no one was killed.[39]

Frenzied Finance

If the Amalgamated Copper management thought that Lawson was safely inside the Amalgamated tent, they misread the mercurial Lawson personality. Lawson fell out with his Amalgamated patrons over his share of profits in manipulating Bay State Gas. Lawson claimed that Henry Rogers owed him $1 million for his acquiescence in the Bay State Gas merger. Rogers said that he had already given Lawson $93,000 as payment in full.[40]

To understand his quixotic fight against Henry Rogers requires an appreciation of the Lawson pride and the Lawson ego. He considered himself Rogers's partner. He had bragged for years that he was the financial genius who had dreamed up Amalgamated Copper and convinced Henry Rogers of the worth of that mammoth undertaking; and now Rogers treated him as a hireling and tried to pay him off with what Lawson regarded as a pittance.

Rogers should have known better. When Rogers bought control of Anaconda Copper, along with Anaconda's mines, mills, and smelters, he also took on Anaconda's mortal enemy, Augustus Heinze. As much sway as they held on Wall Street, the Amalgamated autocrats could not rule Butte, Montana, where the young rival mine owner Augustus Heinze and his lawyer-brother Arthur, along with their obedient judges, harassed the Amalgamated companies with the intricacies of the mining law. Like Lawson, Heinze was a charismatic financial buccaneer. To Anaconda he was a blackmailer who corrupted the courts and sent armed thugs to wage underground warfare in the mine tunnels beneath Butte. To his admirers Heinze was the handsome blue-eyed hero who prevented the Anaconda from squeezing the state of Montana and its workingmen. Heinze had fought first Anaconda and then Amalgamated for years on many fronts, from the courts, to the Montana statehouse, to the dark underground tunnels. The struggle went wildly back and forth, but neither side was able to gain permanent ascendancy. Now Lawson threatened to become every bit as bothersome in New York and Boston as Heinze was in Butte.

When Bay State Gas went into receivership, the manipulators tried to pick a compliant receiver; the judge instead appointed a lawyer of unquestioned integrity, George Wharton Pepper. Pepper discovered that the company held only $26 in its treasury but was owed some $600,000 by Addicks, Lawson, and Rogers. He sued to recover the money.

During the trial Lawson and Henry Rogers contradicted one another on the stand, and Lawson accused Rogers of perjury. His testimony caught the eye of *Everybody's*, a struggling populist magazine. Ida Tarbell's serialized *History of Standard Oil* was then creating a sensation in *McClure's*, and *Everybody's* was looking for a similar exposé. Lawson first refused to speak to *Everybody's*, but after the editors camped out at the door of his office, he relented. Lawson began his serialized inside story of Amalgamated Copper, "Frenzied Finance," in the July 1904 issue of *Everybody's*. It was a wordy and rambling account that stretched into 1906, continually luring the reader on with promises of scandals to be revealed in later installments. Lawson used his convoluted tale to whitewash his disreputable roles in Amalgamated Copper, Bay State Gas, and other machinations.[41]

Parts of Lawson's account were contrary to fact. He repeatedly exalted his favorite, the Butte & Boston Company, and said that Rogers and Rockefeller had cheated the public when they failed to include it in Amalgamated's first stock offering. On the other hand he denigrated Anaconda Copper, the centerpiece of the first stock offering, and said that Marcus Daly had cheated Rogers and Rockefeller into paying too high a price for the Anaconda and the Parrot mines. In fact Anaconda was the world's largest copper producer, while Butte & Boston was much smaller and economically marginal.[42]

As an aside in his series, Lawson revealed the cozy relationships that financiers had cultivated with insurance companies, and told how insurance companies were misinvesting the monies entrusted to them. Lawson's revelation started a storm that led to a congressional investigation, jail sentences for some insurance executives, and regulatory oversight of the industry.

Lawson rode the wave of revulsion from the excesses of the Rockefellers and Morgans. Mainstream financial papers dismissed Lawson as "the frenzifier," but his articles were a sensation, and the first installment in *Everybody's* sold out in three days. A clever marketer sold a card game "Trusts and Busts or Frenzied Finance." The novelist David Graham Phillips published *The Deluge*, whose protagonist was a fictionalized Thomas Lawson. Lawson warmed to his role and cranked out article after article, painting himself as the one honest man fighting to prevent plutocrats from plundering the public. Lawson's serial quickly came out in book form as *Frenzied Finance, the Crime of Amalgamated*.[43]

Public suspicion regarded no event to be beyond the conspiratorial powers of Amalgamated. Montana farmers had sued the Anaconda Copper Company, claiming that smelter smoke had ruined their crops, but unusually high rainfall in 1905 resulted in record good harvests and threatened the farmers' legal case. The farmers angrily charged that Anaconda's parent, the evil Amalgamated, was in cahoots with rainmakers to bring them good harvests and thus ruin their lawsuit.[44]

Mining and financial commentators dismissed Lawson's *Frenzied Finance* as a mishmash of fact and fiction. Horace Stevens, in his authoritative *Copper Handbook*, lambasted Lawson's record of deceit and mismanagement. Clarence Barron's *Boston News Bureau* also missed no opportunity to criticize Lawson and his methods. Lawson charged that Stevens and Barron were blackmailers whom he had refused to pay off.[45]

The Lawson Panic

Lawson defended Amalgamated out of pride of authorship, despite his bitter denunciations of its managers. In November 1904 Lawson advised investors to purchase Amalgamated shares. A week later, however, in one of his breathtaking turnabouts, he bought display advertisements in Boston and New York newspapers proclaiming that Amalgamated was an overvalued swindle. In advertisements on December 6, Lawson predicted that Amalgamated would fall to $33. Actually it slid from $80.75 before his attack, to $58.50 three days later, before recovering to the mid-sixties. The drop in Amalgamated triggered—or some said merely coincided with—a brief sell-down of the stock market that became known by the exaggerated label the "Lawson panic." In ten days Lawson spent $119,000 slamming Amalgamated in newspaper advertisements, including some appearing in London, Paris, and Berlin. Lawson continued his newspaper attacks through February 1905 but with little further effect.[46]

Lawson indignantly denied that he had any financial interest in "bearing" Amalgamated and claimed that his only motive was to warn the public. But despite his denials, he had taken a large short position in Amalgamated and reaped hundreds of thousands of dollars when the price dropped. He then repurchased Amalgamated at bargain prices while still advising the public to sell. He maintained that he bought the shares only to prevent the price from falling too far. It was all supposedly for the public good, but at the end of the "Lawson panic," Lawson was richer, and the public poorer.[47]

The Boston financier gloried in his fame and toured the plains states in his private railcar, giving speeches on Wall Street tyranny to the populist heartland. He spoke himself hoarse in Ottawa, Kansas. In Minneapolis he promised an appreciative audience that he would repay the people for the millions of dollars he had helped Amalgamated swindle from them. In February 1906 he addressed a joint session of the Iowa legislature. A Kansas City reporter voiced a minority view when he found Lawson "loud, vain, and vulgar."[48]

Copper from Clay

In September 1905 Lawson announced that he would graciously accept $10 million—in amounts of not less than $25,000—for a blind pool to profit from an imminent crash in the price of copper. Lawson predicted that the price drop would ruin all copper mining companies, and that his copper pool would enjoy enormous profits by selling short. Word leaked that Lawson's pool was based on his belief that Dr. Shiels of Glasgow, Scotland, had invented a way to create copper from common clay, at a cost of only a few cents per pound.[49]

Lawson learned that he had been gulled by swindlers who had put a copper ingot on his desk and told him that Dr. Shiels had created it from clay. When the story emerged, Dr. Shiels himself denied the ability to alchemize copper. Copper prices prospered, and Lawson's $10 million pool faced ruinous losses.

Doomsday, June 28, 1906

On May 17, 1906, Lawson predicted a "worldwide catastrophe" of an undisclosed nature on or about June 28. Possibly he hoped to induce a panic to allow him to settle his short copper position without too great a loss, but the date came and went without a drop in the market.[50]

A market reverse following the San Francisco earthquake in late 1906 finally drove the copper price down temporarily, and Lawson liquidated his $10 million copper pool at a slight profit.[51]

The Nevada-Utah Stock Game

Lawson's problem in manipulating public opinion was that his credibility suffered when, as often happened, he advised the public to buy stocks that subsequently floundered. He sidestepped the problem neatly in his Nevada-Utah stock game in early 1907 by persuading his public to buy stocks while telling them not to.

Nevada-Utah Mines & Smelters Corporation was occasionally traded on the New York curb market for $4 to $5 in early 1907. When the price rose on rumors that Augustus Heinze was interested in the property, Heinze denied it. When Lawson was rumored to be interested in the company, however, he responded with a large newspaper advertisement that praised Nevada-Utah but warned his followers not to buy until he finished investigating.[52]

Lawson's advertisements appeared on February 27, and shares jumped to over $8. Frantic trading led to fistfights in the curb market, and extra policemen to keep order. Lawson beat the drums every day for two weeks, in large display ads, extolling the greatness of Nevada-Utah while cunningly advising the public not to buy until he gave the word. Lawson advised that Nevada-Utah could soon rocket up to $40 or even $100. Lawson's followers saw that prices had already doubled and that if they waited they would be too late; so heavy buying continued.[53]

Curb traders identified Lawson's campaign as a scheme to manipulate share prices. Nevada-Utah shares made up over three-quarters of the volume traded on the curb market on February 28, 1907. Lawson himself admitted that he had sold his own shares at more than $8, but only, he said, to keep the price from rising out of reach of the small investor. Lawson's charity was duplicated by the Nevada-Utah insiders, who were likewise reported happily selling their large shareholdings at high prices.[54]

After two weeks of teaser advertisements, Lawson announced that he had finished his investigation and could not recommend purchase of Nevada-Utah. On March 12, the morning of Lawson's announcement, the price broke as soon as the stock was called and sent the curb market—not known for dignity even on its best behavior—into pandemonium. The price sank to $2.75 before a common price was established; it closed at $4.50. The public lost about $1 million because of Lawson's publicity, but after all, he had told them not to buy.[55]

Friday the 13th

Lawson's next doomsday chronicle was his serialized novel *Friday the Thirteenth*, which told of a stockbroker stricken with a sudden attack of conscience. Lawson's hero threw his fortune into a bear operation that destroyed all share values. This followed Lawson's theme that share values were only an illusion and a trick by the money powers to cheat small investors.[56]

Lawson advertised that any broker could do the same as his fictional protagonist—completely disrupt the stock market and make a great deal of money without any risk—and he offered $5,000 to anyone who could disprove him. Stockbrokers knew better—all except Albert Appleyard, a Philadelphia broker who lost everything one day in June 1907 in futile attempt to implement Lawson's system. The Philadelphia Exchange booted Appleyard out the next day, after he failed to settle his obligations. Appleyard disappeared to avoid creditors, and other brokers joked that he had gone to Boston to claim his $5,000 from Tom Lawson.[57]

Lawson had been predicting imminent financial apocalypse for more than a year, yet the market was healthy. At first he proclaimed victory of a sort, saying that his bear campaign had forced the Amalgamated managers to spend enormous sums to support share prices. However, the continued failure of the market to obey his dire predictions, and the public's stubborn habit of buying and holding common stocks—despite Lawson's instructions to sell, finally sent him into a tantrum. He announced his withdrawal from the reform movement in newspaper advertisements in December 1907. His *Everybody's* editors begged for an explanation. Lawson sneered, "You talk of what I owe the people. What do I owe the gelatine-spined shrimps?" The public remained bewildered, but Lawson was always an intensely emotional man, and his turnaround may have been brought on by distress at the recent death of his wife.[58]

Lawson allowed the journalist Frank Fayant unrestricted access to his records. Fayant was mesmerized by Lawson, whom he called "one of the brilliant men of his times." Yet when he analyzed Lawson's public stock market predictions from 1904 to 1908, Fayant discovered that Lawson had almost invariably been wrong: Lawson had repeatedly predicted disaster while share prices rose, then predicted boom times just before the crash. Lawson said, "My consistency consists in my inconsistency," and cheerfully admitted to Fayant that he had lied to his followers. "I have caused my following—the public—to lose millions of dirty dollars that in the end they may be prepared for My Remedy."[59]

The Rivals: Thomas Lawson versus Cardenio King

Closely resembling Lawson in flamboyant company promotions and self-glorifying newspaper advertisements was Cardenio King, also of Boston. Both came from humble circumstances. Each recognized in the other a shameless financial mountebank, and pilloried one another in the Boston papers.[60]

King's worst promotion was the King-Crowther Corporation, a fledgling oil company that—so said King—would rival Standard Oil itself. In 1902 he chartered two train cars to wine and flatter newspapermen traveling to and from his operations in Texas. The reporters dutifully responded with puff pieces, which King arranged for the New York *Sun*, the New York *World*, and other papers to print prominently and in full. More knowing publications, such as the *Oil Investor's Journal* in Beaumont, Texas, scoffed at the lies of the company, which owned oil properties of little value. The State of Texas revoked the company charter for violating even the permissive state regulations of that era. King brought the company back in 1906, claiming that it had been "purified by fire." However, the reincarnated King-Crowther was just as much a swindle as the original, and fared no better.[61]

Cardenio King was also notorious for his role in pumping up the price of Orphan Copper, a wildcat Arizona mining company. Under his promotion Orphan Copper shares were introduced on the Boston curb market at $8.38 one Monday morning, but King's spell quickly wore off, and the shares plummeted to $1.50 by Thursday's closing. Within a few years shares could be bought for $0.01.[62]

Although their methods were similar, and the records of their companies were equally dismal, it was Cardenio King who was brought to justice, and not Lawson. King fled the country after being indicted but returned in 1908 and was convicted of fraud the following year. He died in prison in 1913, still blaming his incarceration on a plot he said was orchestrated by Thomas W. Lawson.[63]

The Spell of Yukon Gold

When the Guggenheim brothers formed a company to exploit the golden gravels near Dawson City, Yukon, they chose Lawson to promote the initial public offering. He plunged in with characteristic zeal and claimed that the shares were worth double or triple the selling price of $5. His advertisements created such excitement that New York curb brokers knocked one another over and tore clothing in the rush to buy shares on the first day. Financial and mining periodicals criticized the Guggenheims, who had previously run their mining interests without flamboyant promotion. The Guggenheims denied hiring Lawson and blamed his involvement on their brokerage firm. In truth they had hired him because their own reputation had been tarnished after their Nippising mining company was revealed to be worth much less than advertised.

Despite the first-day excitement, the Lawson name was no longer magic. He sold only half the offering, and share prices drooped to $4.50, which Lawson blamed on "treachery." The *Engineering & Mining Journal* calculated that the shares were worth only a third of their selling price. Yukon Gold was plagued by high operating costs and fights over control. Profits did not materialize, and in the 1920s the company gave up and shipped the gold dredges to Southeast Asia.[64]

The Return of Bay State Gas

"'National Stock' is to be the one absolute power to make and unmake Wall Street prices at will."

—Thomas W. Lawson, August 17, 1908

After two years of forecasting stock market disaster, Lawson suddenly predicted the greatest bull market in history. One stock in particular, he said, would rocket to twenty times its present price. After a few days of buildup, Lawson revealed that the greatest financial colossus of the twentieth century would be none other than the despised Bay State Gas, which he himself had helped drive into bankruptcy. Bay State shares had declined to $0.03 in the dark days of receivership, after Lawson, Addicks, and Rogers had abandoned ship leaving it with large debts and no assets save $80.70 in cash. The receiver George Wharton Pepper had resuscitated the company by suing its despoilers. He was unsuccessful in his lawsuit against brokers Kidder, Peabody & Co., but only after witnesses scattered like cockroaches to avoid subpoenas. Lawson paid back $350,000 in an out-of-court settlement, and Addicks repaid $48,254 and a large block of Bay State shares that he had issued to himself. Pepper assigned the claims against Henry Rogers to a third party for a hefty $1.5 million. Bay State shares had risen to $3 during Lawson's secret campaign to secure control.[65]

Lawson claimed that Bay State Gas, shorn of assets and municipal franchise, was somehow the most valuable stock in the world. The Bay State Gas charter did allow the company to issue new shares at whim and to operate without issuing reports of any kind. The *Wall Street Journal* marveled that the company charter gave it "the right to do anything except manufacture gunpowder and declare war." This made it the perfect vessel for financial piracy but did not guarantee profitability. Nevertheless, Lawson advertised Bay State as a holding company to speculate in the stock market. The name of the company was still Bay State Gas, but Lawson insisted on calling it "National Stock" to give it more importance.[66]

This was Lawson at his most megalomaniacal. He claimed that his overarching goal for the past fourteen years had been to secure control of Bay State Gas. According to Lawson's far-fetched soap opera, Henry Rogers of Standard Oil had seen through Lawson's plan and done everything in his power to keep Bay State Gas out of Lawson's hands. But in the stock panic of 1907, Lawson said he had maneuvered the financial powers into a position where he could dictate terms, and his price was control of Bay State Gas. Lawson never specified the trap in which he claimed to have caught his foes, leaving that to the imaginations of his conspiracy-theory devotees. Lawson claimed that neither himself nor Bay State Gas had any shares to sell but then revealed that he was busy selling Bay State stock and that Bay State Gas held sixteen million unissued shares—more than four times the number of shares outstanding—which it would sell.

Lawson promised that with him at the helm of Bay State Gas, his skill in stock manipulation would enable him to plunder the Wall Street powers. He said that he could take each dollar invested in Bay State Gas and return profits of $4 annually. In newspapers across the country he advertised: "A Fortune for Everyone—Quick."[67]

It was all Lawson moonshine: Bay State Gas sank once more into unmourned obscurity, and Lawson's influence and finances declined inexorably.[68]

The Promoter-Politician

In 1907 Lawson urged a reform fusion ticket of Democrat Woodrow Wilson and Republican Teddy Roosevelt, and pledged all his financial resources. Roosevelt and Wilson, however, were unenthusiastic about joining forces. Lawson himself announced for US senator from Massachusetts in 1912 but lost the Republican primary. He ran for the same seat six years later as an independent but won only five percent of the vote. Despite his own manipulations, he remained a spokesman for reform. The governor of New York invited him to Albany to support bills to regulate stock exchanges. In 1913 Lawson sent congressmen deluxe copies of his latest book, *High Cost of Living*, in an effort to convince them to investigate the stock exchanges.[69]

Throughout his "Frenzied Finance" series, Lawson promised that he would give the Remedy (always capitalized) at the series' conclusion. When the series ended, however, Lawson ducked the question, claiming that the people were not ready. He returned to it in 1912 when he printed "The Remedy" in *Everybody's*. In his rambling fashion, he stretched the series to six installments, later published in book form. The Remedy consisted of federal controls on all corporations and stock exchanges. Much of Lawson's Remedy was enacted following the stock market crash of 1929.[70]

The "Peace Scare"

"War may have its terrors, but they are nothing to those of peace—in Wall Street."

— *Wall Street Journal*, December 23, 1916

When in 1916 President Woodrow Wilson proposed a peaceful resolution of World War I—in which the United States had not yet entered—the stock market reacted violently downward in anticipation of the loss of wartime demand. Investors were jittery after a long bull market, and rumors of European peace had shaken share prices three times in the previous nine days. Wilson's note to the belligerents on December 21 sparked a massive sell-off in what was then the second-largest daily volume in the history of the New York Stock Exchange. The *Wall Street Journal*

"Frenzied Lawson: "Darn your 'Big Stick'! Get a meat AX!!""

Thomas Lawson helped the "Standard Oil crowd" sell watered-down
mining stock but broke with his partners to win a reputation as a reformer.
Here Lawson battles a Rockefeller-headed octopus.
From Denver Post, February 24, 1905.

denounced Wilson's peace effort, writing, "It is time to call the attention
of these irresponsible statesmen to the evil that they can do to security
and commodity markets." The same issue noted that because of the peace
initiative, "Wall Street's Christmas will not be as pleasant as was anticipated. Many a jewelry house has had many a big order cancelled." The
display of Wall Street interest in war, dubbed the "peace scare," appalled
the public; even more appalling were rumors that government insiders
had profited by selling shares short ahead of the announcement.[71]

Lawson grabbed the spotlight by announcing that he had proof that
the peace initiative had been leaked to insiders, who then profited $60
million. He wanted to personally brief President Wilson, and called for an
investigation by leaders of the House, Senate, executive branch, and Supreme Court. Lawson also claimed knowledge of a pro-German financier
who had offered to disrupt the allied financial system, and had secretly
met with the Kaiser, plotting to ship $40 million in US securities back to
Germany in a U-boat.[72]

Lawson met privately with Congressman Henry, head of the House Rules Committee. Henry said afterward that Lawson refused to provide details, but Lawson claimed that Henry wanted only to protect the malefactors. Lawson himself admitted to making hundreds of thousands of dollars from the drop in share prices, although he disavowed any advance knowledge.[73]

As a witness before the House Rules Committee, Lawson at first said that he couldn't reveal what he knew without betraying a confidence, then that the awful truth would undermine the government. Finally, he offered to reveal all, but only as part of a full congressional investigation of the stock exchanges. Lawson relented when the committee pondered contempt charges, and said that Secretary of the Treasury McAdoo had tipped off Wall Street brokers. Lawson could provide no evidence, the men he named disputed the charges, and the matter ended inconclusively, although it provided fodder for Lawson's book *The Leak*.[74]

Lawson's Decline

The law finally caught Lawson in a small way when he and other Boston promoters were charged with illegal stock advertising. Lawson was indignant at being lumped with petty con men and claimed that it was a plot by his enemies. Even under indictment Lawson could not resist the limelight: he tried to testify before a Massachusetts legislative hearing on stock flotations but was ejected after he shouted that one of the committee members was a liar. He pleaded guilty to ten counts, and although the district attorney asked for a jail sentence, the judge let Lawson off with a $1,000 fine.[75]

Lawson's finances declined steadily after his break with Amalgamated, and his publicized stock dealings became infrequent. A financial pinch in 1922 forced him to sell his famous Dreamworld estate, yacht *Trophy*, and art collection. He announced in May 1923 that he was reentering high finance, but the share markets remained quiet.[76]

In October 1923 Thomas Lawson suffered an attack of diabetes. He spent his last year of declining health at his sister's home in Maine. The once-multimillionaire died penniless in February 1925. Thomas W. Lawson today is a historical footnote, most often remembered as a crusading reformer—and for his unique namesake seven-masted sailing ship. Forgotten is the way he plundered those of modest means who put their hopes in him.[77]

Aron Beam and His Gold Process

Mining history is full of "process men," visionaries or crooks who claim secret knowledge to unlock gold from the rock. After the process fails, the inventor usually fades back to obscurity. Aron Beam, however, continued for thirty-five years to find gold where there was none and promote false mining booms in Arkansas, Colorado, Kansas, Missouri, and Oklahoma. His victims always lost money, yet Beam was clever enough to avoid legal problems. As long as his clients had money, he would sell them a dream.

Aron Beam came to prominence at age forty when he promoted the Lost Louisiana mine in Arkansas with false assays. At Bear City, Arkansas, where a spurious rush had created a town of two thousand where there had once been just a pair of farmhouses, Beam learned the swindling techniques of Professor Samuel Aughey and "Professor" R. R. Waitz.

Beam's fraud was exposed in 1888 when he could not replicate his assays under the careful watch of honest assayers. But would-be Arkansas gold miners still believed Beam. He built a house in Bear City and perfected his "Beam Electric Process" to extract gold from barren rock by magnetism. Beam installed his equipment in a failed smelter at Bear and obtained excellent results in test runs. Beam's lies buoyed Bear City miners' hopes for a couple of years, but in the end, his process could produce no gold where there wasn't any.[1]

"Professor" Beam moved to Denver in 1894. He said that his patented ore furnaces could recover more gold and silver than any others. Beam's process was harmless but more expensive than conventional furnaces. The real process that Beam had perfected was his method of using false assays and a worthless process to swindle hopeful miners and investors.[2]

The Golden Jug Handle

A recurring story told how a trickster exposed Beam's fraud by crushing the handle of a clay jug and giving it to Beam for assay. Beam, of course, found that the jug handle contained gold. Some Beam supporters even accepted the story but gave it the twist that the jug handle had been made from clay near the mines of Central City, Colorado, and indeed contained gold. However, the jug-handle assay story was told of assayers long before Beam.[3]

Aron Beam persuaded investors to build a Beam-process mill at Telluride, Colorado, in 1894. The owners advertised that they could recover 125 percent of the fire-assay values by Beam's marvelous process, but the mill shut down later that year.[4]

In 1896 Beam sold the rights to build a Beam-process mill at Empire, Colorado. The Empire mill opened to glowing press reports, but the early optimism proved false. The mill could not recover the promised gold and silver and was never able to run at more than a fraction of capacity. Beam told his unhappy customers that the mill had failed because they had deviated from his specifications. The mill was rebuilt under his supervision but still failed. Beam convinced the owners to add a concentration step, but the Beam mill still could not perform, and it closed in 1898.[5]

His failures would have put a lesser man out of business, but Aron Beam built test mills in Denver and Florence, Colorado. The mills accepted test loads of ore, and with Beam in control his mill results invariably supported the golden dreams of the deluded prospectors. He used his bogus test-mill results to sell more Beam furnaces in Colorado—all failures. Beam also sold rights to his process in other western states. Beam's agents went as far as London to find investors ignorant of his failures. Beam's test mills became shrines to lost causes, and desperate prospectors flocked to them. Beam said that metallurgists denounced his process because it would put the big smelting companies out of business.[6]

Beam arranged favorable write-ups in Denver newspapers. He took out sizable advertisements in the *Mining & Industrial Reporter* and the *Daily Mining Record*, and those publications printed flattering accounts of the Beam process. Out-of-town mining magazines denounced Beam, but favorable coverage in the bribed Denver press gave him credibility.[7]

The South African Veldt and the American Great Plains

The rise of South African gold mining was a boon to mine swindling on the great plains of the United States. South African gold is mined from conglomerate—a sedimentary rock composed of pebbles. The plains states had more sedimentary rock than they had grasshoppers, and to a farmers' eye the grassland of the South African veldt was similar to his own sod haven. The similarities were superficial, and geologists knew that the American plains could not host South African–type gold deposits. But swindlers insisted that since South African sedimentary rocks held gold, those of the American plains must also.

No one was better at swindling plainsmen with visions of South Africa than Aron Beam. He was soft-spoken and confident when he assured people that their land held gold, usually hidden in clay shales of drab appearance. He spoke biblically: "He hides His wonders, until His people are made ready to receive them and draw this treasure from the earth for the good of fellowman." Beam quoted Solomon that "gold is where you find it."

A myth grew in the late 1890s that shales in central Kansas held a great wealth in gold. Beam was there preaching his gospel of non-assayable gold. Would-be gold magnates shipped a few loads of Kansas shale to Beam's plants in Colorado, but in the end Beam was crowded out by other swindlers vying for foolish money. However, the other false prophets disappeared and were not heard from again, while Beam moved to his next fraud.

Dauntless Gold Swindler of the Wichitas

From an early date prospectors looked longingly at the Oklahoma's Wichita Mountains. A "secret" expedition left Texas in 1849 to prospect the Wichitas, and in 1852 an army expedition led by George G. McClellan found some gold-bearing quartz. However, prospectors were frustrated by treaties that recognized the mountains as Indian territory.[8]

In 1876, prospectors trespassing on Indian treaty land found gold deposits in the Black Hills of the Dakota Territory and pressured the government into taking the land from the Sioux. Many saw the Wichita Mountains as another Black Hills, waiting only for the inconvenient Indians (Choctaws in this case) to be removed.

Spanish prospectors explored the Wichita Mountains. Gold production, if any, was minor, but the rumors of lost Spanish bonanzas spurred would-be gold miners to trespass on Choctaw land. A story spread that Indians had killed a prospector who found gold veins in Oklahoma and that the Indians kept the secret location from white men. The gold rumors may have been encouraged by Native Americans having a joke at the expense of credulous and gold-hungry white men.[9]

Prospectors tried to force the issue numerous times from 1880 to 1897, but each wave of trespassers was evicted by US troops.[10]

The federal government finally opened the area in 1901 to homesteaders and miners. Thousands of prospectors dug the hills and gulches. Shafts were sunk on the slightest indication, until mining claims covered nearly every bit of mountain land. The towns of Meers, Oreana, and Wildman sprang up to serve the soon-to-be-discovered bonanzas.[11]

Gold and silver were there, but only in small amounts and of no economic value. An occasional rich specimen sustained the prospectors' hopes, however, and they continued to dig.

Aron Beam Comes to Oklahoma

"The whole excitement appears to be a fraud of the first water."

—*Engineering & Mining Journal*, January 28, 1904

After two years, the Wichita gold searchers had little to sustain them except the eternal and irrational optimism of prospectors. The owners of the Big Four mine near Oreana thought that they had a large body of ore with $100 per ton in gold and silver and sent a wagon load of ore to the Beam smelter in Denver. Aron Beam was too canny to tell the Big Four that their rock was the high grade that they imagined, lest they start shipping it to his smelter and expect him to pay $100 per ton. He told them that his smelter recovered $11.60 per ton, not enough to pay freight to Denver but enough to be economic if they paid him $40,000 to build a Beam smelter at the mine.

The Beam assays electrified wishful Wichita miners. A headline across the front page of the Lawton *News-Republican* declared "The Boom Is On." Hundreds more rushed to the Wichita Mountains, and Beam sent his son to open an assay office in Lawton. Under the expert ministrations of Professor Aron Beam and his son, the same rocks from which legitimate assayers could coax only negligible amounts of gold and silver assayed as valuable ore. The *Engineering & Mining Journal* warned against Beam, but Oklahoma prospectors trusted Beam more than a magazine from New York City.[12]

H. Foster Bain of the US Geological Survey punctured the Wichita mining pretensions in his 1903 report. Charles Gould of the Oklahoma Geological Survey summarized the findings of Bain and other legitimate geologists, all of whom found no basis for optimism. The publications drew a storm of abuse from the Wichita prospectors and boomers, but Beam was unperturbed. He had faced expert critics before and knew that he could triumph through his quiet confidence and supreme indifference to the truth. Beam had the great advantage that the Wichita miners would rather believe his golden lies than disappointing reality.[13]

Beam presented a long refutation of Bain's report to the Roosevelt, Oklahoma, Commercial Club. It was Beam's usual optimistic moonshine and pumped-up assay values, but his audience lapped it up. An owner of Wichita mining prospects persuaded the state chief inspector of mines to print Beam's report verbatim in an official report. The duties of the chief inspector of mines were limited to mine safety, but he gave space—and implied endorsement—in his official report to Aron Beam and the Wichita miners.[14]

Beam's false results gave credibility to other frauds and fools. The Cache Mining Company claimed it had a "mountain of gold ore," enough to keep a thousand miners working for a hundred years. The Lugart Mining Company also had a secret process to extract gold. The company planned to mine ordinary bottomland clay, from which it said it could extract three ounces of gold per ton. Another company claimed to have ore with not only gold and silver but also $60 per ton in platinum.[15]

The *Mount Sheridan Daily Miner* insulted the government geologist Bain as a mere book-expert, empty of practical knowledge. Wichita miners charged that government agencies had been bribed by the "smelter trust" to quash the new district.[16]

E. G. Woodruff, professor of mineralogy at Oklahoma State University, warned the Wichita prospectors that they were being cruelly deceived, but few heeded his warning. Six ore-processing plants were built in the Wichitas. Although Beam and other friendly assayers continued to report that the Wichita mines held ore, none of the smelters could recover metal in paying quantities. The money ran out, and Beam closed his assay office and returned to Denver.[17]

There is no cure for gold fever. Some prospectors lived out their years in their Wichita Mountain cabins, always convinced that they were close to the mother lode, but never finding it. The mining boom towns of the Wichitas have disappeared. The only trace that remains of Wildman, Oklahoma, is the foundation of an ore mill—a mill built to extract gold that never was. It is a historical monument to the optimism and pioneer spirit that built the American West. But it is also a monument to wishful thinking, broken dreams, and the swindles of Aron Beam.

"The Only Gold Mine in Missouri"

"As for Beam, that mild-mannered, soft-spoken gentleman is too old an offender to warrant hopes for reform."

—*Mining & Scientific Press*, April 10, 1909

Beam continued to mesmerize gold-hungry Midwesterners. In 1908 a pair of Denver sharpers named Yager and Reed announced that an abandoned coal mine near New Cambria, Missouri, contained gold ore. They formed the Boone Baldwin Pioneer Mining Company and sold shares. Yager and Reed said that eastern capitalists were eager to buy the gold mine but that they preferred instead to sell shares to residents of New Cambria and nearby prairie towns. They spouted a lot of nonsense that the rock was identical to the ore at Cripple Creek, Colorado, and even spoke of opening a mining stock exchange at the county seat of Macon, Missouri.[18]

To clinch the deal, they called in Aron Beam, who arrived at the mine, took samples, and certified that the clay in the mine shaft was rich in gold. To pay for a Beam ore-treatment mill, the promoters sold shares to locals from January through March 1909. The Missouri state geologist H. A. Buehler tested the clay and found no gold, but Yager and Reed accused him of being in cahoots with the smelter trust. Even after selling shares in the first company without building a mill, they reincorporated as the Pioneer Gold Mining Company in late 1909 and sold still more shares before they absconded back to Denver.[19]

Little was heard from Beam between 1909 and 1919. His gold-shale swindle seemed to have run its course.

Golden Oil Shale

"The Beam discoveries have revolutionized the oil shale business and brightened the future a thousand per cent."

—*Shale Review*, April 1922

High oil prices in the early 1920s started a rush to exploit the oil shale of the western United States. Since Beam had spent a career convincing people that he could recover gold from shale, he applied his unique talent to the oil-shale industry. Beam announced that his process would recover not merely the oil, but also zinc, gold, silver, and platinum, that no one else had been able even to detect.

To turn his lies into cash, Beam formed the American Continuous Retort Company and built a pilot oil shale–processing plant in Denver. The company brochure claimed that there was gold and silver in every oil-shale sample he tested. Shale from Arkansas, Colorado, Idaho, Kansas, Oklahoma, Utah, and Wyoming averaged $3.00 in gold, $1.50 silver, and $4.00 in platinum per ton. Beam announced that the shale in the Santa Monica Mountains near Los Angeles held up to $40 per ton in gold.[20]

The *Shale Review* printed long, illustrated articles praising Beam's shale process. The *Engineering & Mining Journal* criticized Beam and the *Shale Review*, and the *Mining & Scientific Press* called Beam's statements "Another Pipe Dream." The US Bureau of Mines tested oil shale for precious metals and issued two publications refuting Beam's claims.[21]

Despite the findings of the Bureau of Mines, Beam licensed the process rights and sold his Denver pilot plant in 1923. Nothing more was heard of it, as no doubt the equipment ceased recovering precious metals as soon as Beam left. The price of oil declined soon after, and the oil-shale boom went bust. Aron Beam died in 1926, age seventy-eight. He had never suffered any legal penalty for his lifetime of fraud.[22]

"Process men" like Beam are still around, and still popular with prospectors. Prospectors are a peculiar breed: incurable optimists and individual thinkers, always convinced that their ore is worth more than the assayer says. Inventors unencumbered by stifling formal education still emerge from garages and basements with secret processes to recover fantastic amounts of gold, silver, and platinum from worthless rock. Some are sincere, and some are swindlers. They never bring their processes into production, but they are always close to success—at least in their press releases.

Golden Sands of the Adirondacks

"Not all of the Adirondack gold 'miners' are swindlers; some of them are merely stupid."

—*Engineering & Mining Journal*, March 2, 1912

The gold in New York has always been in bank vaults in the city, not in mineral deposits upstate. But gold rushes anywhere in the United States caused mining excitements on Wall Street, which in turn were strangely reflected in spurious gold rushes upstate.

Upstate New York has many legends of gold and silver mines, with the usual dying prospectors, ancient Spaniards, mines known only to Indians, and so on. But however much people searched, no gold or silver mines were found in New York.[1]

The Gold Frenzy of 1880

The years 1879 and 1880 saw mounting speculation in mining stocks on Wall Street, largely in shares from Leadville, Colorado, and Nevada's Comstock lode. The American Mining Stock Exchange opened in New York City in June 1880 to cater to growing interest in mining investments.[2]

Publicity about mines in Colorado and Nevada sparked an epidemic of gold fever in Upstate New York, and in the summer of 1880, prospectors and rumors chased one another through the Adirondack Mountains. The *New York Times* noted: "Within the past year the interest has been constantly growing until at present it is nearly at fever heat, and gold is the one great topic of conversation in these counties." Newspapers told that a rich vein was discovered beneath the Lake Pleasant jail, although why someone was digging beneath the jail was not revealed. Some Fulton and Hamilton county promoters were caught trying to salt their properties.[3]

All the reports of gold came to naught. How many were swindles, how many newspaper hype, how many bad assays, and how many just self-delusion is not known. Supposed discoveries sputtered on for years, but no metal was produced, and interest waned. In spring 1885 the *Engineering & Mining Journal* briefly noted "the usual yearly report of gold in the Adirondack region."[4]

The Golden Sands of Saratoga County

William Bullis just knew that the large banks of sand near the town of Glens Falls contained gold. No reputable chemist could detect any gold in the sand—which only convinced Bullis that his gold must be in a peculiar undetectable form. The gold frenzy of 1880 died and chemists continued to tell him that the sand was worthless, but Bullis kept searching for a process to recover the gold he knew must be there.[5]

John Sutphen of Albany joined Bullis in his quest. Sutphen claimed a degree from the Columbia School of Mines, but some who checked found no record at Columbia. After Bullis's death, Sutphen told investors that the gold was in magnetite grains with a natural coating of Vaseline that frustrated normal assays. Assayers recognized all this as nonsense, which only confirmed Sutphen in his low opinion of scientists.

He announced in 1897 that he had solved the problem by treating the sand with a secret chemical. He formed the Sacandaga Gold Mining and Milling Company in 1897 and sold stock, mostly in Saratoga Springs. The company would mine a bank of sand near Hadley, from which he claimed he could recover enough gold to profit $5 to $100 per ton of sand. The spoilsport assayers, as usual, found no more than a tiny trace of gold, far less than was needed to pay expenses.

Potential investors hired assayer J. C. Minor Jr. to examine the sand deposit near Hadley, and Sutphen's test mill at Glens Falls. Sutphen refused him admittance to the test plant. However, he was allowed to take two samples of the golden sand, one of which he kept; the other was treated by Sutphen's process. Sutphen's process recovered more than five ounces of gold per ton, but Minor later assayed his duplicate sample in his laboratory, and found that it contained less than one-hundredth of an ounce of gold per ton, worth less than $0.20.[6]

Another mining engineer took twenty samples of the sand, nineteen of which he carefully guarded from tampering; the twentieth he left unguarded before leaving the mine. Analyzed later in New York City, the unguarded sample showed nearly two ounces of gold per ton, while the other nineteen had only traces of gold.

Sutphen and his supporters replied to unfavorable assays with an alchemical argument known as the "green gold" myth: normal processes could not detect the gold because it was only "half-formed," and all methods save that of Sutphen destroyed the incipient gold.

The Saratoga County Gold Rush

The Sutphen process was publicized at the same time newspapers were filled with discoveries of fabulously rich gold-bearing sands along the Klondike River in Canada. If there was gold in the Yukon, why not in Upstate New York? News of Klondike gold reached Seattle in July 1897. By September gold fever had seized the southern Adirondacks: Argonauts rushed north to their imagined Klondike along the Hudson and Sacandaga Rivers, and all the ground between Saratoga and Hadley, twenty-two miles apart, was soon claimed, as were lands in the vicinity of Greenfield, Wilton, and Corinth. In previous years, Upstate New York gold "discoveries" had been of hardrock lode deposits; now they were almost entirely of placer sands, like the Klondike deposits in the news. The excitement moved west to Northville and Gloversville in neighboring Fulton County.[7]

New York law reserved all precious metal deposits to the state. A prospector could gain the right to mine his discovery, no matter under whose land it lay, by filing a claim with the government in Albany. When gold prospectors began crawling over farms and estates, landowners protected themselves by filing mining claims on their own properties.[8]

The Sacandaga Gold Mining and Milling Company built its $22,000 mill next to the deposit of unremarkable-looking sand from which they said they could mine gold. John Sutphen promised to be producing gold by January 1, 1898, but that year came and went with not a glint of gold.

Other Companies Join In

A boom in mining stocks on the Boston Stock Exchange in 1899 prompted more unscrupulous promoters to promise great fortunes from the gold-bearing sands of the Adirondacks. The following year Boston newspapers carried advertisements for stock in Adirondack gold companies. Prominent among them was the Boston-Newton Investment Company, which promised that its sand near Belford contained between twenty-five and a hundred million ounces of gold, far more than was ever mined from California's Mother Lode.[9]

The dentist Charles Bellows, from Gloversville, and his relative E. P. Bellows promoted the Boston-Croghan Company to Boston investors, claiming one to two million ounces of gold in its sands at Croghan. The Bellowses had been promoting gold mines based on barren sand piles in Upstate New York for more than ten years, but their failure to produce gold did not deter them from further lies. Their advertisement in the Boston *Herald* announced: "The Greatest Metallurgical Problem of the 19th Century Solved. The Golden Placers of the Adirondacks to be made to yield their Millions of Treasure."[10]

E. P. Bellows convinced Philadelphians to build a hundred-ton-per-day mill near Croghan. Their Keystone Reduction Company used an "electro-amalgamation process" and claimed to be in successful commercial operation in 1905 but never produced any gold. The Philadelphians tried to entice other investors to take it off their hands four years later, but mining engineers found that the mill somehow recovered $3.40 in gold per ton of sand when the promoters were there but no gold when the promoters were away.[11]

None of the companies could spin gold from the straw-colored sand, and they eventually ceased the pretense. In 1900 the Sacandaga Gold Mining Company property was sold for $700. The buyer sold the land to a farmer and carted off what machinery he could move to a woolen factory in Warrensburg.[12]

The Return of the Adirondack Gold Sands

The failure of Adirondack gold sands put an end to the gold rush, but inventors tinkered with new alchemy to replace the discredited Sutphen process. The Waverly process had its brief fame but also failed. Chicago promoters used clairvoyants to convince the gullible to buy shares in worthless New York gold mines. The law took a narrow-minded view of the practice and jailed two officers of the Chicago Adirondack Gold Mining Company. The psychic financial adviser apparently escaped.[13]

An inventor built two experimental gold-extraction factories near Lowville and looked for investors to build a full-scale version to treat one hundred tons of supposed gold-bearing sand per day. The Gulf Mines Company formed in 1914 to mine gold from sands near Salisbury. Stock for a company to mine New York's supposed golden sands was being sold in California in 1915, promising that the electro-amalgamation process could recover twice as much gold as could be detected by fire assay. The Ambraw Milling Company sprang up in 1916 and was selling shares to Illinois farmers in its gold-sand deposit near Gloversville until the New York state geologist learned of the swindle and sent the farmers his standard no-gold-in-the-sand letter. Another company with a secret process sold shares in another supposed gold mine near Gloversville in 1925. In the 1920s the prospectus for the Reservation Gold Dredging Company claimed that it owned "the richest deposit of platiniferous gravel in the world" on the Cattaraugus Indian Reservation in western New York, but no gold or platinum was ever produced.[14]

Later alchemists promoting Adirondack sands took advantage of public fascination with radioactivity and claimed that they used "radiomagnetic" substances to transform the waste into gold. The Los Angeles *Oil*

& Mining Bulletin noted in March 1915: "The gold-in-sands myth of the Adirondack region is like the seventeen-year locusts, colds, and interest on the mortgage, in that it shows up unannounced every so often." The Adirondack gold sands have been quiet for years, but the lure of hidden gold, and its potential for swindle, is still there.[15]

15

Mining Gold from the Sea

In 1866 a French chemist announced that ocean water contained dissolved gold. The concentration was small, but quick arithmetic showed that the sea contained more gold than had ever been mined throughout history. There was no practical way to recover the gold, but the possibility inflamed the imaginations of inventors, investors, and swindlers.

The Formerly Reverend Prescott Jernegan

Prescott Ford Jernegan was the son of a Yankee sea captain from Martha's Vineyard, Massachusetts. Prescott raced through a four-year course at Phillips Academy at Andover in two years then graduated Phi Beta Kappa from Brown College. He was tall, athletic, and popular. His father urged him into commerce, but young Jernegan had imbibed the utopian socialism of Edward Bellamy's novel *Looking Backward,* and by his own account, "despised business."

He became a Baptist minister and assumed a pastorship in Middletown, Connecticut. To his prosperous congregation, however, the brilliant young preacher seemed disproportionately interested in social reform. Jernegan later wrote: "I had hammered dishonest capitalists from my pulpit....They laughed at my thunderings." After a year in Middletown he offered his resignation, which, to his surprise, they accepted.

Jernegan found a new post with an even wealthier congregation in DeLand, Florida, and this time took care to preach sermons less unsettling to his flock. But he was again dissatisfied with the meager bounty that his congregation shared with the less fortunate.

On vacation back home in Edgartown, Massachusetts, in the summer of 1896, Jernegan fell delirious with typhoid fever. During six bedridden weeks he despaired of religion and of persuading the wealthy to share with the poor. His nurse read a news item to him about the chemist Edward Sonstadt who had found gold in ocean water, and Jernegan seized the idea of a gold-from-seawater swindle to part the upper crust from their bread.

When at Newton Theological Seminary.

As he appears today

As an Evangelist in 1896.

Reverend Prescott Ford Jernegan sold a phony process to extract gold from seawater. From the Boston Sunday Globe, *August 7, 1898.*

As soon as Jernegan recovered, his wife told him that she had met a man she preferred to him, and asked for a divorce. By the time Jernegan returned to DeLand, Florida, he was determined to carry out his swindle.

A boyhood friend, Charlie Fischer, met him in Florida. Fischer, also the son of a sea captain, had left home for adventure, which he had found in the American and British cavalries and as a private detective. He had just been fired from his job as a floorwalker in a New York department store when he met Jernegan in Florida. They briefly considered joining a socialist colony in Mexico but preferred the thrill and satisfaction of defrauding a few overly wealthy capitalists of their spoils. "My theology was bankrupt," wrote Jernegan, "my family broken up and I felt a driving force within me not to be resisted." He added, "Now, I would strike back."[1]

God's Secret Process

Jernegan resigned from his congregation and moved to Boston. He announced in November 1897 that a vision from heaven had revealed to him the secret of extracting gold from seawater. Rather than patent his process, Jernegan kept the heavenly secret to himself. The apparatus consisted of electrodes and liquid mercury in a tank through which seawater circulated.[2]

To prove his process, Jernegan instructed potential investors how to work his "gold accumulator" submerged at the end of the wharf on Narragansett Bay. While investors prepared for the test, Jernegan made his own secret preparations. He outfitted his confederate, Charlie Fischer, with underwater diving gear to add the gold to the submerged gold accumulator during the test. However, Fischer nearly drowned during a test dive and refused to have anything more to do with diving equipment.[3]

The test went on as scheduled. Jernegan brought the gold accumulator to the end of the wharf, and the investors added their own chemicals then stood vigil in a shanty on the wharf through a New England winter's night. At sunrise they removed the mercury and found that it contained gold. Jernegan had introduced the gold by sleight of hand the previous day while he showed them how the accumulator worked.

Jernegan had no problem organizing the Electrolytic Marine Salts Company to exploit his secret process. The company attracted respectable officers from Connecticut and Massachusetts and sold ten million shares of stock to New Englanders. Although Jernegan had wanted to rob from the rich, most investors were middle-class like himself, including many ministers and their widows, who trusted a fellow preacher.

Ministers-turned-swindlers were not unknown at the time. Besides Reverend Jernegan, ministers who quit gathering souls to gather greenbacks in shady mining deals included the formerly reverend Richard Flower, Samuel Aughey, W. P. Fife, C. E. Nylin, W. R. Price, and Charles McCrossan.[4]

Prescott Jernegan was riding high on success and even persuaded his wife to rejoin him. He knew, however, that he would soon have to flee.

The Plant at North Lubec

Narragansett Bay was convenient to investors but was not ideal for other than a demonstration plant. The full-sized plant was built in an abandoned grist mill at North Lubec, Maine, and gold accumulators were installed. The gristmill had been built to harness tidal currents. Each high tide would bring in fresh seawater through open gates.

Jernegan could not always be at the plant, so he relied on his friend Charles Fischer, and on William Phelan, a private detective friend of Fischer's. With Fischer and Phelan in charge, the plant began pouring gold bullion. At the same time, sealed packages labeled "platinum wire" were delivered to the building, which was evidently how the gold entered the plant.

The Sea-Gold Fever Spreads

The federal assay office in New York received its first ninety-two ounces of gold and silver bullion from the Electrolytic Marine Salts Company on March 16, 1898, after which it received an average of one bullion shipment per week from the company. Despite the convincing proofs, the *Engineering & Mining Journal* warned: "A mining enterprise 'run' by a minister of the gospel, of any sect or church, is a good thing to avoid. We cannot recall a single instance where it was not either an outright swindle or a bungling failure."[5]

But visionary investors expected just this sort of obstructionism from a mouthpiece of the old gold-mining industry about to be swept away by God's and Reverend Jernegan's process. Share prices began at $0.50 and rose to $1.50 on the favorable publicity. The company received $954,000 from stock sales, out of which Jernegan took $340,000.[6]

Jernegan's success inspired other gold-from-seawater schemes. The promoters of the Neptune Gold Extracting Company said that their secret process was even better than Jernegan's, and prepared to sell stock. However, the Electrolytic Marine Salts Company was not standing still, and it licensed the secret process to investors in Springfield, Massachusetts.[7]

The sea-gold euphoria peaked in June 1898. The Electrolytic Marine Salts Company had sent $22,000 in bullion to the New York assay office and announced that its first dividend would be paid on July 15. The company was building a second gold-extraction plant a mile and a half down the coast from the first.[8]

Jernegan Disappears

Now that the stock was sold, Jernegan had to end the charade before dividends used up all his profits. In July a man identifying himself as Frank Thompson began paying cash for large amounts of US government bonds at New York banks. Mr. Thompson kept returning for more bonds, and the banks watched his movements. Investigators saw that Mr. Thompson, who was registered at a hotel as "A. C. Spencer," went about town with Prescott Jernegan. Bankers confronted Thompson and Jernegan. "Mr. Thompson" turned out to be Reverend Jernegan's younger brother, Marcus.

The good reverend explained that he needed the bonds to pay European manufacturers for new machinery and preferred to transact business quietly. Although Jernegan was within the law, the banks watched him. An agent for the Fourth National Bank photographed Jernegan on July 23, 1898, boarding the steamship *Navarre* with his wife and son. Without informing his directors, Jernegan sailed for Europe under the pseudonym

Louis Sinclair, carrying an estimated $200,000 in US government bonds and an unknown sum of cash.

Company officials at first derided suspicions. Even though Jernegan had left without warning, they accepted his explanation that he went to Europe to buy equipment.[9]

His accomplice Charlie Fischer destroyed part of the gold accumulators and fled to New Zealand, carrying an estimated $100,000 in his suitcase but leaving his wife and son behind. Without Jernegan or Fischer, no one at the company knew how to operate the apparatus, and company officers and shareholders began to realize that they had been swindled. They stopped sales of capital stock and laid off the six hundred workers at North Lubec.[10]

Just before he left, Jernegan had mailed a letter to Electrolytic Marine Salts president W. B. Ryan. Jernegan wrote that his associate Charlie Fisher had wrecked the gold-concentrating equipment and had run off with part of the secret process. "He may have deceived me," Jernegan warned, "even in the experimental state, since, now I see back, I see he was situated where he could do so in every experiment that has been successful." However, Jernegan insisted that his gold accumulator was no fraud and advised a strict scientific test to establish the fact. He promised to return soon. "I give you my word of honor." However, few were fooled by Jernegan's attempt to push the blame on his pal Charlie Fisher.[11]

The unbending honesty of the Jernegan family would not permit them to excuse Prescott's fraud. Before leaving, Prescott Jernegan had given his brother Marcus $45,000 and apparently an additional amount with which to reimburse friends and relatives for their investments. But when Marcus realized that his brother had likely defrauded investors, he turned all the money over to the company. His father Jared Jernegan admitted, "My boy Prescott has gone wrong." The old sea captain said that he had never believed in the ocean gold. "I had sailed over too many miles of salt water to be taken in." Jared Jernegan died six months later.[12]

Prescott Jernegan was indicted for obtaining money under false pretenses, but as it turned out, this was not an extraditable offense, and the Jernegans disembarked unmolested when the ship arrived in France. The victims sought to have Jernegan indicted for fraud, for which he could be extradited. In the meantime, French police followed the Jernegans, but Jernegan eluded them and dropped from sight. The directors discovered that it was unlikely that Jernegan could be jailed for fraud if he returned to the United States.[13]

An Electrolytic Salts Company chemist was arrested, apparently because he was the highest authority left behind at the North Lubec plant when the others fled. He objected that his only duty was to prepare the bullion for shipment and that he hadn't been privy to the fraud. The *New York Herald* persuaded the third conspirator, William Phelan, to reveal how the swindle had been perpetrated, and he told the story then disappeared like Jernegan and Fisher.[14]

Keeping the Faith

Electrolytic Marine Salts stock sank to $0.10, and investors sued company officials. The directors held a company meeting but discovered that since Jernegan and Fischer owned most of the stock, they had no quorum. Shareholders applied to the courts to declare Jernegan's and Fischer's stock void.[15]

The corporation president still believed that Jernegan had solved the problem of extracting gold from seawater. He convinced the shareholders to authorize a committee to investigate the apparatus. A minority unsuccessfully sued to stop him from spending money on worthless experiments.[16]

The experiments dragged into 1899 without recovering any gold. A professor examined the equipment and reported to the directors that the process was worthless. Most directors still refused to believe that they had been duped, however. The directors voted to keep the text of the professor's report secret, but dissidents leaked the conclusions to the public.[17]

Jernegan Returns

Prescott Jernegan settled in Belgium but in October wrote to his father that he intended to return to the United States. He began long-distance overtures to settle legal actions in exchange for partial repayment. In December 1899, while living in Vienna, he sent $75,000 to his victims and authorized them to recover another $35,000 in Jernegan's name in a US bank. Investors recovered about $0.25 for each $1.00 invested.[18]

Jernegan's fortune had shrunk to $6,000. His wife and son returned to the United States, while he joined the English chemist and inventor Edward Sonstadt, whose experiments had first inspired him to fraud. He had repented and was trying to rehabilitate himself by helping Sonstadt discover a real way to recover gold from seawater, but the chemist achieved no success.

Prescott Jernegan quietly returned to the United States to join the Nome, Alaska, gold rush but returned with only $200 in gold dust. He worked as a laborer in Canada and then Washington State before one of his former teachers obtained for him a teaching job in the Philippines in 1901. He never changed his name, and was embarrassed two years later when his identity as the famous swindler was publicized. He continued teaching in the Philippines until 1910 then moved to California where he wrote and lectured on the Philippines. After a year without success, he took a job as a high school principal in Hawaii, where he taught until his retirement in 1924.[19]

Jernegan had not seen his wife since he left for Alaska in 1901, and she obtained a divorce in 1906. He retired to California and died in 1942.[20]

His accomplice Charlie Fischer worked on a steamship between New Zealand and Australia. Although his stern sea captain father declared him dead, even publishing a false obituary notice in 1902 and putting up a gravestone, the family continued to receive word of Fischer's activities for some years.

Gold from Seawater: The Dream Survives

Inventors the world over continued trying to extract gold from the sea. At least four gold-from-seawater patents were issued in the United States, Britain, and France between 1899 and 1903. Rather than deter others, Reverend Jernegan's fraud inspired imitators.[21]

In 1899 a German chemist solicited funds for his secret gold-from-seawater process. His literature even cited the Electrolytic Marine Salts Company as a success. In 1901, London promoters floated Sea Gold Limited, with a capitalization of £300,000, to remove gold from ocean water. Sea Gold sank with no survivors.[22]

The Duped Nobel Laureate

"There is no doubt that Mr. Snell has proved that gold can be profitably obtained from sea-water on a large scale."

—Sir William Ramsay, Nobel Prize winner in chemistry

In 1904 the Englishman Henry Snell said he had four processes to recover gold from seawater. The first process recovered the gold but was too expensive. The second process was a failure. The third process, he said, was so toxic that his partner was poisoned to death and Snell himself fell ill for eight months. It was his fourth process that Snell claimed as a success and demonstrated to the London company promoter F. L. Ramson. Ramson had a history of promoting visionary but impractical companies, but he convinced capitalists to fund Snell's demonstration plant on the Isle of Wight.[23]

The demonstration plant consisted of concrete tanks that filled and emptied with the tide. The cautious investors hired Sir William Ramsay to investigate Snell's process. Ramsay was a professor of chemistry at University College of London, a fellow of the Royal Society, president of the Society of Chemical Industry, and a member of the French Legion of Honor, and earlier that year he had won the Nobel Prize in Chemistry. He seemed the perfect choice to test Snell's startling claims.

Ramsay visited Snell's seaside laboratory, watched the tank fill with seawater, watched Snell add his chemicals, and collected the residue left after the seawater drained out. Ramsay found gold in the residue and pronounced Snell's process an unqualified success. He admitted that the seawater tanks could not be shut off from tampering hands, and so his tests were not absolutely protected from fraud but dismissed that possibility based on his confidence in Snell, saying: "During my stay at Hayling Island I was brought into close contact with Mr. Snell for nearly a week, and I am confident that his integrity is to be relied upon."

Ramsay's assistant replicated the experiment in Snell's apparatus and obtained the gold from seawater. Ramsay was completely convinced, and the story hit the press in January 1905.

Mining men had seen it all before. The *Mining World* in Chicago observed that Ramsay, although an outstanding scientist, seemed "particularly gullible," and the *Engineering & Mining Journal* recommended that would-be investors visit Prescott Jernegan's deserted gold factory in Maine. The *Australian Mining Standard* urged that Ramsay "for his name's sake, if nothing else," leave sea-gold schemes alone. The *Financial Times* called it a joke and compared it to Jonathan Swift's idea of extracting sunbeams from cucumbers. But Sir William Ramsay's endorsement attracted investment to the blandly named Industrial and Engineering Trust, Ltd., which had been formed to extract the gold. Among the initial shareholders were three members of Parliament and an ex-governor of the Bank of England.[24]

The Industrial and Engineering Trust bought a property on England's southern coast, consisting of two water reservoirs, between four hundred and five hundred acres of low-lying ground, and easements preventing industrial developments along the adjacent 50 miles of shoreline. The land was to be flooded with seawater, chemicals added, and the process repeated daily. Each acre would yield, it was claimed, £20,000 worth of gold (approximately 4,700 troy ounces) annually.[25]

Success was so certain and the cost was to be so small that the trust's only worry was driving the gold price down through overproduction. The firm's chief consulting engineer calculated that the market could absorb more than seventy million ounces of gold per year, which the Industrial and Engineering Trust would cheerfully provide. A newspaper speculated that the Industrial and Engineering Trust, by controlling the world's gold supply, could "forbid Governments to make war and dictate terms of peace, arrange international treaties and literally control the destinies of mankind." The trust was so confident that they searched for other suitable sites and settled on three properties in the south of Ireland.[26]

Further experiments in 1905 no doubt showed that Snell's process was worthless, because the Industrial and Engineering Trust suddenly disappeared from the news. The company died so quietly that one senses the acute embarrassment of its sponsors.[27]

More Processes—More Failures

Henry Snell convinced Umberto Ciantar and H. C. Ciantar with a demonstration at his laboratory on Hayling Island in 1905, and the brothers licensed Snell's gold-recovery processes for a seawater gold factory on the Mediterranean island of Malta. The Ciantars later claimed that they had invented an improved gold-recovery process and formed the Atomised Gold Recovery, Ltd., to put their own process into production.

Ciantars and Snell sued and countersued. A London jury decided that the Ciantars had stolen Snell's process, but judged it to be worth only a farthing (one-fourth of one penny), which they ordered the Ciantars to pay. More seriously the jury also ordered the Ciantar brothers to repay Snell £25 for damage to a shed at Snell's laboratory. Whoever stole whose process, no gold was ever produced.[28]

In 1904 A. Argles began to build a plant near Byron Bay, New South Wales, Australia, for his patented process to extract gold from seawater. Argles spent more than $10,000 of his own money on the project but produced no gold.[29]

In October 1911 Dr. Oskar Nagel lectured a meeting of the American Chemical Society on his process to precipitate gold from seawater. Chemical Society members criticized Nagel's work and voted not to accept the paper in the society's transactions. Nagel was undeterred and searched for a suitable location along the East Coast to erect his gold-extraction plant. He must never have found the right spot, for his gold factory was never built.[30]

Between 1912 and 1938 at least six more seawater-gold processes were patented in Britain, France, Germany, and the United States. A German company was planning a seawater gold plant for the southeastern US coast in 1920 but never built it.[31]

In the early 1920s Dr. Fritz Haber, who had won the Nobel Prize in Chemistry, devoted himself to recovering gold from seawater as a way to pay the crushing German World War I reparations. Haber discovered after ten years of careful work that seawater contained much less gold than previously reported. He concluded that economic recovery was not possible and published his disappointing results. When highly respected Haber gave up the quest, most chemists concluded that it was pointless to investigate further.[32]

The Soviet Union, like Germany desperate for hard currency, also directed its scientists to find a way to draw gold from ocean water. They likewise had no success.

The hope of extracting gold from seawater faded, but the dream of discovering gold dissolved in inland waters persisted. Into the twentieth century Colorado prospectors searched for a legendary mountain stream so rich in dissolved gold that animals drinking from it developed gold-plated teeth. In the 1920s Mono Lake, in California's Sierra Nevada, was falsely promoted as holding enormous quantities of dissolved gold and platinum.[33]

No Trick—Real Gold Finally Recovered from Seawater

The gold in seawater was so dilute that its metallic gleam always flickered at the edge of the chemist's ability to detect it. With better tests allowing chemists to detect tinier and tinier concentrations of gold, the amount of gold thought to be dissolved in ocean water became smaller and smaller, like a disappearing dream. This was not, as someone hypothesized in 1924, because sunspots were causing gold to disappear from ocean water but rather because earlier chemists had been mistaken. In 1886 a chemist had erroneously announced that he had found sixty-five milligrams of gold in each ton of seawater, but by 1927, chemists with better tests had reduced that amount to no more than 0.044 milligrams per ton.[34]

In the early 1960s experimenters for Dow Chemical Company extracted the gold from fifteen tons of seawater. The total recovered was 0.09 milligrams, less than one three-hundred-thousandth of a troy ounce, and worth only a small fraction of one cent. This is the only metallic gold known to have been recovered from seawater.[35]

16

The Spoilers

"This is, perhaps, the most detestable scandal that has ever besmirched the record of any administration. It is without one redeeming trait, for it is sordid and mean from first to last."

— *Washington Post*, February 19, 1901

Every gold rush has its winners and losers. When the gold-bearing gravels of Nome, Alaska, were discovered, the biggest winners were a few Scandinavians who found and claimed for their own the richest tracts. Left out, as usual, were those who arrived too late, cursed their bad luck, and cursed the fortunate few.

Jafet Lindeberg arrived from Norway on a US government-sponsored expedition to introduce reindeer into Alaska. Erik Lindblom was a tailor from Sweden, naturalized as an American citizen, who came north on a whaling ship (according to one account, he had been shanghaied while drunk in San Francisco), and jumped ship in Alaska. John Brynteson, who had moved to the United States from Sweden when he was sixteen and was also a naturalized citizen, first came to Alaska to mine coal. All three were bitten by the gold bug. They met at Council City, Alaska, in August 1898 and partnered to prospect uncharted Alaska. The next month they discovered gold on Anvil Creek. Word first reached fellow Scandinavians, including some reindeer herders, 120 miles away in Saint Michael, and by the time the outside world arrived the best ground had been staked, mostly by Laplanders, Finns, Swedes, and Norwegians.

Foreigners had joined in every American gold rush since the California forty-niners, and federal mining law opened the mining claim process to anyone who was an American citizen or who had declared an intent to become one. Lindeberg knew this, and two months before he found gold at Anvil Creek he had declared, before the nearest federal government official, the commissioner at Saint Michael, Alaska, his intent to gain American citizenship. Lindeberg did not know, however, that the commissioner was not empowered to receive such a declaration.

Claim Jumpers in the Land of the Midnight Sun

Prospectors and hustlers crowded into Anvil City, soon renamed Nome, which was at first just a mass of tents. The latecomers enviously called the discoverers "the three lucky Swedes," and those without the gold tried to figure a way to get it for themselves. The schemers decided that foreigners should not be allowed to own the best mines of Nome, and in June 1899 claimed the properties held by the Scandinavians: in the language of prospectors, they "jumped" the claims, by filing mining claims to the same ground with the district recorder. But the Scandinavians still held physical possession of their mines.[1]

In July 1899 the claim jumpers, perversely calling themselves the "Law and Order League," held a mass meeting to declare invalid mining claims made by foreigners. Since those without valid claims outnumbered those with, they knew that they could carry the vote. They would light a bonfire when the measure passed, to signal their confederates poised in outlying areas, who would then seize the disputed mines. The whole procedure was illegal and would certainly result in violence, but there was little law in Nome and a fortune in gold to be taken. However, when word of the scheme reached the army base Fort Get There at Saint Michael, a lieutenant traveled to Nome with three soldiers. When one of the schemers proposed the antiforeigner resolution to the meeting, the officer marched his men to the front and ordered the jumpers to withdraw the resolution. When the leaders balked, the soldiers cleared the tent at bayonet point and broke up the scheme.[2]

The claim jumpers looked for a more sophisticated way to steal the claims: lawyers. But when a judge traveled hundreds of miles from Sitka, Alaska, to hear the cases, he ruled flatly against the claim jumpers. The original discoverers' citizenship was irrelevant, for it was a principle long affirmed by the US Supreme Court that a "first in time" mining claim, even if invalid, could be vacated only by government initiative, and not in a suit by later claim jumpers.

The Nome law firm of Hubbard, Beeman, and Hume agreed to take the claim jumpers' cases in exchange for an interest in their claims and soon had half-interests in about one hundred mining claims declared worthless by the court. The lawyers knew that the law was squarely against them, so they set out to change the law. With the richest gold mines in Alaska at stake, in the fall of 1899 they sent one of the partners, Oliver Hubbard, to see if the theft could be arranged in the halls of Congress.[3]

Claim Jumpers in the Land of the Midnight Deal

Hubbard had made valuable contacts when he clerked in the Justice Department of the Democratic administration of Grover Cleveland. Democrats were now out of power, but Hubbard knew that money could cross party lines. In New York City, he and one of his clients who had been arrested for claim jumping met with the political fixer Alexander McKenzie, the Republican national committeeman from North Dakota.

Alexander McKenzie was big, tough, and inclined to action. He had gone west from his home in Canada as a boy, and after stints as a scout for Custer's cavalry, a railroad laborer, a manufacturer of soda pop, and then a frontier sheriff, he had made his fortune as a receiver for the financially embarrassed Northern Pacific Railroad. Although he could barely read and write, he had a keen mind for politics. He preferred to stay behind the scenes and didn't run for office. He and his allies bribed, stuffed ballot boxes, and did not shrink from intimidation and beatings. McKenzie was called the "senator-maker" and held great sway with the Republican delegations from Montana, North Dakota, and Minnesota.[4]

To cheat reindeer herders out of their gold mines seemed easy to McKenzie. He had been riding over Scandinavian agricultural interests in North Dakota for years, and he thought them no match. "Give me a barnyard of Swedes," he bragged, "and I'll drive them like sheep."

The most straightforward way to confiscate the mines was to change the law to declare invalid those claims that noncitizens had staked in Alaska. A slight turn of phrase might slip through Congress and could hand the best gold mines of Nome over to the claim jumpers. To this end, McKenzie conferred with Senators Henry Clay Hansborough of North Dakota and Thomas Carter of Montana.

A routine bill before the Senate dealt with the legal code for Alaska Territory, but, as had the previous code, the bill guaranteed the rights of foreigners to own and convey property. On April 4, 1900, Senator Hansborough quietly proposed an amendment invalidating Nome mining claims previously staked by foreigners. The retroactive provision would be constitutionally suspect, but the placers would probably be largely exhausted and the gold safely in the hands of the jumpers by the time a court could declare it invalid.

The amendment might have slid through unnoticed but for the alertness of the California mining capitalist Charles Lane. Some of the reindeer herders who had followed Brynteson, Lindblom, and Lindeberg had sold their gold mines to Lane, so that his title rested on the validity of the Scandinavian claims. Lane also had influential friends in Washington. He was the chairman of the National Free Silver Committee, which gave him the ear of politicians who favored the remonetization of silver.

Senators William Stewart, elected on the Silver Party ticket in Nevada, and Henry Teller, a "silver Republican" from Colorado, spared no adjectives in opposing the Hansborough measure. Senator Teller thundered, "I have lived in a mining country for nearly forty years, and of all the contemptible creatures who ever saw the light of the sun it is the man who jumps another man's claim." The Scandinavians had declared their intentions to become citizens before the American commissioner at Saint Michael, not knowing that the official was not empowered to receive such declarations. Opponents saw nothing but endless litigation from the measure. Lawsuits over mining titles had long been the curse of American mining, and the Hansborough amendment would make the situation even worse for mining in Alaska.[5]

Senator Hansborough described his amendment as protecting American citizens. He and Senator Carter defended the claim jumpers of the Law and Order League as honest Americans and introduced a letter from their lawyer Hubbard that denigrated the original locators as "alien Laplanders." In fact, some of the jumpers themselves were noncitizens, and others of the jumpers were working on grubstake agreements from an English company, which would make the English company part-owner of the jumped claims.[6]

Rhetoric grew more strident when the measure was taken up by the Senate two weeks later. Each side hinted through the syrupy insincerity of senatorial courtesy that the other was in the pocket of special interests. Hansborough and Carter said that opposition to their amendment came from wealthy speculators. The opposition hinted that the amendment was bought and paid for by lawyers for the claim jumpers. Senator Teller continued to hammer the claim jumpers: "These people," he told the Senate, "are blackmailers or thieves, and since mining was begun in this country they have been the curse of every mining camp. They are among the worst scoundrels that ever went unsung."[7]

McKenzie's Senate allies Hansborough and Carter twice proposed substitute amendments, but both Democrats and Republicans opposed putting such a controversial provision on a routine bill. When it became clear that they had no support, Hansborough and Carter dropped the amendment.[8]

The Poor Swede

McKenzie had not risen to wealth and power through lack of determination and so devised another tactic: he would arrange to steal the richest gold placers of Nome using the federal courts. McKenzie incorporated the Alaska Gold Company, with himself as the president and general manager.

The claim-jumper attorney Oliver Hubbard swapped his half-interest in the jumper claims for shares of the Alaska Gold Company and became the company secretary.

McKenzie then secured the appointment of Arthur Noyes as one of three federal judges for Alaska. He knew Noyes as a man who would take orders. Noyes had recently been defeated in the polls for a state judicial post in Minnesota and would now have the federal district that included Nome. McKenzie also won the appointment of other cronies, including C. L. Vawter as US marshal for Nome, and Joseph Wood as Nome district attorney. With a fortune in gold at stake, McKenzie headed north to lead his gang of crooks in person. McKenzie, Noyes, and their bunch arrived at Nome on the same boat, and their plot to steal the gold mines soon became obvious.

Another judicial appointee, James Wickersham, who was not part of the conspiracy, sailed on the same ship to Alaska. Judge Wickersham heard McKenzie brag that he had secured the judgeship for Noyes and that he had big mining deals in Nome that would make him rich. Wickersham remembered Judge Noyes as "an agreeable man, though he seemed to be immoderately fond of the bottle."[9]

McKenzie debarked from the SS *Senator* as soon as it arrived at Nome on July 19, 1900. While Noyes spent his first few days on the ship anchored off the beach, McKenzie and Hubbard met with Hubbard's legal partners, Beeman and Hume, and told them of the agreement made in Washington. When Beeman and Hume balked at splitting their plunder, McKenzie told them that Judge Noyes would do his bidding. "This judge," he told them, "is weak and vacillating and uncertain." McKenzie threatened that if the attorneys didn't sign over half of their interest in exchange for Alaska Gold Company stock, their jumped claims would be worth nothing. In addition McKenzie demanded and received a silent one-fourth interest in the law firm to hold in trust for Judge Noyes. His crony, the new US district attorney Joseph Wood, received another one-quarter. McKenzie assured everyone that he had the federal judiciary in his pocket, and if the lawyers cooperated they would all become rich. Beeman and Hume agreed, and the papers were drawn up and signed. Then three stenographers spent the next four days drawing up the petitions in five lawsuits while McKenzie urged them to hurry.

On July 23 Judge Noyes had not yet opened his court, but McKenzie and Hume came to his hotel room at 6 p.m., and gave him four petitions from the claim jumpers to put the gold mines in receivership until the court cases were settled—with Alexander McKenzie's own Alaska Gold Company named as the receiver. The claim-jumping conspirators knew that their legal case was worthless but also that they could drag it out,

The weak-willed Arthur Noyes was given a federal judgeship in Alaska to help steal the best gold mines at Nome. From Appleton's Booklovers Magazine, *February 1906.*

possibly for years. The best parts of the gold placers could be worked out in a few seasons, and by the time the claim jumpers had exhausted their legal options, the receiver would likely have exhausted the gold in the creek gravels.

Hume began to read one of the petitions aloud to Noyes, as was customary, but the judge told him it was not necessary and signed immediately, without notifying the defendants or even reading the orders he was signing. The papers ordered the mine owners to surrender their properties to the receiver, McKenzie, immediately. McKenzie was required to eventually post bond of $5,000 for each property, a wholly inadequate amount for the rich placers, one of which by itself was producing $15,000 worth of gold each day. As soon as Noyes signed, McKenzie rushed off in waiting wagons and headed to the claims on the creeks, where he ordered the rightful owners off their properties.

McKenzie found Jafet Lindeberg asleep in his tent on the mining claim and ran him off. However, some of Lindeberg's men had the presence of mind to distract McKenzie while others sneaked in the back window of the mine office and recovered most of the nuggets in the safe before McKenzie could grab the gold. By midnight McKenzie had all the gold mines in his possession.[10]

The Poor Swede

A square-headed, hard-working Swede,
    Propelled by inordinate greed,
    Mucked around in the cold
    Till he found some coarse gold,
    And then came to town at full speed.
A lawyer with galvanized jaw,
    Whose mode of procedure was raw,
    Sent a thief out to jump
    The rich claim of the chump
    And stake it "according to law."
Now the Swede is now stretched on the rack
    And trying to get his claim back,
    While the Court takes its time,
    To consider the crime
    Till the receiver fills up his long sack.

— SAM DUNHAM[11]

McKenzie was a master at enriching himself through receivership, and with his flunkies being the only law in Nome, the job must have looked easy. He began producing gold as quickly as possible. He knew that his receivership was temporary, so he hired every able-bodied man available to work the claims around the clock. The rightful owners were excluded from the properties and could not even learn how much gold he was taking from the gravel.

The brash seizure of the Scandinavians' properties halted serious work on other mining claims around Nome. Even though the mining season was short, owners dared not work their ground for fear that if they found gold Noyes and McKenzie would steal the property through some bogus legal quirk. Many would-be mine owners figured that Judge Noyes must know the law, and so they jumped every valuable claim around, hoping to prove later that it had not originally been staked to the exact letter of the law. Other claim jumpers jumped the first claim jumpers, until some ground had as many as twelve people claiming ownership.

Some of the crew who arrived with Noyes proved to be as conscienceless and crooked as McKenzie. Archie Wheeler, the court stenographer who arrived with Noyes, was soon supplementing his government salary by hiring himself out as a lawyer and promising favorable rulings from his boss, Judge Noyes.

When McKenzie learned of a rich beach placer sixty miles away at Topuk that was owned by the Black Chief Mining Company, he straightaway shipped his mining machinery down the coast. While the machinery was en route, he had Noyes appoint one of his cronies as receiver of the claims, for no good legal reason. The receiver was ordered to put up $10,000 bond, equal to only one good day's gold production. The receiver removed the gold-washing machinery already on the property and installed McKenzie's equipment. The Black Chief owners protested to Noyes, but Noyes only suggested that they hire his secretary, Archie Wheeler, to represent them. Wheeler told the claim owners that he could clear their title, but only in exchange for a large interest in the company, an offer they refused.

The owners of the Black Chief Mining Company managed to bring their case to trial after several months, but after the jury restored their property, the receiver reported that he had recovered only $30,000 of gold, while his expenses, inflated by a charge of $29,000 for McKenzie's redundant mining equipment, had totaled $35,000 so that the property had operated at a loss. The owners appealed again that the properties must have produced at least $200,000 worth of gold during that time. Even Noyes backed away a bit from McKenzie's raw theft, appointing a referee to examine the books. The referee found that even by the receiver's own account, the receiver had taken $100,000 in gold, so that even with overstated expenses, the receiver still owed the original owners $65,000. The receiver simply forfeited the $10,000 bond and walked away with all the gold.[12]

The Ninth Circuit Court of Appeals to the Rescue

Noyes and McKenzie triumphed easily in their kangaroo court, but the Scandinavians did not give up. The "three lucky Swedes" had combined their properties into the Pioneer Mining Company, and Jafet Lindeberg, although at twenty-four the youngest of the trio, emerged as their leader.

Also in Nome to fight for his gold mines was the veteran California mine owner Charles D. Lane, who had organized his Nome mines into the Wild Goose Mining Company. Lane had defended his rights in Washington and was now back in Alaska to oppose McKenzie again. Lane was a tall, old miner, and although his beard was pure white, as was the hair that clung in a stubborn fringe around his head, he was still strong, quick-witted, and every bit a match for Alexander McKenzie. Lane had gone west from Missouri as a young man, risen to the ownership of mines in California and Nevada, and learned how to protect his mines from schemers.[13]

Charles Lane and Jafet Lindeberg did not expect the claim jumpers would just give up after their defeat in Congress and so had hired a pair of top San Francisco lawyers, Samuel Knight and William Metson, to come to Nome ready to act. Knight and Metson petitioned Noyes for a hearing the morning following appointment of the receiver, but Noyes put them off repeatedly. Noyes finally heard their petitions on August 10 and dismissed them all.

The lawyers filed for the right to appeal on August 14. The next day, Noyes not only denied the petition for appeal but punished the recalcitrant defendants by authorizing the receiver to seize their personal property on the mining claims, including tents, beds, money, and clothing. "I am going to tie your people up all around," he told a defense attorney. Since most miners lived out on their placer mines, this order left them literally out in the cold, without money or even a change of clothes. The lawyers for Lane and Lindeberg knew that Noyes would never give them a fair hearing, but they had now a court record of the gross unfairness of the proceedings. With this legal ammunition, they took the next ship south to the outside world.

The defense lawyers sent certified copies of the Noyes court proceedings to US Attorney General Griggs in Washington and petitioned Griggs to remove Noyes for incompetence. Griggs ignored the petition.

The defense lawyers traveled to San Francisco to petition the federal Ninth Circuit Court of Appeals, where Judge William Morrow, a former Republican congressman, read the legal papers in disbelief. The record showed flagrant disregard for both judicial procedure and the Scandinavian locators' most basic rights. There had been not the slightest pretense of due process, and the appointment of a receiver violated both statute law and common sense. He called Noyes's actions "high-handed and grossly illegal." McKenzie and Noyes no doubt expected higher courts eventually to undo their illegalities and depended on the snail's pace of the law to give them time to rob the gold mines. But this time the law sped. On

August 29 Morrow revoked the receivership orders and directed Noyes to return the properties to the original locators.[14]

> *"The order so made was so arbitrary and so unwarranted in law as to baffle the mind in its effort to comprehend how it could have issued from a court of justice."*
>
> —Ninth Circuit Court of Appeals, *In re Noyes*, January 6, 1902

The lawyers for Lane and the Scandinavians raced back to Nome with certified copies of Judge Morrow's orders. Their ship arrived at Nome, but a storm prevented them from landing with the papers. McKenzie learned of the writs while the ship was still stormbound off the beach, and he went to the Alaska Bank & Safe Deposit Company to retrieve his ill-gotten gold. However, the original claim owners were watching his movements and surrounded the bank entrance with armed and angry men. McKenzie was forced to leave the building empty-handed through the rain and the crowd.[15]

Finally, Charles Lane found boatmen foolhardy enough to row out to the ship, and the old man himself went with them into the storm and brought back the writs, which his lawyers presented to Judge Noyes and Alexander McKenzie on September 14.

Judge Noyes wavered among inconsistent positions, none of which complied with the order. He successively held that he had no duty to enforce the appeals court writ, that the writs might not be genuine, and that they required no more than for him to maintain the status quo of the receivership. He finally drew up an order rescinding his receivership orders of the previous July—but filed the papers away without issuing them. Instead, he ordered military authorities not to enforce the appeals court writs.

Alexander McKenzie also ignored the federal court of appeals. His defiance was urged by his coconspirators, the US district attorney for Nome Joseph Wood, the assistant US district attorney C. A. S. Frost, and Dudley DuBose, attorney for the claim jumpers. Far from enforcing the order of the appeals court, US District Attorney Wood and Assistant District Attorney Frost ordered the US marshal in Nome to swear in a posse to prevent anyone from trying to enforce the writs.

Frost also hired three private detectives at government expense to watch the movements of Charles Lane, the Scandinavians, and their lawyers, and to report their findings to Alexander McKenzie. One detective was told to get a job with the lawyer Metson, the better to overhear defense strategy. McKenzie unsuccessfully offered William Metson $1 million to throw over his clients. When the defense attorney Samuel Knight tried to gather evidence of the conspiracy, Noyes ordered Knight jailed for suborning perjury. Knight evaded arrest by smuggling himself out to a ship leaving Nome, perhaps reflecting that this was not the sort of jurisprudence he had learned at Yale Law School.[16]

When Judge Morrow in San Francisco learned that the receiver, the US district attorney, the assistant US district attorney, and even a federal judge were flouting his court orders, he sent two federal marshals to Nome to enforce the writs and arrest Alexander McKenzie. It was already October 1, and Nome would soon be locked in ice until the following summer, so the marshals left San Francisco for Nome a few days later.

The federal marshals arrived in Nome and arrested McKenzie while he ate breakfast. McKenzie at first insisted that no "son of a bitch" could arrest him, but the marshals would not be intimidated. That afternoon McKenzie summoned Noyes and told the judge to issue a writ to set him free. However, Noyes knew that to undo the order of a superior court would be indefensible, and he refused.

Judge Noyes was distancing himself from McKenzie. A few days before McKenzie's arrest, Noyes had finally returned the mines to the original owners, but McKenzie had convinced him to enjoin the owners from removing any gold from his jurisdiction, hoping that the writs could somehow be reversed. McKenzie was still trying to protect his plunder of the previous ten weeks: five hundred pounds of gold nuggets and gold dust, then worth $150,000, in safe-deposit boxes in a Nome bank.[17]

McKenzie gave the only keys to the bank boxes, along with some confidential papers he did not want to fall into unfriendly hands, to US District Attorney Joseph Wood. The marshals demanded that Wood give them the keys, but the US District Attorney refused even to admit that he had the keys. The marshals had their orders, however, and smashed into the safe-deposit boxes with sledgehammers while the bank manager pleaded with Wood to give them the keys. The marshals handed the gold over to Charles Lane and the Scandinavians and left with Alexander McKenzie in their custody on the last ship leaving Nome ahead of the grip of winter ice.

The Ninth Circuit Court in San Francisco found Alexander McKenzie guilty of contempt of court on February 11, 1901, and sentenced him to one year in the Oakland, California, jail. The three judges agreed that McKenzie merited far graver punishment, but since he was on trial only for contempt of court, the main punishment would have to await a trial for conspiracy.

The facts of the case laid out in the unanimous opinion of the Ninth Circuit startled Washington and prompted a look at McKenzie's conspiracy. The Washington *Post* led its editorial section with the headline

"Remove Judge Noyes." The *Post* urged Congress to impeach Noyes, unless the president acted first to "remove him under every circumstance of disgrace."[18]

Noyes postured as the outraged innocent. He wrote Senator Hansborough: "I never did a dishonorable thing in my life, and never spent a dishonorable dollar in my life and I never expect to. The golden opinions of my fellow men are treasures far too rich to be swapped for gold dust." He charged that all his troubles were the fault of the owners of the Pioneer and Wild Goose Mining companies, who had slandered him after he had spurned their bribes.

Some senators demanded to impeach Noyes, but the judge's defenders accused the Ninth Circuit Court of Appeals of bias. In the end Congress and the president joined in doing nothing and left Noyes sitting uneasily on the Nome bench.[19]

McKenzie immediately applied to the US Supreme Court; that tribunal lost no time in taking up the case but denied McKenzie's plea in March 1901. McKenzie remained in jail.[19]

The federal appellate judges wrote in their opinions of the obvious evidence of a criminal conspiracy to use the federal judicial system for the theft of gold mines. But those who waited for an official investigation of the conspiracy waited in vain. The McKinley administration was not about to prosecute a conspiracy that reached back to fellow Republicans in Washington.[20]

Neither would President William McKinley let his friend and political ally Alexander McKenzie suffer a year in jail. McKinley commuted McKenzie's sentence on May 23, 1901. McKenzie was released the following day, after serving a bit more than three months of his twelve-month sentence. The president justified his pardon by saying that McKenzie was in such ill health that he might not survive the jail term. Also, the president said, McKenzie had finally given back another $9,000 in gold that he had stolen and shipped to Seattle. McKenzie's health bloomed anew, and a few hours after his release he bounded down the railway platform in Oakland to catch a train. He remained a power in the Republican Party until his retirement in 1908.

Without McKenzie to give him backbone, Judge Noyes weakened further. He often conferred privately with lawyers representing clients before his court, leading to suspicion of more corruption. He boozed and issued contradictory rulings favoring whichever lawyer was the latest to catch his ear. In a belated attempt to mend fences with the Scandinavians at Nome, he appointed the Swedish American Lars Gunderson as American commissioner and claim recorder of the newly formed Kusatriem mining district. Jafet Lindeberg, adapting to the bribery he had found in

America, even became a partner with Judge Noyes in financing a prospecting expedition, in which Lindeberg generously paid Noyes's share of the investment.[21]

The Ninth Circuit Court of Appeals had jailed McKenzie and now turned to his coconspirators. Although McKenzie had already been released from jail by the president, in July 1901 the court ordered Judge Noyes, US District Attorney Wood, Assistant US District Attorney Frost, and two of the claim-jumper attorneys to appear in San Francisco for trial for contempt of court.

Summoning Judge Noyes to San Francisco solved a problem for the Alaska judiciary. His incompetence had caused so much trouble in Nome that Territorial Governor Brady asked Washington to transfer Noyes to Juneau, but the Juneau bar vigorously protested any plan to saddle them with Noyes.

An honest anomaly in Noyes's entourage was his chief law clerk, George Borchsenius. Noyes had hired young Borchsenius only under pressure from two US senators and had trouble with him from the start. Borchsenius was of Scandinavian descent himself and sympathized with Jafet Lindeberg and the Scandinavians. When Noyes tried to handpick a grand jury, Borchsenius refused to approve the list and insisted that the jurors instead be chosen by lot from the jury pool, as the law required. However unhappy he might be with his head law clerk, Noyes was reluctant to discharge a young man with friends in the Senate. But Noyes finally fired Borchsenius on July 13, 1901, when his own days on the bench were numbered.[22]

Judge Arthur Noyes left Nome on the evening of August 13, 1901. He stumbled drunk onto the deck of the SS *Queen*. In his intoxicated condition, he dropped a heavy package onto the deck, where it remained undisturbed while Noyes sequestered himself in his cabin all the next day and consumed more liquor. Lawyers with cases pending, some of which were due to come up the very next day, boated out to the *Queen* before it sailed, to urge Noyes to sign last-minute orders favoring their clients. The following night someone examined the package Noyes had dropped upon boarding: there abandoned on the ship's deck were two sacks of gold nuggets, more than one hundred pounds of gold. On the second day of the voyage, Noyes sobered up enough to report the gold stolen. How a federal judge of modest means came in possession of such a treasure was a puzzlement to some, but not to those familiar with Noyes and his methods.[23]

A few days after Judge Noyes sailed into the sunset, fifty-two lawyers at a meeting of the Nome bar chipped in $2 each to send a petition to President McKinley asking for Noyes's removal. The petition called

Noyes "weak," "vacillating," "dilatory," "careless," "negligent," "partial," and "absolutely incompetent." There were 127 lawyers in Nome, and the large number in the small frontier town speaks volumes about the litigious state of mining titles in Nome.[24]

In January 1902 the federal appeals court in San Francisco found Judge Arthur Noyes guilty of contempt of court. At the same time, US District Attorney Joseph Wood and Assistant US District Attorney C. A. S. Frost were found guilty of urging Alexander McKenzie to defy the court of appeals writs and, in the case of Wood, also for defying orders from federal marshals. The claim-jumper attorney Dudley DuBose had been found guilty of contempt months earlier. The court handed the three crooked lawyers sentences of between four and twelve months. Crooked judge Arthur Noyes was fined $1,000 but spared a jail sentence. Although one of the appeals court judges recommended that Noyes spend eighteen months in jail, the other two considered a jail sentence tantamount to removing the judge from office, a measure they regarded as the prerogative of the president.

Attorney General Knox investigated the judicial fitness of Judge Noyes in February 1902, a move that had been urged in Alaska and Washington, DC, nearly a year earlier when Alexander McKenzie had first been convicted. The attorney general's investigation began shortly after senators asked on the floor of the Senate why the president had not removed Noyes. Senator William Stewart recounted at length the history of the swindle, and according to reports he broke with Senate decorum by alluding to the participation of some of his colleagues; but the next day Stewart agreed to substitute milder language in the printed version for the *Congressional Record*. After a brief investigation, Attorney General Knox announced Noyes's removal.[25]

Noyes never returned to Nome, and his successor found a court docket clogged by the year of inaction that followed Noyes's vacating his post. Many cases were tangled by the hasty and contradictory orders Noyes signed while aboard the SS *Queen* before it sailed. Affairs were further confused by the fact that in more than a year on the bench, Noyes had made no written opinions.

Arthur Noyes returned to his former home in Wisconsin, where he died a few months later. Contemporaries wrote that the theft at Nome would have been far more successful had Noyes shown the cleverness and courage of his boss, Alexander McKenzie. The writer Rex Beach exulted of Alexander McKenzie: "What a superb villain he is!" but held Noyes in contempt. The Nome *Gold Digger* memorialized Noyes as "too honorable to be a scoundrel, too much of a scoundrel to be an honorable man;

too strong to be a pliant tool; too weak to protect his own honor; brilliant and brainy, with many of the mental graces of a Chesterfield, but a fool."

Meanwhile, Back at the Ranch

Noyes had refused to return some of the mines to their owners, and then sailed off—leaving Nome without judicial authority. The rightful owners finally despaired of the legal system, and after more than a year of patience righted the wrong themselves. Claim owners on Ivan Creek and Gold Run Creek banded together as vigilantes to run off jumpers.

In his drunken haste to leave Nome, Noyes had signed rulings giving three different parties possession of a valuable gold property along Glacier Creek. On the night after Noyes left Nome, the original claim owners moved against the jumpers that Noyes had installed on the mine, which had originally been owned by the Pioneer Mining Company. Seventy-five vigilantes, among them Jafet Lindeberg, armed and masked themselves and forced the claim jumpers off the property. Some of the claim jumpers were injured, one seriously wounded by a bullet through his thighs. Soldiers from Fort Davis moved in and kept both sides from mining the disputed properties.

The assassination of President McKinley in September 1901 placed his vice president, Theodore Roosevelt, in the White House. Roosevelt was more reform-minded than McKinley but showed no inclination to go after corruption high in his own party. None of the three small-fry lawyers serving jail time in Oakland were pardoned (as their influential boss Alexander McKenzie had been), yet US District Attorney Wood and Assistant US District Attorney Frost, both convicted, kept drawing government salaries even while they lolled in the Oakland jail. Senator Tillman of South Carolina suspected hush money and charged, "You have to pull somebody's leg in Washington in order to accomplish what these men are doing."26

Since President McKinley had already pardoned his friend Alexander McKenzie, few had considered it useful to charge McKenzie with criminal conspiracy while McKinley was still president. However, a few weeks after President McKinley's death, a grand jury in Nome considered conspiracy indictments against Noyes, McKenzie, Wood, and others, including even some US senators. The power of prominent politicians still protected the conspirators, however. The new judge and the new acting US district attorney for Nome persuaded the grand jury that the conspiracy was ended and that further punishments were not needed. The matter died, to the relief of many in Washington.

Instead of pursuing the conspirators, the US district attorney obtained indictments against the vigilantes who had retaken their own properties by force following Noyes's departure, although only seven of the vigilantes could be identified. A grand juror slipped word to Jafet Lindeberg that he had been indicted, and Lindeberg fled to California. Another vigilante went to Mexico, and the court lacked the funds or the will to extradite them. The others were tried, but Nome juries refused to convict them. It was even said that the first jury panel itself, selected from among respectable Nome businessmen, contained three of the Glacier Creek vigilantes.[27]

The Spoilers

"And I've got the biggest scheme that ever came north, backed by the biggest men in Washington."

— "Alexander Macnamara," a fictionalized Alexander McKenzie, in Rex Beach's novel *The Spoilers*

The writer Rex Beach, who had been in Nome during the steal, published a fictionalized version of the events in his 1905 best-selling novel *The Spoilers*. He followed with the nonfiction details in a magazine series *The Looting of Alaska*. The novel and articles caused a sensation, especially among Washington politicians who would rather forget their complicity. The scandal aided reform candidates in North Dakota, who ousted the McKenzie machine in the 1906 elections.[28]

The Spoilers was made into a stage play, then five times made into movies. The most famous movie version starred John Wayne, Randolph Scott, and Marlene Dietrich in 1942. Alas, the roles played by Wayne and Dietrich were the inventions of the author Rex Beach and later screenwriters. However, Randolph Scott played Alexander Macnamara, based on the real-life spoiler Alexander McKenzie.

In the movie the sneaky villainy of politicians was neatly stopped by John Wayne's fists, but the real-life ending was not so cinematic. Alexander McKenzie's scheme was halted by smart legal footwork, not brawls. The crooked judge Arthur Noyes was stopped only by judges more honest. The malefactors, in fact, suffered little. McKenzie spent only three months in jail. Judge Noyes left Nome with a hefty mass of gold and never served a day in jail.

The "three lucky Swedes" went back to mining gold until they took their Pioneer Mining Company public in 1902, when they sold their interests. John Brynteson and Erik Lindblom moved to California. Brynteson moved back to Sweden in 1906 but never lost the prospecting fever; he later discovered gold in Siberia. Jafet Lindeberg gained his American citizenship in 1914. Although he made his fortune in gold mining, he never forgot his reindeer, and as late as 1928 was promoting Alaska reindeer meat as a staple for the American dinner table.[29]

17

Whitaker Wright's House of Cards

On January 26, 1904, all eyes watched a portly well-dressed man in his late fifties enter the London court. J. Whitaker Wright, one of the wealthiest men in England, a friend of royalty, and the owner of a fabulous country estate, was there to hear his fate decided by a jury of twelve. Everyone knew of his rise to wealth, the collapse of his mining empire, and charges that he had defrauded investors. Newspaper headlines informed the world of his international flight from arrest. Parliament had debated Wright's prosecution, and his extradition had been contested in the United States Supreme Court.

All this everyone knew. But as Wright entered the courtroom for the final day of his trial, he knew something that no one else could: in his pants pocket were cyanide capsules and a loaded revolver.

To Seek His Fortune in America

"How did I get my start? Why, I went West, and as I made a little money, I saved it and bought a few shares of a mine that looked as if it would be profitable."

—Whitaker Wright, in Ludlow Street Jail, New York, March 1903,
New York Times, January 27, 1904, p. 2

John Whitaker Wright was born in Cheshire, England, in 1845, and sailed for America in 1868. He started in Philadelphia as a trader in grain and that new commodity, petroleum. Philadelphia could not contain the young man's ambition, however, and he moved west to find gold and silver. By Wright's own account, he barely escaped being killed by Indians while prospecting in Idaho.

In the winter of 1879 Wright joined the silver boom at Leadville, Colorado, where he and some friends moved into a shack with little to live on but determination. The temperature hung at thirty degrees below zero and water froze each night inside the cabin as Wright shivered in all his clothing. A Leadville contemporary described him as tall and athletic, with dark eyes and hair. He kept well groomed, won friends with his engaging personality, and was known for his shrewd business acumen.[1]

Wright and two associates bought a mining claim in December 1879, then two more claims the following month. If Wright and his partners were as poor as he later described, then they bluffed their way into some expensive properties. The price of two of the claims alone was reported as $325,000, although the down payment was probably more modest. They incorporated the claims as the Denver City Consolidated Mining Company and sold shares in New York and Philadelphia.[2]

Leadville papers breathlessly reported an immense and valuable ore body at the bottom of the initial shaft, and Denver City Consolidated built a surface plant and hoisting apparatus. But instead of developing his supposed ore body, Wright started sinking another shaft only 800 feet away, at which he built another surface plant and hoisting machinery. By installing redundant surface facilities, he delayed the revelation that he had discovered little paying ore. Denver City Consolidated finally started shipping ore in late 1881 and appeared on the brink of wiping out its debts and paying dividends. Paradoxically, share prices declined.[3]

Despite Wright's puffery Denver City Consolidated had found only low-grade ore and could not pay its debts. But before this became known, Wright effected a reorganization of the sort that would baffle investors in later years. He split off the Lee Basin Company as a separate corporation and sold shares to pay the debts of Denver City Consolidated. Even the Leadville *Herald,* loath to criticize local mining companies, called the plan "ridiculous." Wright was buying time. In April 1882 it became obvious that the ore was too low-grade, and the Denver City Consolidated mine shut down. The companies strung investors along with assurances that an expensive ore-concentration plant would bring the long-promised dividends. Those hopes eventually faded as well, and neither Denver City Consolidated nor the Lee Basin ever paid a dividend.[4]

Wright later said he lost $1 million at Leadville, but any million dollars he lost clearly wasn't his own. On the contrary, shortly after his Denver City Consolidated promotion he bought a four-story stone house in Philadelphia. Wright left Leadville with capital to invest in other mining properties.

Lake Valley—Another Fortune

In 1881 Wright led Philadelphia investors in buying a controlling interest in a group of mining claims on low grass-covered hills at the foot of the Black Range in New Mexico. Although the claims were more efficiently operated as a single unit, Wright split them into four separate companies for maximum promotional value. He floated the Sierra Plata, Sierra Madre, Sierra Apache, and Sierra Grande companies on the Philadelphia Mining Exchange.[5]

The first months were disappointing, as shafts sunk to the pay horizon showed no ore rich enough to mine. But no discouraging words clouded the public reports from the mine superintendent, George Daley, who telegraphed that the shares were worth much more than their current price. Daley joined a posse against the Apaches, but the Apaches ambushed their pursuers in the hills west of the Lake Valley mines, killing Daley. Meanwhile, the remaining miners had discovered a large cavity lined with silver minerals on the line between the Sierra Grande and Sierra Plata. The discovery of the fabulous "bridal chamber" posthumously gave truth to Daley's exaggerations—at least about the value of the Sierra Grande and Sierra Plata companies.[6]

George D. Roberts, the notorious mining sharp who had sold the property to Wright, still owned an interest. Thomas Ewing, who had helped Roberts in his swindles at the State Line mines in Nevada and the Robinson mine in Colorado, became superintendent of the Lake Valley companies in January 1881. Mining and financial periodicals disagreed on the value of Lake Valley. The opinion-for-hire New York *Stock Report* first blasted Lake Valley then reversed itself.[7]

The ore body in the Sierra Grande and Sierra Plata was limited in size but very rich. Wright and Roberts led a party of eastern investors to New Mexico, where the visitors were literally dazzled by the metallic gleam of silver ore lining the bridal chamber. The group included the ever-optimistic Yale professor Benjamin Silliman Jr. Silliman grossly exaggerated ore in sight, and endorsement by the popular professor had the desired effect on the excursionists, even though Silliman's notorious reports on the Emma mine in Utah had compromised his credibility among mining men. The *Engineering & Mining Journal* complained, "In the past, Professor Silliman has displayed a lack of judgement which should suffice to render impossible any attempt to float mining property on his unsupported testimony." Wright and Roberts arranged for the Sierra Grande to absorb the Sierra Plata, and floated the combination in New York. The *Engineering & Mining Journal* warned its readers away, but investors snapped up shares as soon as they appeared.[8]

The Sierra Grande company paid dividends for a few months. The dividends were not enough to repay investors, but the temporary profits sent share prices of all Lake Valley companies sky-high. Prices retreated the following year, when the ore body was exhausted, but by that time Wright had sold his shares and become a millionaire at age thirty-six.[9]

Wright's Lake Valley companies issued false press releases that magnified small ore bodies into great bonanzas, but none of the companies ever returned to profitability. The Sierra Grande Company was reorganized in 1887 and turned over to tribute miners to pick the bones.[10]

The Philadelphia Broker

"The people want to be skinned, and I am going to skin them."

—Whitaker Wright, to a Philadelphia friend, *Philadelphia Press,*
January 27, 1904 p.1

Wright used his knowledge and connections in the western mines to attract Philadelphia investors. He formed the Security Land, Mining and Improvement Company to develop and sell Leadville mining properties. Wright always attracted money but was less successful at making profits. A Philadelphian had him arrested in 1888 for misusing investor money, but the case was dropped.[11]

Wright was elected president of the Philadelphia Mining Stock Exchange in April 1881. He lived in West Philadelphia, then upscaled to fashionable Haverford. He bought a luxurious summer home at Long Branch, New Jersey, and expensive yachts.[12]

The Lure of Western Australia

Wright returned to England in 1889, intending to retire. He was a naturalized American citizen with a Philadelphia-born wife and three children. However, the financial panic of 1893 deflated a boom in steel companies, and the consequent failure of the Gunnison Coal & Iron Company took away most of Wright's fortune.

Spectacular gold discoveries in the Western Australia desert electrified London investors in 1894. Wright bought options on claims in the Coolgardie district; he formed the West Australian Exploring and Finance Corporation in September 1894 and offered it to investors as an exploration and promotion company. Wright kept promotion profits reasonable, drawing praise from the *Financial Times.* Critics, however, spurned West Australian Exploring as a blind pool and noted that Wright had a bad record on both sides of the Atlantic. Nevertheless, the public offering was a big success.[13]

West Australian Exploring promoted new mining companies, including Hannan's Golden Dream, Paddington Consols, Victorian Gold Estates, and Wealth of Nations. None ever paid a dividend. Wright then floated another blind pool, the Austin Friars Finance Syndicate, which floated mining companies in New Zealand and Australia.[14]

Wright organized the International Mining Corporation to mine the Pacific island of New Caledonia. The company spun off Fernhill Gold, International Copper, and International Nickel, but the company and its progeny did not fare well. International Nickel was sold in 1897 to another Wright entity, the Nickel Corporation. Fernhill Gold and International Copper were merged in 1899 into Caledonia Mining Corporation in a deal designed to generate profits for Wright's London & Globe company but which left the merged company without sufficient capital. Wright then organized the International Corporation, which took over the remaining assets of International Mining Corporation.[15]

The London & Globe

Wright floated another holding and promotion company, the London & Globe Co. Ltd. The London & Globe prospectuses omitted careful estimates of ore values, but the investing public was too enamored of Wright flotations to care. Ivanhoe Gold and Lake View Consols, London & Globe flotations in Kalgoorlie, Western Australia, were very profitable. In 1897 Wright merged his sagging West Australian Exploring into the London & Globe Finance Corporation.[16]

The British America Corporation

"Was there ever such a preposterous document as the prospectus of the British America Corporation?"

—British Columbia *Mining Record,* February 1898

In 1897 I. N. Peyton and US Senator George Turner sailed for England to sell the Le Roi Mining and Smelting Company. Le Roi, in the Canadian province of British Columbia, was the jewel of the Rossland mines, having already paid more than £130,000 in dividends to its American owners. Turner and Peyton spent fruitless months in London discovering that the English profoundly distrusted companies, such as theirs, incorporated in the United States. They finally contacted Whitaker Wright, who agreed to buy the company—but only if he could reincorporate it in Great Britain.[17]

Wright decided to make Le Roi the start of a systematic promotion of western Canadian mines. His London & Globe Finance floated the British America Corporation as a promotion and holding company in December 1897, with the understanding that the first acquisitions would be the Le Roi mine and some Klondike gold properties. As usual, investors were invited to entrust their money to Wright to do with as he pleased. Many in the financial community scorned the blind pool and warned investors that they were better off investing directly rather than through watered schemes. The public ignored the experts and clamored for shares.[18]

The British America Corporation, BAC for short, issued no reports to shareholders until it had been in existence for two years, and then gave its owners only a vague accounting.[19]

Le Roi Mining Company

The much-awaited prospectus of the Le Roi Mining Company was over-loaded with superlatives but devoid of specific facts. Even those eager to praise the Le Roi mine admitted that the document was completely un-businesslike. The British Columbia *Mining Record,* although always sanguine about the Rossland mines, said that those who bought on the basis of the prospectus were forced to invest "blindly." The *British Columbia Review,* a weekly devoted to encouraging investment in western Canada, complained that the prospectus of the province's best-known mine resembled a "collection of choice panegyrics on a new soap." Such quibbles did not deter the share-buying public. Subscription books opened in December 1898, and within three days there were two and a half times more applications than available shares.[20]

Wright knew the tendency of mineral veins to disobey the promises of promoters, but too much was at stake to allow Le Roi shares to suffer the vagaries of nature. Wright replaced the mine manager with Bernard MacDonald, on whom he could depend for optimistic publicity. In 1879 MacDonald had been arrested, but acquitted, for murder in a gunfight with claim jumpers at Bodie, California. He had first worked for Wright at the Lake Valley mines of New Mexico in 1881.

Despite MacDonald's constant promises of profits, the fact was that the Americans had thoughtfully removed the best ore before they sold. Le Roi paid no more dividends and stopped ore shipments in March 1900. MacDonald blamed the interruption on inadequate hoisting equipment, but the stoppage shook shareholder confidence.[21]

Since the Le Roi flotation garnered high earnings, British America and London & Globe joined for the next initial public offering, a collection of Rossland mining claims unimaginatively named Le Roi No. 2. British America and London & Globe pocketed £550,000 out of the initial capital of £600,000, leaving the new concern only £50,000 in working capital.[22]

The public recognized that Le Roi No. 2 had nothing but its name in common with the Le Roi Company, which itself had yet to pay a dividend to the new investors. Previous flotations had been offered preferentially to shareholders in London & Globe and British America Corporation, but

Wright's usual investors were not as confident as before, so the subscription books for Le Roi No. 2 were open to all. Wright announced that the flotation was a complete success, but in reality many Le Roi No. 2 shares had to be bought by Wright's own agents.[23]

Bernard MacDonald also managed Le Roi No. 2 and lied that the company was already on a paying basis, with enormous tonnages blocked out and ready to mine. Despite a quick dividend in June 1900, ore shipments fell far below MacDonald's confident promises, and the guinea pig directors grew uneasy. The 1900 annual meeting was postponed until January 1901 when the auditors decided that they needed to inspect the books in Rossland. The chairman and two directors resigned.[24]

London & Globe and British America next jointly floated the Rossland Great Western Mines, Ltd. Bernard MacDonald wrote in the prospectus that the mines were ready to provide twenty-five percent annual returns (£125,000) to shareholders. However, after more than a year of mining at a loss, the remaining ore had a net worth of only £27,000.[25]

To pump up profits of the ailing London & Globe at the expense of the British America Corporation, Wright had the London & Globe organize and float the Kootenay Mining Company, then directed the British America Corporation to buy the entire initial offering. The ore in the Kootenay property averaged about $9 per ton, only a fraction of that claimed in the prospectus.

Guinea Pig Directors

Wright knew that British investors required prominent guinea pig directors, a rotten old tradition of London finance. Wright's prize guinea pig was Frederick Blackwood, the Marquis of Dufferin and Ava, the former governor-general of Canada and India, and a former ambassador to France, Turkey, Russia, Italy, and Egypt. Lord Dufferin became chairman of the London & Globe Financial Corporation and Le Roi Mines. Other guinea pig directors included the master of the queen's household, and a groom-in-waiting to the king. A colonel of dragoons later testified that he had fulfilled his entire duty as a director by signing stock certificates. Some suspected that titled directors also shielded Wright from legal prosecution.[26]

Bribing the Gentlemen of the Press

Wright hewed to the tradition of corrupting the financial press. During the 1898 bankruptcy trial of the company promoter Ernest Terah Hooley, names on Hooley's canceled checks included many connected with financial journals. Some were well-known mercenaries such as Harry Marks

Whitaker Wright's prize guinea-pig director Frederick Blackwood, Marquess of Dufferin and Ava. Lord Dufferin capped a distinguished diplomatic career with the disgrace of his association with the Whitaker Wright companies. Courtesy Library of Congress.

of the *Financial News.* Hooley said that he had given him £17,000 because Marks was a friend. In fact, Marks had attacked Hooley's companies in print, up until the date on the first check. Other objects of Hooley charity were the wife of an editor of the *Financial Times,* and an editor of *Rialto.* Hooley later complained, "I have promoted companies that I have not made a single penny out of, because the newspapers took all the profit."27

After the city editor of the Westminster *Gazette* died, his employers discovered that he had taken bribes from Whitaker Wright. Bribes for favorable notice of the Le Roi flotation had cost, it was reported, £1,500 for two newspapers alone.28

Wright later named editors, publishers, and reporters of the *Daily Mail, Financial Times, Financial News, Truth,* and *Citizen* as recipients, though he insisted that the payments had not been bribes but press calls. Wright had sold shares to journalists then bought the same shares back the next day at a higher price, an immediate profit of £0.50 per share for the journalists. Wright said that his journalistic subsidies had exceeded $43,000 for each stock flotation. The press denied that the press calls had influenced them in any way.[29]

The Grand Life

Wright employed nearly a thousand laborers to build a two-thousand-acre estate at Godalming, Surrey. The grounds included gardens of palm trees, antique fountains and statuary from Italy, summer houses, and lakes stocked with trout. When a hill blocked a view, that hill was leveled. The stables housed fifty horses, each in a stall with a decorated ceiling and electric light. An underground passage from the thirty-bedroom mansion led to a glass-walled billiard room at the bottom of a garden lake. At one end of the ballroom was a stage, and at the other end a huge pipe organ. Above the ballroom was an observatory with one of England's largest telescopes. Critics sneered that the architecture was an awkward jumble of whims, but the public was awed.[30]

In 1898 Wright bought a luxurious steam-powered yacht, said to be the fastest in Britain. He renamed it the *Sybarita* and used it to entertain the Prince of Wales as well as other nobility, cabinet members, and ambassadors. The triumph of the *Sybarita* over the German kaiser's *Meteor,* as well as Wright's donations of prizes for yachting contests, won him memberships in six exclusive yacht clubs.

In addition to his country estate, Wright kept a house on Park Lane in London, where he spent four days each week. His son was admitted to Eton. Wright had entered the social and financial elite of Britain and the world.

The Fall in Mining Shares

The Western Australia mining boom went flat in the late 1890s. Wright knew booms and busts and must have known that it would end sooner or later. Wright had lost a fortune before but rose again with the next boom. This time he was determined not to lose his great estate, his position, and his privileged friends.

As investors realized that none of the companies promoted by Wright's Western Australian Exploring was likely ever to pay a dividend, share prices drooped. Wright performed his financial juggling by consolidating thirteen of his West Australian companies and refloating them as the Standard Exploration Company in February 1898. How mines that could not profit individually might pay after merging was not explained, but London was enthralled by merger mania. The public eagerly subscribed to shares in Standard Exploration. Unbeknownst to investors, Standard Exploration was burdened by a secret agreement to acquire the assets of Wright's worthless Austin Friars Syndicate. Like its constituent companies, Standard Exploration never paid a dividend.

Westralian shares continued to slump, and Standard Exploration and London & Globe lost money. Wright hid the losses with a frantic shell game, shuffling assets and liabilities between his companies as the annual balance sheet of each in turn was propped up using the assets of the others. Each company was a separate entity with different shareholders, but Wright swapped assets as if they were all his personal property. When investors tired of waiting for dividends from his Victorian Gold Estates, Wright split it in two—Lodden Valley, and Moorlort Goldfields— and made even more money from the promotion. Wright needed all the promoters' profits and more to fight bear attacks on his hollow financial colossus.

Fighting the Bears

To maintain the value of his holding companies, Wright had to prop up the shares that comprised their assets. But in 1900, stock market operators recognized that share prices were unrealistically high, and they thought to make money by selling short and bearing the shares. Chief among the bears was Charles Kaufman, an American mining engineer whose bad reputation dated from his participation in the State Line swindle in Nevada. He worked for Wright's Australian ventures but left Wright in 1897 to work with Horatio Bottomley and other manipulators.[31]

The bears first attacked Le Roi No. 2, selling large blocks of shares in the hope of inducing panic selling. The bears underestimated public confidence in Wright. Shareholders refused to sell at depressed prices, and the bears lost heavily when they had to buy shares at £20 to fill their £5 short contracts.[32]

Despite heavy losses in Le Roi No. 2, the bears attacked Rossland Great Western. Again, shareholders weren't interested in selling at a loss, share prices rose from £5 to £10, and the bears had to pay high prices to fill their contracts.

The London & Globe branched out by building a London commuter train, the Baker & Waterloo Railroad, organized in November 1900. The plan was solid, but construction drained cash from the Wright companies at a time when they needed every penny to support sagging mining shares.

The Ivanhoe Mining Company had performed admirably for Wright. The Western Australia mine had profited £140,000 in 1898 and £286,000 in 1899. By mid-1900, however, directors were desperate to justify the overblown share price. When the mine's highly regarded manager, Thomas Hewitson, refused to exaggerate ore reserves, the London directors fired him and brought over the more cooperative manager of Lake View Consolidated. The mining engineer Charles Kaufman used Hewitson's firing to try to convince shareholders to install Kaufman's own slate of directors.[33]

A Corner in Lake View Consols

One of the flotations of the London & Globe was the Lake View Consolidated Mining Company, at Kalgoorlie, West Australia. It was a very profitable mine, and its shares, Lake View Consols, were a staple of the London stock market. In early 1899 the mine tapped an extremely rich gold vein. Miners called the ore body the "duck pond" and gathered its golden eggs as the Lake View mine poured out more than thirty thousand ounces of gold each month. In November 1899, however, the miners exhausted the rich oxidized ore and found that the deeper sulfide and telluride ores were leaner and more difficult to treat. Production dropped suddenly to twelve thousand ounces per month.[34]

According to one account, the mine manager hid the condition of the mine from Wright, but told Charles Kaufman, who sold Lake Views short. When the bears attacked, Wright thought that he would punish them as he had with Le Roi No. 2 and Rossland, by cornering the market in Lake View Consols. He ordered the London & Globe to buy heavily.

A corner, however, is a slippery thing. When bullion shipments from the Lake View mine plummeted, Lake View Consols dropped sharply, costing the London & Globe £1 million. Wright hid the magnitude of the loss by juggling assets among his companies. Just before the annual meeting of each company, Wright would shift assets from the others into its treasury to keep the appearance of an invulnerable economic powerhouse.[35]

The following year Wright's engineers reported favorable new developments in the Lake View mine. Although his finances stretched thinner this time, he formed a syndicate to corner the market on Lake View Consols and gambled all he could borrow on the scheme. He was deceived a second time. The supposed great new ore bodies did not materialize, Lake Views declined further, and Wright's London & Globe lost £750,000 ($3.6 million). On Friday, December 28, 1900, the London & Globe Corporation once again prevented Lake View shares from sliding downward, by ordering its brokers to buy at the high prices. The bears had been defeated once again, or so it seemed until the brokers tried to cash their checks from the London & Globe but discovered that the bank accounts of the great company were empty. The dishonored checks threw thirteen brokerages into failure.[36]

The next day was pandemonium at the stock exchange. All day the bears hammered shares connected with Whitaker Wright. London & Globe, which closed at £0.70 on Thursday, sold for £0.30 on Saturday. British America declined moderately, from £0.66 on Thursday to £0.52 on Saturday, but its progeny fared much worse. Le Roi, flagship of the British America flotations, sank to £3.25, down from £7.12 on Thursday. The bears drove Rossland Great Western from £8.00 down to £1.50. Le Roi No. 2, which had maintained its £23.25 price only by the determined support of Wright, was left a wreck at £2.00.[37]

Juggling the Books

Desperate to prop up London & Globe, Wright ordered his accountants to shift almost £1 million in liabilities to the British America Corporation. The accountants protested but Wright bullied them into signing off on the transfer. The liability was fatal to the British America Corporation. Wright's shell game did not save London & Globe but only pulled BAC. down into bankruptcy as well.

At an emergency meeting of London & Globe shareholders in January 1901, the company president, Lord Dufferin, rose to a chorus of boos. Lord Dufferin confessed his ignorance of the company he headed, and his forthright manner won sympathy from the audience. Whitaker Wright then told the shareholders that he was finishing arrangements with the creditors and that the London & Globe would soon be healthy again. Heartened by Wright's promises, shareholders authorized a voluntary liquidation, which was granted by the court five days later.

Voluntary liquidation allowed Wright to keep control. The accountants appointed to liquidate the London & Globe were the same firm that for years had audited London & Globe's accounts without finding anything amiss. The accountants continued obediently—and falsely—to report nothing irregular.[38]

Wright was an experienced and tenacious infighter, and many in the City expected him back on his feet with fists swinging. The London Stock Exchange no longer had blind confidence in Wright, however, and few placed their money in Wright's reconstruction scheme. Two months after the fiasco, the London correspondent to the *Engineering & Mining Journal* observed, "The collapse of the Whitaker Wright group is beginning to look as if it were to be more than temporary."[39]

False Balance Sheets

"It is the sort of statement that ninety-nine directors out a hundred will make at a stockholders' meeting."

—Whitaker Wright at his trial, on false balance sheets,
Times (London), January 27, 1904, p. 7

Wright applied for voluntary liquidation of Standard Exploration and the British America Corporation, but shareholders had heard too much about Wright's management to be as trusting as those of the London & Globe had been a few months earlier. By insisting on involuntary liquidation, the shareholders secured independent management. The new management of the British America reported that the company had been systematically robbed to benefit the London & Globe. In their joint operations the London & Globe shared the profits but allowed British America to suffer all the losses.[40]

Wright tried to bring Standard Exploration and British America back under his control by proposing that they be merged with the London & Globe. It was Wright's old shell game—pretending to create a valuable company by merging worthless ones. Wright lost control of London & Globe in 1902 when creditors obtained a court order for compulsory liquidation. The new receiver studied the books and found the same sort of irregularities in the London & Globe.

Dissident Le Roi shareholders forced a special meeting in August 1901. Whitaker Wright, Lord Dufferin, and all but two of the other directors resigned. Although Wright did not dare attend, his backers were there armed with proxies. However, the dissidents defeated Wright's choice for chairman. The new chairman denounced Wright, appointed new directors, and sent a mining engineer to British Columbia to inspect the mine. The mining engineer sent by the new directors found gross mismanagement at the mine and fired Le Roi's manager, Bernard MacDonald.[41]

The House of Cards

Some of Wright's guinea pigs revolted and tried—usually too late—to protect the companies they headed. Directors of the Nickel Corporation sued the London & Globe to reverse some of Wright's financial sleights of hand that had bled the Nickel Corporation.[42]

Creditors of the London & Globe Finance Corporation received five cents on the dollar, and shareholders lost their entire investment. The directors put the blame squarely on Wright. Lord Dufferin wrote that he would have been "highly criminal" had he knowingly approved Wright's stock speculations; fortunately for his sense of honor, he had always rested in blissful ignorance of his companies.[43]

Wright tried to shift blame onto a group of London financiers who had agreed to support his stock operations in Lake View shares. According to Wright, it was their failure to live up to the agreement that had sent Lake Views, then the whole Wright empire, crashing. Wright sued his partners for damages.

The London & Globe's lawsuit for £1 million damages against its former financial backers came to trial in June 1902. As he neared the end of his resources in November 1900, Wright had made a last-minute agreement for the financiers to provide cash for his purchase of Lake View shares. In return he gave them large blocks of the Lake View shares he was buying. When the silent partners saw that Wright was losing the battle, they liquidated their holdings, and Wright found himself buying back his own shares. Wright's solvency had depended on maintaining Lake Views at £17, and he testified that, in addition to the written contract, he also had an unwritten agreement that the financiers would not sell their Lake Views for less than £17. But the jury took only fifteen minutes to find against Wright, and they foreclosed his last chance to recoup his fortune and good name.[44]

"Will Anyone Get Up and Say That a Man Ought to be Prosecuted for Issuing a False Balance-Sheet?"

"The English are not good sports. They don't know how to lose gracefully."

—Whitaker Wright, *Philadelphia Press*, January 31, 1904, p. 7

Despite Wright's obvious falsification of company balance sheets, British attorney general Sir Robert Finlay concluded that Wright had broken no law. In parliamentary debate Solicitor General Sir Edward Carson noted that misleading financial statements were common practice in the City, and challenged the opposition: "Will anyone get up and say that a man ought to be prosecuted for issuing a false balance-sheet?"[45]

Wright's solace was short-lived. The Companies Act allowed plaintiffs to instigate criminal charges at their own expense, but only if they could convince the court of the probability of their case. Investors were determined not to allow Wright to sit comfortably at Godalming and asked the courts to allow prosecution.

Justice Buckly ruled in March 1903 that those harmed by the London & Globe failure had the right to prosecute Wright. The next day a policeman called on Wright's London residence with an arrest warrant, but Wright's wife told him that her husband had left for Egypt, for his health.[46]

While London police searched for Wright, the crew of the New York–bound ocean liner *La Lorraine* puzzled over two passengers who had boarded on March 7 at Le Havre, France, under assumed names and occupied three of the liner's finest suites. M. Andreone and Mlle. Andreone, once out of port, informed the crew that they were J. W. Wright and his niece, Florence Browne. Wright told the ship's officers that the tickets were issued in the Andreones' names by mistake.

Before Wright boarded *La Lorraine,* Anna Wright had cabled her husband in Le Havre that "everything looks bad." Wright wired his wife to send £500 with his niece to Le Havre. When his niece arrived the next day, they boarded the ocean liner. Nearing New York, Wright tried to persuade the crew to hale a tugboat or other small ship so that he might leave the *Lorraine* before it docked, but the ship's officers refused.

As soon as Wright's flight was discovered on March 11, a dozen Scotland Yard detectives began tracking him. They followed a trail of banknote serial numbers through Paris to Le Havre, where a man of Wright's description had boarded the US-bound *La Lorraine* the previous week. On March 14 Scotland Yard sent Wright's description and requested his arrest via transatlantic cable to US law enforcement agencies.[47]

The next day the New York papers reported that Wright was thought to be in Egypt, but law officers knew better. Waiting for the *Lorraine* that morning were two Pinkerton detectives, two US marshals, and two New York police detectives. City police detectives were the first law officers to board the *Lorraine,* and they arrested Wright in his stateroom.[48]

The Pinkertons were a few steps behind the police and were chagrined to learn that their prize had already been captured. The US marshals on the pier tried to arrest Wright as he left the ship, but the city detectives told them that Wright was in police custody.

Extradition: Were False Balance Sheets a Crime in the United States?

Age and prosperity had taken away the athletic body of Wright's youth and padded it with the corpulence associated with men of substance. Dark eyes now peered out through pince-nez glasses from a fleshy face with heavy jowls and a double chin. His black hair had receded and was streaked with gray on the sides. The Scotland Yard fugitive circular described him as five feet ten or eleven inches tall and 250 pounds, with a florid complexion, large head, small eyes, and slight American accent.

Wright fought extradition but said that he would return voluntarily to England if the British government would withdraw its extradition request. The British authorities had seen Wright flee once and were not willing to risk losing him to some nation without an extradition treaty.[49]

Whitaker Wright built a mining financial colossus but was destroyed when traders learned the true value of his companies, and his paper empire crumpled. From British Columbia Review, *January 12, 1901.*

Extradition depends on mutual criminality: the act must be a crime in both countries. Wright's lawyers argued that the attorney general of Great Britain had determined that there were no grounds to prosecute Wright, so he had demonstrably not committed a crime in Great Britain, and furthermore, false balance sheets were no crime in the United States.

The US Supreme Court ruled against Wright in June 1903. Chief Justice Fuller wrote for the majority that the Companies Act of Great Britain had its equivalent in the penal code of New York, where Wright was arrested. Wright's lawyers fought for four more months before he realized that he could delay but not avoid extradition. He waived further appeals and returned to England escorted by police detectives. On shipboard back to England, Wright made comments to one of the escorts that prompted the detective to warn his English counterparts that Wright might attempt suicide.

The Trial

On his return to Britain, Wright was released on £50,000 ($242,000) bail. Police detectives shadowed Wright to see that he did not again flee the country. Wright's health was visibly failing.[50]

Wright was brought to trial in January 1904 on the charge that by falsifying the 1899 and 1900 balance sheets of the London & Globe, he had defrauded stockholders and creditors. The prosecutor was Rufus Isaacs, who had questioned Wright in the various bankruptcy inquiries.

Wright was the sole witness for the defense. His barrister led him through the financial tangles of 1899 and 1900, and the judge repeatedly asked for clarification as Wright described asset transfers and retransfers. Wright told the court that all his actions had been approved by the directors, and he denied that he ever misled stockholders. He blamed his troubles on a group of stock market wreckers. The defense argued that Wright himself had lost a fortune trying to prevent the collapse and that he should not take the blame while his guinea pig directors went free.[51]

The prosecutor Isaacs then put Wright through three days of cross-examination. Wright parried questions from Isaacs and bulldoggedly defended each transaction in the complicated juggle among his companies. He insisted that Lord Dufferin had approved and signed the 1900 London & Globe balance sheet, even though the signature page was missing and Dufferin had been in Ireland on that date. He said that changing the format of the 1900 balance sheet was not to disguise losses but was merely the decision of a bookkeeper. He dismissed his misrepresentations at shareholders' meetings as "slips of the tongue."[52]

A Loaded Revolver and Cyanide Capsules

"I failed to accomplish the impossible. I gave up a fortune in the attempt, but I could not stand alone against the entire London Stock Exchange."

—Whitaker Wright

Unknown to those who watched him enter the court on January 26, 1904, Whitaker Wright carried in his pants pocket a loaded revolver and capsules filled with potassium cyanide, a chemical used by his companies to treat gold and silver ores.[53]

The jury returned after only an hour and announced that they had found Whitaker Wright guilty on all but two counts. The judge passed sentence immediately, said that he could not conceive of a worse violation of the Larceny Act, and imposed the maximum punishment of seven years' imprisonment. Wright first appeared agitated, but then calmed. He began to proclaim his innocence but was silenced by court officers, who led him away.

There are various accounts—conjectures really—of how he took the poison. According to one, immediately after the verdict he took the cyanide capsules out of his pocket, concealing them in a handkerchief, and while wiping the handkerchief over his mouth, he swallowed the poison there in the courtroom. Another supposed that he swallowed the capsules in the lavatory after leaving the courtroom.[54]

Wright went to a consultation room in the court building to spend his last moments with his lawyers and associates, as he waited for the capsules to dissolve in his stomach. The jailers waited outside to take him to Brixton Prison. He drank a whiskey and water as a chaser for the cyanide and gave his pocket watch to a friend, saying that he would have no use for it where he was going. Fifteen minutes after he left the courtroom, as he leaned forward to light a cigar, Wright pitched forward, unconscious. His solicitor and a court attendant caught him, laid him back in his chair, and ran for a doctor.

A doctor tried to revive Wright but listened to his heart as the beating weakened and stopped. Police found the loaded and cocked revolver in his pocket. The doctor presumed that he had suffered a heart attack, but the coroner detected the strong scent of prussic acid from Wright's corpse and knew immediately that he had died of cyanide. The autopsy found partially dissolved capsules of potassium cyanide in Wright's stomach.

The Legacy of Suicide

Four days later his family stood in pouring rain to bury John Whitaker Wright in Whitley Parish churchyard, near his country estate. Sympathy went out to Wright's grieving widow and children.

His suicide accomplished what his living arguments could not: a rehabilitation, at least among some, of Wright's reputation. Previously, it had been said that Wright exploited the titled class to shield himself from the law, but some now considered the opposite to be true, believing that he had protected many—including members of the royal family—who had enriched themselves by his illegal manipulations. The Westminster *Gazette* claimed that many distinguished people were relieved that Wright had gone to his grave in silence. Harry Marks of the *Financial News* wrote:

> Wright might have made disclosures which would have seriously discomfited some people, but as he was loyal, as well as courageous, he

carried his secrets to the grave. To protect others from divulging what he chose to conceal, he destroyed many papers, thus completing his sacrifice.[55]

Wright received less sympathy from the Australian *Mining Standard*, which condemned the injury to Australian mining companies caused by Wright's managing them as a roulette game rather than as serious industrial concerns.[56]

The mines that Wright puffed up to unreasonably high values retreated to more pedestrian share prices and began orderly exploitation. First, however, they had to disentangle themselves from the financial wreckage. Le Roi stumbled through a series of superintendents who mistakenly reported profits while the company lost money. Standard Exploration shareholders sued a pair of guinea pig directors for the false prospectus that went out under their names; the lord and the honorable lieutenant-general were further shocked when the court held them responsible.[57]

Responding to the systematic robbery of shareholders, Parliament had passed a new Companies Act in 1900. The act made explicitly illegal such common practices as false balance sheets and false prospectuses. As a result, by 1904 nearly two-thirds of new British companies were floated without a prospectus. To avoid the new requirements, many chose to incorporate "offshore"—on the Isle of Guernsey, where the 1900 Companies Act did not apply.[58]

American newspapers noted that Wright's Wall Street counterparts continued to escape legal judgment. Wright's American lawyer Samuel Untermeyer ventured that Wright would have escaped punishment had he done the same misdeeds in New York—and judging by Untermeyer's own success in promoting the Harney Peak tin bubble, perhaps it is so. Even in London, Wright was not nearly as bad as such financial clowns as Terah Hooley, who seemed to grow richer each time he declared bankruptcy, or Horatio Bottomley, who wrecked numerous Australian mining companies for his own profit and then was elected to Parliament.[59]

George Graham Rice's
Adventures with Your Money

After each fraud conviction, George Graham Rice emerged from prison to sell more overpriced or worthless mining stock. He was a conscienceless swindler, but to some investors Rice was a financial genius battling Wall Street. Rice was an incurable gambler who lost at cards and lost at horses. With mining stocks, however, he rigged the game so that he always won.

Alias Joseph Hart

George Graham Rice was born Jacob Herzig in New York and from an early age was an avid but unskilled gambler. The teenager was more skillful at forging signatures on checks from his father's furrier business to pay his gaming debts. His father at first paid the drafts, but after young Herzig cashed $40,000 in forged checks, the elder Herzig preferred charges against his son, who at age nineteen was convicted of larceny and spent two years at the youth reformatory.

After his release Herzig, now calling himself Joseph Hart, again indulged his passion for gambling but unfortunately with no more skill. He again paid his gambling debts by forging the signatures of his father and brother on checks for thousands of dollars before he was arrested in Boston in 1894. The young defendant declined to testify at his trial, and the jury quickly convicted him. But he had somehow charmed the jury—or perhaps it was the beautiful young woman with him in the courtroom—and they recommended leniency. The judge was unswayed, however, and noted that "young men so intelligent as he may not expect from this court mercy or consideration on account of their social position." The judge sentenced him to six and a half years in prison.[1]

Herzig married Theramutis Ivey on the train to prison, reportedly to defy his family by marrying outside his faith. The bridegroom made his vows while handcuffed to a deputy sheriff. Years later the woman declared that Herzig had married her thinking that she could make life in prison easier for him—and that he had declared the marriage invalid when he learned that she could not help.

Alias Graham Rice

Out of prison in 1899, Jacob Herzig, a.k.a. Joseph Hart, now took the name of a fellow inmate, Graham Rice, initially as a pen name for a story he wrote for the *Youth's Companion*. Rice had found his true talent in writing. The New Orleans *Times-Democrat* hired him in 1900 to report on the disastrous flood at Galveston, Texas, but once on the scene he switched to the higher-paying New York *Tribune*.[2]

The following year found him jobless in New York, where he spent his last $7 to place a newspaper ad for the Maxim & Gay Company—which existed only in his imagination—promising to supply winning race-track tips. Rice was astounded when gamblers eager for the inside track lined up around the block to pay Rice to tell them which horses to bet on. Money poured in to Maxim & Gay, but Rice admitted that he lost it almost as quickly—by betting on his own horseracing tips.

Though a fashionable dresser who sometimes changed suits four times daily, Herzig was unimpressive physically. He was thin and unathletic—he himself jokingly described his physique as "cadaverous." Photographs reveal dark hair and an innocent clean-shaven face with spectacles. A gregarious sort, he was quick to offer one of the fine cigars he carried whenever he could afford them. He was a persuasive dynamo, especially on the printed page.[3]

Rice expanded Maxim & Gay into a bet-by-mail scheme and bought a New York sporting newspaper, the *Daily America,* to tout the company. Despite huge profits from his bet-by-mail business, he was soon beset on all sides. The *Daily America* lost money by the bucket and was sued for libel for printing dubious gossip about the socially prominent. On the complaint of a Kansas City man, New York City police arrested Rice in June 1903 for running an illegal bet-by-mail scheme. In December postal authorities issued fraud orders to halt mail delivery to the Maxim & Gay offices in New Orleans and New York.[4]

Rice also found himself in a domestic tangle that plagued him for years. As he left the criminal courts building, his wife served him with divorce papers. She had read in the newspapers that Jacob Herzig had made a great deal of money under the name Graham Rice, and she thought that she deserved some of it. Complicating the lawsuit were the facts that both Rice and his wife had since remarried: she had married a former prison friend of his; and Rice had married Frances Drake, a San Francisco actress and journalist. The former Theramutis Ivey was at the same time separating from her second spouse, and a New York lawyer found himself representing two different husbands defending themselves in simultaneous lawsuits for divorce brought by the same woman.[5]

Alias Jack Hornaday

Rice eluded punishment, but in April 1904 the post office halted mail delivery to Graham Rice himself, for mail fraud and for running an illegal lottery. Maxim & Gay had profited $1.5 million in its three years of operation, but Rice had lost most of it at the track. He took a few months to gamble away what was left, then put his last $200 in his pocket, left New York and $6,000 in gambling debts behind, and headed west. In California he worked on a farm for a few weeks then decided to see San Francisco.

By chance Rice met a friend from New York, William J. Arkell, the former owner of *Leslie's Weekly* magazine, in San Francisco in late 1904. The pair soon were publishing a West Coast racetrack tip sheet under the pseudonym "Jack Hornaday."

The Lure of Goldfield, Nevada

When Jack Hornaday's tips lost favor with San Francisco bettors, Rice and Arkell traveled to Tonapah, Nevada, where they promptly lost their money at a roulette wheel. Arkell was broke, but he represented himself as an agent of wealthy eastern investors and bluffed his way into an option on the Tonapah Home mine. Arkell was familiar with mining promotion, having been a founding director of the disastrous Joseph Ladue Gold Mining Company during the Klondike boom.[6]

Mine promoters believed the charade and took Arkell and Rice down to Goldfield, Nevada, to sell them a mining property. Arkell was unimpressed and promptly cashed a bad check to pay his fare back to San Francisco. Rice, however, was enthralled by the desert boomtown and stayed in Goldfield. He became a reporter for the Goldfield *News* but was fired in less than a week because of his ignorance of mining, *News* readers' principal interest.

Alias George Graham Rice

Rice, who by this time was calling himself George Graham Rice, started an agency to place advertisements for Goldfield penny mining stocks in newspapers throughout the United States. He ran the advertising agency along with the *Nevada News Bureau* in the back room of his wife's soda fountain in Goldfield, supplying news items to keep Goldfield before the public eye. Newspapers published his news items to keep receiving his paid advertisements. His ignorance of mining was no bar to success; on the contrary, his enthusiasm unrestrained by knowledge was very effective with the investors his advertisements targeted.[7]

Promoters organized about two hundred Goldfield mining companies: some good, many worthless, and all hungry for investment capital. Local newspapers demanded money from mining companies for printing uncritical puff pieces about their properties. The news columns of the Goldfield *Tribune* were for sale at $0.60 per line, and in one issue all but two inches were reportedly paid for by mining stock promoters.[8]

Goldfield attracted many men with shady pasts who were hoping for either a clean start or more opportunities for illegality. The town was tolerant. Goldfield became the temporary residence of New York mine swindler Dr. J. Grant Lyman. Sydney Flower, who hid his criminal past under the pseudonym Parmeter Kent, edited the Goldfield *Gossip*. "Judge" Francis Burton left behind Colorado mining swindles and a prison record. Even while in a Boston prison cell, Burton had coolly swindled the warden and some guards out of $15,000 for shares in a fake mine. He arrived in Nevada to found a Goldfield bank but fled the state in 1905, leaving only $21 in the vault.[9]

To stimulate Rice's journalistic enthusiasm for the newly laid out town site of Rhyolite, Nevada, the town's promoter, C. H. Elliott, gave him seven lots in the new town, and ten thousand shares in each of two Rhyolite mining stocks. Rice's news bureau obligingly sent out wildly optimistic news items about the mines of the area, publicity that he said enticed Charles Schwab into losing millions of dollars in the Montgomery-Shoshone mine. Rice later admitted that none of the mines of the district ever paid a dividend but that the sale of the stock and town lots gained him $20,000.

One night in Goldfield he lost thousands of dollars—all the money he had—playing faro. The next day he borrowed a small sum and set out for a new mining strike he had heard of called Manhattan. Arriving in Manhattan, Nevada, at night, he slept in the open, and the next morning bought a mining claim for $5,000, giving a $100 down payment with a bad check drawn on a Goldfield bank. Rice knew the dangers of passing bad checks and hurried to beat his worthless draft back to Goldfield. He quickly sold his claim for $20,000 on the basis of the ore specimens he brought back, covered his check, and pocketed the profit.

Rice realized how much money he was making for others by promoting their mining companies. When some Goldfield promoters made the mistake of telling Rice that they were negotiating for a certain Manhattan mining property, Rice quickly bought the property himself and promoted it as the Blizzard Mining Company. To disguise his own involvement, the promotion was done in the name of a fiscal agent, Rudolph Gnekow, Rice's nineteen-year-old stepson and office boy.[10]

Haunted by His Past

Rice's ex–tip sheet partner, William Arkell, arranged a merger of mining properties as the Tonopah Home Consolidated Mines. Arkell advertised both US senators from Nevada as directors and began selling and manipulating the shares. However, after they read the lies that Arkell had published in the prospectus, US Senators George Nixon and T. L. Oddie both denied being company directors. The San Francisco and Tonopah stock exchanges were tolerant of manipulations, but when Arkell began cheating brokers, they booted him off both exchanges. Arkell's finances crashed along with the share prices of Tonopah Home. In May 1905 the San Francisco *Chronicle* exposed the shady pasts of Arkell and his ex-partner Rice. When Rice learned of the article, he quickly offered $1 each for copies of the *Chronicle,* depriving Goldfield of that issue. The Goldfield papers had no interest in giving bad publicity to a civic booster such as Rice and so ignored the story, although the Goldfield *News* stopped selling him advertising. Rice's fast action kept his record quiet in Goldfield—for six months.[11]

Rice's past was forcefully brought to public notice by the Denver *Daily Mining Record* in November 1905. The newspaper had been paying large advertising commissions to Rice's *Nevada News Bureau,* and according to Rice the *Mining Record* attacked him after he refused to split his commissions with them. For whatever reason, the *Mining Record* went after Rice in article after article. Rice indignantly denied the facts and sued for libel.[12]

Wing Allen, the Goldfield correspondent for the *Mining Record,* suffered the wrath of a town dependent on doubtful promotions. Major W. A. Stanton, the mining engineer and old Civil War veteran in charge of Rice's companies, went to Allen's home one night, wrecked the furnishings, and wrecked Allen enough to send him to the doctor. Allen ended the beating only by pointing a revolver at his assailant.[13]

Wing Allen got no sympathy from Goldfielders, newly ex-friends, or the doctor who sewed his wounds. Rice was never one to skimp on legal fees and hired a former Colorado governor as his attorney to file a libel suit against the newspaper. Rice secured a restraining order against further articles about him and bragged that the libel judgment would soon make him the newspaper's owner. A Denver court dismissed the restraining order, but the case dragged on for almost a year.[14]

In the end, Goldfielders accepted Rice with his criminal record. Rice was performing a needed service in booming the town; in the excitement of the gold rush, his legal lapses seemed quibbling matters. Rice himself belittled his criminal record with the odd euphemism "I had a youthful past."

Shanghai Larry Sullivan

In February 1906 Rice joined "Shanghai Larry" Sullivan to form the L. M. Sullivan Trust Company stock brokerage. The gregarious Larry Sullivan became president, and Peter Grant became treasurer. Rice kept a low profile. The Tonopah *Miner*, as recipient of much advertising from Sullivan Trust, praised Sullivan's "extensive experience covering many years in the West."[15]

In fact Sullivan's experience was mostly in running a Portland, Oregon, boardinghouse and a gambling saloon, both catering to sailors. Sullivan's boardinghouse was notorious for delivering its roomers—sometimes willing, but mostly not—over to sea captains eager for crews. He also shanghaied unfortunates by promising a party on board ship or by adding knockout drops to their drinks in his saloon. According to one story, spread by his bitter rival "Bunco" Kelly, Sullivan once put so much dope in the drinks of a Finnish sailor that, hours later, the Finn still would not wake up to convince the sea captain that he was able-bodied. Sullivan was in a hurry to collect his fee, so he doused the man's boots with kerosene and set them afire. Sullivan later paid the court a $1,000 fine for the incident; the sailor died of his burns.[16]

Sullivan's partner Peter Grant was another former dealer in reluctant oceangoing crews, who had also run boardinghouses in Portland and Astoria, Oregon, as fronts for shanghai kidnappings. His criminal activity in Astoria was protected by his brother, Chief of Police Ignatius Grant. In 1905 Larry Sullivan and Peter Grant handed the shanghai business over to another Grant brother and left together for Goldfield, Nevada. There they added to their mining expertise by running a popular gambling house and saloon. They would run their new stock brokerage in the same manner, except that the customers in their casino—or their boardinghouses—faced better odds than investors in the mining companies promoted by the L. M. Sullivan Trust Company.

Rice and his wife owned half the company, while the other half was split between the saloon keepers Larry Sullivan, Peter Grant, and another Grant brother. Sullivan and the Grant brothers supplied the capital, and Rice the promotional genius. They floated new stock issues and brokered and speculated in shares—Rice's wife was said to be an avid speculator who rarely missed a session of the Goldfield Stock Exchange.

Sullivan and Grant also contributed political savvy. Back home Sullivan had been a powerful ward boss, and he bragged that he was "the law in Portland." Together with Grant he had engineered Oregon's "Larry Sullivan Law," which gave them a practical monopoly of sailors' boardinghouses in Portland. Sullivan had been a stalwart Republican in Oregon, but in Nevada he allied himself with the Democrats. Sitting Nevada governor "Honest John" Sparks served as president for many L. M. Sullivan Trust mining incorporations. In return Sullivan Trust bankrolled Sparks's reelection campaign.[17]

The Sullivan Trust Company quickly put out three new mining companies—Jumping Jack, Stray Dog, and Indian Camp—at the rate of one per month and sold shares across the country. Sullivan Trust bought good press by buying large display ads in southern Nevada newspapers, including the Goldfield *News,* whose new proprietor, Charles Sprague, had none of the prejudices against Rice's methods that had caused the previous owner to deny him advertising space. Their ads also bought them good press in New York. Larry Sullivan served as an officer or director of each of the new companies, but Rice kept discreetly off the list of directors.[18]

The gambler-newspaperman Rice and the boardinghouse- and saloonkeeper Sullivan had few qualifications for developing mining property. Rice had been fired from the Goldfield *News* because of his ignorance of mining. Larry Sullivan, who had heard the term *winze* in connection with mines, lied to a group of investors: "Right now I've got a whole carload of winzes coming in to rush development work on half a dozen properties." Sullivan, the officer and director of mining companies, didn't grasp that a winze is a type of mine shaft—a hole in the ground. Despite Sullivan's ignorance Nevada newspapers receiving his advertising printed long and serious interviews of his views on the mining share market.[19]

The feud between Rice and the *Daily Mining Record* continued, and Wing Allen was again the target. Rice and his partner, Larry Sullivan, swore out charges of extortion, testifying that Allen had demanded that they advertise in the *Daily Mining Record* in return for stopping newspaper attacks on Rice. Sullivan accompanied a Goldfield policeman to Allen's house to arrest him one midnight. But Allen explained the situation to the policeman, who then declined to place him under arrest. Sullivan cursed the policeman, and the officer responded by punching Sullivan to the ground. Allen showed up at court the next morning to post bond.[20]

Rice finally dropped his extortion charge against Wing Allen, and his libel suit against the *Daily Mining Record,* and in return the newspaper stopped its attacks. Rice's aggressive legal tactics had cowed an opponent into silence.

One of the scoundrels in Nevada to make a quick dollar was John Grant Lyman, who was trying to recover from his worthless International Zinc promotion. Despite his notoriety and the abysmal record of his mining companies, southern Nevada gushed and fawned over Lyman. Rice and the local newspapers later insisted that they had known nothing unfavorable about Lyman, which is difficult to believe.

Lyman bought the Rush property at Bullfrog, Nevada. Rice arranged for his company to buy and promote Bullfrog Rush shares. Things went swimmingly until the miners discovered that the mineral vein at the Rush came to an abrupt halt. Lyman saw no reason to stop the promotion just because they knew for certain that the property was worthless, but Rice didn't want to ruin his future promotions and so offered to buy back shares in Bullfrog Rush—he wouldn't pay back cash, of course, but swapped Bullfrog Rush stock for shares in his other dubious promotions so that it actually cost him nothing.[21]

Greenwater, California, "Monumental Mining-Stock Swindle of the Century"

Sullivan Trust returned to new flotations with a vengeance in the fall of 1906. One of the brief booms that quickly appeared and disappeared was that of the aspiring copper district of Greenwater, California, near Death Valley. Despite its verdant name, water had to be hauled into the desert town of Greenwater and was so scarce that when Chuckwalla Charley's saloon caught fire, onlookers debated whether it would be less expensive to let it burn and then rebuild it, or to use costly water to douse the flames. The saloon burned down before the townspeople could agree.[22]

Freshly staked Greenwater mining claims were incorporated, and shares sold nationwide in 1906 and 1907, using exaggerated promotions. At the start of 1907, a thousand men were prospecting and developing the copper veins in the district. The number quickly dwindled when no ore in commercial quantities was found, despite good surface indications. The whole town of Greenwater even picked up and moved a few miles to a new location in January 1907. In the spring of 1907, lots in the town of Greenwater were selling for $5,000, but within a few years the town site and the surrounding camp were deserted.[23]

Rice himself called Greenwater the "monumental mining-stock swindle of the century," but he, too, tried to cash in. Rice's L. M. Sullivan Trust Company incorporated a Greenwater property as the Furnace Creek South Extension Copper Company, but when the boom went flat, Sullivan Trust and Rice were left holding a bag of unmarketable shares.[24]

Fall of the Sullivan Trust Company

Sullivan Trust Company advertisements at the end of 1906 bragged that out of eleven flotations, the shares of all but Bullfrog Rush were selling above the initial price. The prices reflected not any intrinsic value, however, but only a boom in southern Nevada mining shares, and market manipulation by Rice. Mining properties at Goldfield, Tonopah, Rhyolite, and Bullfrog had captured investors' interest, and share prices doubled and tripled in late 1906. Brokers scrambled to set up offices in Goldfield. A seat on the Goldfield stock exchange sold for $1,800 in November 1906, and two more stock exchanges were planned for the town. Goldfield brokers knew better than to list the L. M. Sullivan companies on the Goldfield exchange, so Rice and Sullivan took their shares to San Francisco, where the San Francisco Stock and Exchange Board—deservedly notorious for tolerating worthless mining shares—put the companies on the board. Some legitimate brokers complained, but the commissions were too lucrative for the brokers to pass up.[25]

Rice and Sullivan secured bank loans using their own shares as collateral, but the bank was ready to recall the loans if the share price declined. As the boom softened, the Sullivan Trust Company was forced to buy back its own shares to firm up prices. High finance was unfamiliar territory to Rice and Sullivan, and they badly overextended themselves.

Wildcat Nevada mining stocks were losing public favor, and further disaffection threatened in the person of Lindsay Dennison, of *Ridgeway's* magazine, who arrived in Goldfield and began writing a series criticizing the overblown mine promotions. Larry Sullivan took exception to the articles and went looking for the magazine writer, who had stepped out. However, Sullivan so terrified Dennison's wife that the writer moved his temporary residence to Tonopah. The articles in *Ridgeway's* came out in November and December 1906 and signaled the end of blind infatuation with Nevada mining shares.

L. M. Sullivan Trust Company—less than one year old—did not have the funds to pay drafts to a San Francisco stockbroker on December 26 and 27, 1906. Oblivious to the danger, George Graham Rice left for a holiday trip back east and stopped on his way through Salt Lake City to buy $100,000 worth of shares. The bank draft to the Salt Lake stockbroker was also returned for insufficient funds, and San Francisco stockbrokers realized that the Sullivan Trust Company was in trouble. Nevertheless, Larry Sullivan left Goldfield on New Year's Day to attend a prize fight in Tonopah and was not heard from for nearly a week—rumor had it that he had injured a man in a New Year's Eve brawl.

Without purchases to prop up prices on the San Francisco exchange, the overvalued Sullivan stocks crashed on January 5, 1907. Sullivan Trust's treasurer, Peter Grant, beat up a broker on the floor of the stock exchange but found that his methods that had been so useful with drunken sailors in Portland were ineffective in the San Francisco financial district.

Although Goldfield papers tried to put a good face on it—blaming the trouble first on slow mail service, and then on Sullivan and Rice's unwisely being out of town for the holidays—the L. M. Sullivan Trust Company was bankrupt because of its flimsy financing of properties inflated far beyond their real values. Rice and Sullivan had taken all the investor money, so that all the Sullivan mines closed soon after the failure. By one account the company's most coveted asset was not its mining properties but its "sucker list" of two hundred thousand gullible investors, which rival promoters tried to seize. The Sullivan Trust crash in turn caused the failure of the Goldfield State Bank and Trust Company, which had foolishly lent a great deal of money to Sullivan Trust with mining shares as collateral. When creditors reorganized the Sullivan Trust Company, Rice, Sullivan, and the Grant brothers were not invited to remain.[26]

Rice characteristically blamed his troubles on false rumors that he charged had been spread by the Republican US senator from Nevada, George Nixon, whom he said was trying to ruin the L. M. Sullivan Trust Company because it had funded Democratic candidates. Rice promised to pay all his creditors in full but then hurriedly left Goldfield and returned east.

Rice stayed in New York only a few weeks before realizing that Nevada was still his land of opportunity. Sullivan Trust creditors made Goldfield too hot for him, so he moved to Reno, Nevada. Larry Sullivan had also gone east, but after he was arrested in New York City on a complaint of a man who had lost money in Bullfrog Rush stock, Sullivan, too, returned to Nevada. Rice and Sullivan again started dealing in stocks as Sullivan & Rice, Inc., and Larry Sullivan even announced his candidacy for US senator.

Sullivan and Rice quarreled after only a few months, and Sullivan, who despite top billing was the minority owner, tried to take control of the company. While Rice was home ill one day, Sullivan went to the office and began giving orders. When clerks disobeyed Sullivan, he fired them, but they refused to leave. Rice's secretary drew a revolver on Sullivan to keep from being physically thrown out of the office, but Rice's cousin Irwin Herzig was unarmed, and Sullivan beat him severely. Sullivan was arrested for assault with intent to kill, misspending funds, and stealing the firm's books. However, since Sullivan held the lease on the office, he booted Rice out.[27]

Sullivan tried to promote mining companies on his own and announced that his next Nevada mining-stock flotation would feature New York congressman Tim Sullivan (no relation) as president, but Tim Sullivan was a notorious Tammany Hall crook and failed to inspire investor confidence. Without Rice, Larry Sullivan's stock promotions fell flat. Sullivan later tried to promote Mexican mines, then turned up in Los Angeles as an investigator for radicals accused in a dynamite plot. He returned to Oregon ten years later, but after an arrest for bootlegging, he was reduced to laboring in a Portland shipyard, until his death in 1918.[28]

In July 1907 Rice took over as editor of the Nevada *Mining News* and filled the publication with his usual outrageous exaggerations. The master of aliases wrote under different names to make it appear that he had an extensive network of correspondents in the mining camps. He brought Merrill A. Teague to Reno as coeditor. Teague had some credentials as a financial moralist based on articles he had written for *Everybody's Magazine* exposing bucket shops. In Reno, however, he cooperated with Rice's obvious lies.

The Nevada *Mining News* criticized the mine owner George Wingfield, until Merrill Teague met the object of his barbs in a Reno stock brokerage. The ex-cowboy Wingfield punched him then followed him out onto the sidewalk where he thrashed Teague and promised to do the same to George Graham Rice. Rice hurried out of the state and returned with an armed bodyguard at his side. George Wingfield was sometimes called "the owner and operator of the state of Nevada," and his retribution could extend far beyond the swing of his fists; Rice had made a powerful enemy.[29]

The post office sent an inspector to Rice's office to investigate possible mail fraud, but Rice had US senator from Nevada Francis Newlands intercede to stop the investigation. The postal service was not an effective anti-fraud agency due to this sort of political meddling. Rice blamed the investigation on his enemies George Wingfield and US senator from Nevada George Nixon.

Nat Goodwin Goes from Comedian to Stockbroker

"Although I made but little money at Goldfield, I was very greatly attracted by its life; the utter abandon, the manhood, the disregard of municipal laws, the semblance of honor which fooled so many, the codes of right and wrong, the tremendous chances that were being taken with the dice box."

—Nat C. Goodwin, *Nat Goodwin's Book*, p. 297

Rice's next partner was Nat Goodwin, one of America's most popular comic actors. Goodwin had met a free-spender named William J. Brewer, the owner of Goldfield mining properties, in the spa town of Glenwood Springs, Colorado. Goodwin was intrigued by the tales of fortunes being made in the desert, so when Brewer invited him to attend a prize fight in Goldfield, he accepted. Like Rice, Goodwin caught gold fever and decided to stay in Goldfield.[30]

With his new friend Brewer and an Englishman named Kennedy, Goodwin bought a large number of $0.15 shares in the Triangle Mining Company. While Goodwin was en route to New York to tell his friends about the wonderful property, Brewer and Kennedy were salting the rock on the hundred-foot level. Share prices jumped to $1.50 when the phony ore was discovered, and unsuspecting Goodwin bought still more shares.

Brewer and Kennedy sold their shares, then knowing that the salting would soon be discovered, sold short. Share prices continued to rise, however, and the pair were unable to cover their short contracts. They borrowed a hundred thousand shares from Goodwin then promptly went broke. Despite being suckered by con men, losing most of his stock by lending it to them, and even cosigning a $10,000 promissory note from one of the defaulters, Goodwin came out of the affair $20,000 ahead. A federal grand jury later indicted Brewer for mail fraud in another swindle.[31]

Goodwin was impressed with his beginners' luck in mining investments, and, like Rice, thought that his brief stay in Nevada made him an expert judge of mining property. He met Rice in Reno and the two hit it off famously. The comedian lent his capital and famous name to form the Nat C. Goodwin Company, promoters of mining stock. He and partner Dan Edwards served as company officers, but although the company denied it at the time, the real brains of the firm was George Graham Rice. Rice concealed his partnership while he promoted Goodwin's mining stocks through the Nevada *Mining News*.[32]

Nevada's latest mining sensations were the instant boom towns (now ghost towns) of Wonder and Fairview. Rice and Goodwin brought out the Reliance Nevada Fairview Mining Company in October 1907 and the promotions were off and running.[33]

Rice again secured the services of his prize guinea pig director, Governor "Honest John" Sparks, who should have known better after Rice's disastrous L. M. Sullivan Trust promotions. Sparks fronted not only for Rice but also for other mine swindlers such as the notorious Burr Brothers, Shelton and Eugene, who promoted numerous bogus mining and oil stocks.

In addition to new company promotions, Rice kept up his unprofitable feud with George Wingfield. Rice tried to conduct bear campaigns on the

San Francisco exchange against companies associated with Wingfield, and Wingfield likewise tried to use the share markets to ruin Rice financially.[34]

Wonder and Fairview were succeeded in 1908 by Rawhide, Nevada, as a magnet for stock frauds. The Goodwin Company joined E. W. King to promote Rawhide Balloon, and then Rawhide Coalition. Rice continued searching his thesaurus for superlatives as he filled the Nevada *Mining News* and its successor, *Mining Financial News,* with ridiculous praise for his Goodwin Company promotions. A headline named Rawhide's Balloon Hill the "GREATEST MOUNTAIN OF RICH ORE WORLD HAS EVER SEEN." Another headline across the top of page one proclaimed "RAWHIDE COALITION MINING PRODIGY OF THE HEMISPHERE." Goodwin and Rice had optioned blocks of Rawhide Coalition shares at between $0.12 and $0.15 per share and sold them to customers for $1.50.[35]

Rice used his news bureau to keep Rawhide in the public eye by making up Wild West news items that appealed to big-city editors. On a slow news day, one of his writers fabricated a tale that robbers had stolen a $47,000 payroll of the Rawhide Coalition mines and were being chased by a posse. Out-of-state papers printed the story and clamored for follow-up. Rice was used to writing fiction and no doubt appreciated that his employee had cleverly given Rawhide Coalition legitimacy by wildly inflating the payroll for a company that employed only six miners. However, Rice considered robberies bad for stock sales and ordered his employee: "Kill those robbers and be quick; do it tonight so that you choke off the demand for more copy, or you're a goner!" The copywriter obediently sent out an item that the posse had chased the robbers into Walker Lake, where the outlaws drowned. Nevadans laughed at the implausible story, but out-of-state papers printed the neat end of the tale without question.[36]

Rice's biggest publicity coup was persuading the writer Elinor Glyn to visit Rawhide. Glyn, whose beauty and scandalous novels made her a celebrity, traveled by train to Goldfield, and after a few days and a mine tour there, arrived in Rawhide for a Wild West adventure scripted by Rice. At Tex Rickard's saloon, she saw a play-acted poker game with $1,000 chips. She didn't realize that it was all being staged for her amusement but watched as the gamblers quarreled, drew pistols, and fired; two of them fell apparently dead. Saloon owner Rickard let her win $1,000 in a rigged faro game, and during a tour of the town she witnessed the heroics of the volunteer fire department's fighting a fire in some deserted shacks outside town—another event planned by Rice. Rice made sure that the national spotlight shone on Glyn's one day in Rawhide. Newspapers across the country printed stories about the beautiful and controversial novelist's visit to the rough boomtown.[37]

Rawhide's balloon rose as it filled with Rice's hot air, then sank. Rawhide Coalition shares were introduced on the New York curb market at $0.25, and marched up past $1.00 on Rice and Goodwin's lies that they had a workforce of more than two hundred men and were spending $400,000 per year to develop the mine. Rice and Goodwin cashed in their options at the high price, making a reported $500,000. Rawhide Coalition had little in the way of intrinsic value to interfere with Rice's gyrating to push the shares up, and curb brokers trying to stampede it downward. After a high of $1.50, the stock inevitably crashed, and Rice and Goodwin blamed the loss on unscrupulous traders. Despite Rice's lies about great ore discoveries, and Goodwin's confident predictions that the price would soon shoot to $5, Rawhide Coalition shares stayed well below $1.[38]

Rice and Goodwin made a fortune on their Rawhide Coalition options, but the company itself struggled to find money and leased its best tracts to tributers. In truth, the Rawhide camp and its prospects were outrageously overpromoted, and real mining people knew it. Few of Rice's suckers ever read the legitimate mining press, however, and they threw their savings into his worthless Rawhide companies.[39]

B. H. Scheftels & Company, Stockbrokers

B. H. Scheftels & Company, a Chicago stock brokerage, had dealt heavily in Nevada mining stocks since opening a Reno office in 1907 and had acted as the Goodwin Company's Chicago agent. Scheftels combined with the Nat C. Goodwin Company in January 1909. Bernard Scheftels became the president, and Goodwin the vice president, but George Graham Rice continued to be the real moving force. In January 1909 Rice moved back to New York City to oversee promotions for B. H. Scheftels & Company.[40]

It was a triumphant homecoming for Rice. He had left New York as an ex-convict and failed racetrack tipster and now returned as a financial genius. His father accepted him back into the family. The Scheftels' offices in New York hired more than a hundred employees to handle correspondence. Tour guides were soon pointing out Rice's offices overlooking the Broad Street curb market.[41]

Rice continued to use the *Nevada Mining News* to boom his projects. B. H. Scheftels moved the publication to New York and changed the name to the *Mining Financial News*. Among its writers the paper boasted "Red Letter" Sullivan, a brilliant stock analyst when sober, but one whose weakness for liquor led him to associate with swindlers such as Rice and "Yellow Kid" Weil.[42]

Nat C. Goodwin & Co. had bought a controlling interest in the Ely Central Company in Nevada, which lay between two lucrative copper properties—the Ruth to the east, and the Eureka to the west. The trouble with the Ely Central tract was that the bedrock was hidden under lava flows and no one knew if there was copper ore beneath them. Ely Central had been incorporated in 1906 but the original promoters were too busy selling stock to seriously investigate if it really held any copper ore, and they shut down work on the property at the end of 1907.[43]

Rice, Goodwin, and Scheftels bought seventy-five thousand shares of Ely Central at $0.50 each and optioned other large blocks at similar prices. Rice's promotional abilities swung into action. He wrote to prospective investors that he would soon have a hot stock tip that would give them fabulous returns. After a buildup, he telegraphed "Buy Ely Central" and began selling his shares at great profit. Rice's campaign was assisted by rumors—which of course he did not deny—that Lewisohn copper interests were trying to buy the company.[44]

Rice hired Colonel William Farish to examine the property. Farish was a mining engineer whose good name had survived his unfortunate association in the 1880s with the famous swindles of the State Line mines in Nevada, and the Silver Cliff mine in Colorado. Ely Central press releases trumpeted Farish's conclusion that the property was a good prospect but ignored his observation that the company had not yet found any commercial bodies of copper ore.[45]

Most of the copper at Ely was then, and still is, mined from large low-grade ore bodies in big open pits. Because the southern part of the Ely Central property lay between two of these pits, the possibility existed of a similar, large, low-grade ore body, although at greater depth than its neighbors. Exploration for this possible ore would take time, however. Since Rice wanted an immediate splash for stock sales, exploration concentrated on the northern part of the Ely Central property, where high-grade but small copper veins had little chance of developing commercial tonnages but made for dramatic headlines in Rice's *Mining Financial News* and other compliant publications.[46]

By November 6, 1909, Rice's promotion had pushed prices up to $4.25, although some said that these were wash sales. Large blocks of new shares were issued, but only a small percentage of the proceeds went to the mining company. The great bulk of the money—millions of dollars—went to B. H. Scheftels & Co. That weekend the *Engineering & Mining Journal* came out with a scathing five-page damnation of Ely Central that detailed Rice's criminal past. The article caused a sensation and Ely

Central opened on Monday sharply down at $1.50. Rice spent a fortune buying Ely Central shares to support the share price. Both the Ely Central Copper Company and B. H. Scheftels & Company sued the *Journal* for libel, but the magazine would not back down and the suits went nowhere. Some Nevada papers told readers that the *Engineering & Mining Journal* was controlled by the Guggenheims and was part of a conspiracy to grab control of Ely Central.[47]

Rumbles of possible legal trouble, including an attempted shakedown by a New York City police captain for protection money, prompted Nat Goodwin to resign from the Scheftels Company. He returned to acting but always insisted that his mine promotions had been honest and that his friend George Graham Rice was a courageous man ruined by the big Wall Street interests.[48]

In distress, Rice appealed to his brother Charles Herzig. The two had not seen one another in fifteen years. Rice had been in and out of reformatories and prisons since his youth and had identified himself with wildcat mining stocks. Charles Herzig had meanwhile earned an engineering degree at Columbia University and established a reputation as a mining expert of solid judgment. Rice had to arrange a meeting with his brother through a third party. He asked his brother to investigate and report on the Ely Central. Charles Herzig accepted warily, on the condition that he was permitted to make an oral report; he wished to prevent misuse of his report for stock manipulation, as Rice had done with Farish's report.

Herzig concluded, as Farish had, that the property was promising but unproven. Rice persuaded his brother to serve as the vice president and general manager of Ely Central, and the two began a long association. Charles Herzig performed an ethical balancing act at Ely Central and other Rice promotions. Those parts of his reports that were made public indicate that Herzig was honest in his evaluations, but he had to know that his boss and brother George Graham Rice was issuing blatant lies about the same properties and misusing the mining reports by taking portions out of context.

Ely Central hired an expert, Walter Weed, to make yet another report. Weed was less optimistic than Farish or Herzig about the possibility of finding high-grade veins—the fruitless search for which had consumed large amounts of investor money—but agreed with them that Ely Central probably contained a large body of low-grade ore at depth. As before, B. H. Scheftels & Co. tried to misinterpret Weed's report as proving the value of Ely Central, which it did not.[49]

Herzig hired H. C. Wilmot as mine manager, but Wilmot quit after only five days, unhappy about the way the company twisted his reports to mislead investors. Herzig defended Ely Central management and ventured that the criticism was part of a plot by the adjacent Nevada Consolidated, which coveted the property. The *Mining & Scientific Press* noted of Herzig: "His many friends regret greatly to see him entangled in any way in the scandal, as a result of a natural though probably unwise effort to assist a relative."[50]

Charles Herzig examined Rawhide Coalition properties for a week. He found little that he could call ore reserves, because of the shortsighted policy of allowing tributers to run the properties, the absence of systematic development, and a lack of ore-processing facilities. Herzig's report showed that the Rawhide Coalition Company had very little to show for the fortune invested by shareholders. However, Herzig saw reason for optimism if the prospects were run on a more businesslike basis, and Rice's newspaper allies seized on Herzig's hope for the future rather than wonder what had happened to the rich ore reserves Rice claimed to exist.[51]

B. H. Scheftels & Co. floated the South Quincey Copper Company in March 1910 to develop some promising tracts in northern Michigan. The lands had genuine value, but South Quincey Copper investors learned too late that their company owned no interest whatsoever in the copper prospects it promised to develop.[52]

Despite the great amount of money arriving daily from investors, B. H. Scheftels & Co. had difficulty paying its broker accounts in August 1910; after a near-riot of curb brokers, it had to borrow money from another shady stock brokerage, Charles Stoneham & Company. Scheftels no longer had the funds to support its stock manipulations, and its pet stocks sagged. Rawhide Coalition, once a Nat Goodwin Company "buy" recommendation at $1.50, was selling for $0.10. Ely Central, highly touted by Rice at its high of $4.25, had sunk to $0.51.[53]

The US Postal Service and the Reform Movement

The only organization policing fraud on a national level in the United States was the post office, whose enforcement was spotty at best. The most common punishment was to halt mail service to the offender, which, as Rice knew firsthand, was easily evaded by changing the company name. Postal inspectors were few and susceptible to financial and personal appeals at both local and national levels. Swindlers learned that the local postal inspector was a good man to cultivate socially and that a friendly politician in Washington could keep the post office from investigating.

Post office fraud prosecutions increased in 1906, and the era of timid prosecutions was definitely over when President Taft appointed Frank Hitchcock as postmaster general in 1909. Hitchcock made mail fraud his number one priority. He disdained merely stopping mail service and so assembled an aggressive team of inspectors to put swindlers in prison. He shuffled his inspectors to different cities to end cozy local relationships with swindlers. Now the B. H. Scheftels Company would be his first big case.

The Raid

In September 1910, Justice Department agents simultaneously raided B. H. Scheftels offices in Boston, Chicago, Detroit, Milwaukee, New York, Philadelphia, and Providence. They sealed the premises, seized records, and arrested seven employees, including Bernard Scheftels.[54]

During the raid Rice happened to be on the street supervising his brokers in the curb market. There was a warrant for his arrest, however, and the following morning, after he had arranged for immediate bail, he surrendered and was released. Rice was no stranger to post office investigators. His New York racetrack tip sheet had been stopped by a fraud order from the postal service, and the post office had begun investigating his Goldfield mining promotions.

A federal grand jury indicted Rice, Scheftels, and three others. The Ely Central Copper Company was forced to halt operations, for Rice and Scheftels had pocketed almost all the proceeds from stock sales. The company was in fact heavily in debt to Rice and Scheftels, who were funding Ely Central operations for their own stock manipulations. Rice and Scheftels had some of their employees form a "reorganization committee," which proposed a $0.25 assessment, more than the current share price.[55]

In fact the Ely Central property had merit, although Rice grossly exaggerated the value. The Ely Central was merged with the adjoining properties in 1913 to form the Consolidated Copper Mines Company. For many years the Ely Central property lay unmined between two great open pits. It was finally mined out when the pits were merged into one, but not until many decades later.[56]

President E. W. King of Rawhide Coalition told reporters that Rice and Scheftels's arrest was of no consequence to the mining company. In reality Rice had left nothing in the Rawhide Coalition treasury and was funding the company hand to mouth. After the raid on B. H. Scheftels & Co., Rawhide Coalition tried to save itself by stopping the free-spending ways that Rice and Scheftels had imposed upon it. The board of directors axed a host of figurehead officers and employees but had to shut down exploration a few months after the raid on Scheftels & Co.[57]

The postal service did not stop with Scheftels & Company but followed two months later with a raid on another nest of thieves posing as stock brokers, the Burr Brothers company. Sheldon and Eugene Burr had for many years sold shares in their own mining companies, which ranged from worthless to completely phony. The postal service finally ended their career.

Nevada Politics

Rice had a mutually profitable relationship with Nevada Democratic politicians. He bankrolled their campaigns; in exchange they provided him with guinea pig directors and, according to Rice, political protection. Rice tried to increase his influence in the election year of 1910 by weighing in with a heavy wallet in Nevada's elections for governor and Congress.

Charles Sprague, owner and editor of the Goldfield *News,* had been a fair-weather friend to Rice. He had taken Rice's advertisements and given his promotions uncritical publicity but took to denouncing Rice in 1909. However, when Sprague sprouted congressional ambitions the following year, he suddenly realized that George Graham Rice wasn't a bad sort. Sprague and Denver Dickerson, who was running for reelection as governor, met Charles Herzig at a Reno Hotel in July 1910.

George Graham Rice accompanied his brother Charles Herzig to Goldfield later that month to examine some of Rice's mining promotions. Rice was warmly written up in the Goldfield *News* by Sprague, who less than a year earlier had written an editorial titled "Propaganda of Lies," which said "many people are being duped by the Rice crowd." Now, Democratic congressional candidate Charles Sprague wrote nostalgically of the easy-money days of the L. M. Sullivan Trust Company and headlined "RICE'S RETURN IS LOOKED UPON WITH MUCH FAVOR." On his return to New York, Rice told an employee that Sprague had promised to give Rice control of the Goldfield *News* while Sprague served in Congress. Rice's *Mining Financial News* endorsed Sprague and Dickerson.[58]

Rice's arrest in September gave the Nevada Republican papers an issue to hammer Democrats. A Republican paper ran a cartoon of Democratic candidates Charles Sprague and Denver Dickerson taking orders from Charles Herzig in exchange for cash. The Goldfield *Tribune* accused Rice and Sprague of conspiring to rig the courts to steal the Goldfield Consolidated Mining Company property in an apex lawsuit. Both Sprague and Dickerson lost their elections. As he had done many times before, Rice had put his money on losers, although this time his favorites were handicapped by their association with him.[59]

Far from gaining influence, Rice had only further antagonized Nevada's incumbents and the Republican administration in Washington. Rice's hopes of having the political influence to rein in the postal inspectors died on election day in 1910.

The Goldfield, Nevada, Tribune *celebrated the arrest of the mine swindler George Graham Rice and used it to tar Rice's political allies in Nevada. From* Goldfield Tribune, *October 20, 1910.*

"My Adventures with Your Money"

While under indictment Rice stole a page from a fellow mine swindler, Thomas Lawson, and cast himself as a reformer persecuted by Wall Street barons. He wrote his serialized account of his career, *My Adventures with Your Money,* which was published in *Adventure* magazine in 1911, and later published as a book. Rice said that he was a victim of crooked Wall Street powers intent on quashing competition and blamed his prosecution on a who's who of mining and finance, including J. P. Morgan, Charles Schwab, Bernard Baruch, the Guggenheims, and the Lewisohns.[60]

Rice's second wife sued him for divorce and alimony and named the chorus girl Bessie La Fell as corespondent. Miss La Fell's kind offer to pawn her jewels and her French lingerie for George Graham Rice's defense fund did little to reconcile Mrs. Rice to her husband. The former Frances Drake said that she had not known of his criminal past when she married Rice, and alleged that he had tricked her into signing property over to him and had committed her to an insane asylum to force her to surrender her remaining assets. She said, "I am only one of the innumerable dupes of George Graham Rice." Rice's curious defense against paying alimony was that the marriage was invalid because he was still married to his first wife when he married his second. The court granted divorce to Frances Drake Rice without opposition.[61]

B. H. Scheftels & Co. was out of business, but Rice reportedly planned more mining promotions. The *Mining Financial News* had folded after the raid, but Rice's favorite editor, the pseudo-reformer Merrill Teague, found another job as editor for the *Copper Curb and Mining Outlook*. Some said that the *Copper Curb* was a swindle sheet run by Charles A. Stoneham, although Teague denied that Stoneham had any connection with the paper. Stoneham, like Rice, was a mine promoter and bucket-shop operator who had used and discarded aliases as smoothly as a snake molts and slithers out of his old skin. Stoneham was tied financially to B. H. Scheftels & Co., and after the postal service raid, Stoneham had tried to bribe a postal inspector to turn over Rice's sucker list of Ely Central investors. Now Stoneham's employee Merrill Teague was promoting a wildcat mining company in the Porcupine mining district of Ontario, and Rice was rumored in on the steal.[62]

The Trial

The trial began in October 1911, a year after the raid. The defendants were Rice, Scheftels, and three employees, all charged with mail fraud. Former employees testified that the firm had pocketed investor money rather than buy the shares. When some investors had insisted that the company deliver the actual certificates, Rice eventually, but grudgingly, bought a few shares. Most investors, however, had trusted B. H. Scheftels & Co. to hold their shares—shares that it had not in fact even bought. The government also showed that the Scheftels Company had repeatedly secured options on large blocks of shares then boomed the companies to its investors while pretending to be a disinterested broker.[63]

B. H. Scheftels & Co. had also illegally drained the cash from the companies it controlled, most notably Ely Central. Bernard Scheftels pocketed at least $60,000 in Ely Central funds, without leaving an accounting.[64]

Rice's idea from the start was to drag out the trial. The government played into the strategy by bringing in thousands of exhibits and hundreds of witnesses, many offering mind-numbingly repetitive testimony. Rice's lawyers challenged each bit of evidence and line of testimony. After several months prosecution witnesses—under subpoena to remain in New York ready to testify—protested en masse that they were being impoverished by having to live in New York on the government stipend of $1.50 per day. The prosecution realized that the length of the trial was out of control and wound up the government case after two months.[65]

Bribing Jurors

In December 1911 a vaudeville theatrical agent approached one of the jurors and offered a $1,000 bribe for a "not guilty" verdict. Police arrested the agent and three others. The judge remained close-mouthed about the bribery investigation; however, a few days later he revoked Rice's bail, but not the bail of the other defendants, "in the interest of the administration of justice." The lead defense lawyer had protested that he knew nothing of the bribery and would withdraw from the case if his clients were responsible. After the judge yanked Rice's bail, the lawyer Abram Rose withdrew from Rice's defense but continued to represent the other defendants. Rose refused to explain, but everyone remembered his promise that he would not represent clients attempting to bribe jurors.[66]

Rice fired his new court-appointed lawyer and declared that he would represent himself. Prosecuting attorneys regarded it as a desperate move by Rice and thought that he was trying to establish grounds for appeal through his incompetence in conducting his own defense.[67]

Rice showed the jury photos of the fire that burned the Rawhide, Nevada, business district, and tried to blame the crash of Rawhide Coalition share prices on the fire—but the government observed that the fire did not harm Rawhide Coalition property and at any rate occurred years before the fall in share prices. Rice called Tex Rickard to testify to the value of Rawhide Coalition and Ely Central; but Rickard was a saloon keeper and fight promoter, not a mining expert. Rice also called Harry Hendrick, but the court ruled that being a newspaper reporter in Rice's employ did not qualify him as a mining expert. A name conspicuously absent from the list of defense witnesses was that of the mining engineer Charles Herzig.[68]

The defense dragged out for another three months. The case was backlogging the docket, and jurors protested that their businesses were suffering from their long absences. Amid speculation that the defense case could last months more, the prosecution negotiated a plea bargain. The stalling tactics of Rice and the defense lawyers had paid off. Rice and

Scheftels agreed to plead guilty. In return, Scheftels was given a suspended sentence and charges against his three other employees were dropped. Rice was sentenced to one year at the penitentiary at Blackwell's Island, the sentence to be measured from the previous December.[69]

From prison Rice announced that he would continue fighting the Wall Street powers. He told reporters that two unnamed civic organizations had asked him to lead a reform effort to prevent the organized stock exchanges from fleecing investors. "All the evil in this country is centered in the Stock Exchange," he declared. Rice also announced his support of the Democrat Woodrow Wilson for president—indeed, it would have been odd to endorse the Republican administration that had just put him behind bars.[70]

Rice left prison in October 1912 and bought the New York *Mining Age* to publicize his new promotions, starting with the International Mines & Development Company. Rice wanted to be known as the "Napoleon of finance," and even displayed in his office a photograph of himself posing as Bonaparte in an antique French army uniform, hand tucked in his jacket.[71]

Swindling Returns to the Emma Mine, Utah

When Rice traveled to Salt Lake City in 1915, the Salt Lake Stock and Mining Exchange delayed its afternoon session so that brokers hungry for a stock boom could have a leisurely feast with the convicted swindler George Graham Rice. An observer complained, "Every parasite of mining here is toasting and running after this man. It is disgusting." Rice left Salt Lake with an agreement to sell stock in the Emma Copper Company and the Old Emma Leasing Company.[72]

The Emma mine at Alta, Utah, had lain mostly idle since the 1870s, when its rich but limited ore body had lured British investors into a swindle. Its famous failure had prompted congressional hearings, propelled the US minister to London into hasty retirement, and sent the company, its investors, and its promoters into an interminable tangle of lawsuits. However, some always believed that the Emma had mined only a piece of a larger ore body and that the rest of the bonanza still lay undiscovered under the mountain.[73]

Rice hired respectability for Emma Copper in the person of William Ridgely, a former comptroller of the currency in the Theodore Roosevelt administration. Rice sent his brother Charles Herzig to oversee work at the mines. While his brother was laboring in Utah, Rice's New York *Mining Age* printed ecstatic accounts of new discoveries in the Emma mine.[74]

The New York curb market was exactly that. Trades were made in the open air on Broad Street, in near anarchy. The curb was a haven to those not welcome in the organized exchanges, such as Jewish and Irish brokers, and, in 1908, the first African American stockbroker. The only real rule was not to cheat fellow brokers, and the only punishment was that other brokers could refuse to deal with the transgressor. Cheating customers, however, was an ancient and approved practice, as brokers reported fictitiously high prices to the buyer and fictitiously low prices to the seller and split the difference between themselves. Curb brokers were known for "pushing, shoving, luck, double-dealing, muscle, cheating, cleverness, or a combination of these qualities."[75]

The curb brokers knew that Rice specialized in worthless companies, so many of them pocketed their clients' money and "bucketed" orders for Emma shares. They considered Emma Copper highly overpriced at $0.80, and planned to buy the shares later for a few pennies each.

Unfortunately for the bucketers, Rice's publicity sent Emma Copper shares above $3, and their clients began to demand that the brokers hand over the stock certificates. Unable to admit that they had—illegally—not executed the orders, and unable to afford Emma Copper shares at the higher prices, the brokers organized "bear raids" on the stock, trying to panic share prices down by staging phony wash sales among themselves at lower prices.

Rice was too adept at manipulation to allow the bears to panic his stock. While the low-price wash sales were going on a short distance down the street, Rice coolly offered to buy any shares genuinely offered at the high current price, and it became clear that the low-price sales on the curb were phony. Some brokers approached Rice offering to split the difference and return the share prices to their former level so that the bucketers would not face jail or bankruptcy. Rice refused.[76]

Rice went to the district attorney's office, not—as he was accustomed—to answer charges, but to make them against other curb brokers. Rice claimed that the brokers were short 1.5 million shares, even though the total number of Emma Copper shares was only 1 million. Rice claimed that he, Ridgely, and a few other company insiders controlled all but 181,000 shares.[77]

On October 6, 1916, the bucketers were frantic to buy shares, and tempers rose. One of Rice's brokers got into a fistfight over some Emma shares for sale. The brokers were again swinging and dodging fists the following day on the Broad Street sidewalk, and police stepped in.[78]

Rice incorporated Emma Consolidated as a holding company to operate the Old Emma and the Emma Copper companies. Brother Charles Herzig was president, Rice's former son-in-law Rudolph Gnekow was vice president, and William Ridgley was treasurer. The new company was listed on the Salt Lake Stock & Mining Exchange to great hoopla amid the *Mining Age*'s predictions of fabulous profits. Within a couple of months, however, the Salt Lake stock exchange suspended trading in Emma shares because of company misrepresentations. Charles Herzig quit his position as the president of Emma Consolidated and stopped working for his brother.[79]

Rice's *Mining Age* lies never materialized, and the company treasury could not stand the expense of Rice's manipulations. The mine shipped a few thousand tons of ore in 1917, but the Emma Consolidated Mining Company went into receivership the following year. Rice had used the famous Emma mine to milk another generation of suckers, then abandoned the financial wreckage to plan his next swindle.[80]

The Rice Oil Company

During World War I, millions of patriotic Americans had made their first investments by buying Liberty Bonds. Many promoters had volunteered their staff for the Liberty Loan campaigns, and after the war they used their lists of Liberty Bond buyers to advise these unsophisticated investors to gain higher returns by using Liberty Bonds as collateral for stock purchases. Rice was never less clever than his fellows and organized new companies to entice Liberty Bonds holders.[81]

Rice branched out into oil stock. When he was an obscure criminal, he had hidden behind front men, but now that he was a famous criminal, he named the company after himself, because his notoriety was magic with investors who believed Rice's tale that he was persecuted for being a friend to the small investor. Rice named his first petroleum venture the Rice Oil Company. Rice retained his Emma guinea pig, the former federal comptroller William Ridgely, to lend innocence by association.

The company never produced oil. Despite a handsome capitalization of $5 million, Rice Oil drilled only two holes, for which it did not even pay the driller. In 1918 Rice was arrested and charged once more with using the mail to defraud. He was released on $7,500 bail.[82]

By the time the authorities cracked down on Rice Oil, the company had ceased to exist, having been absorbed by Rice's next creation, Appalachian Oil. Rice never forgot the value of journalism, and touted the new oil stock as well as other pet promotions, such as Lampazos Silver Mines, in *George Graham Rice's Industrial and Mining Age*. Apparently Rice literally didn't consider the Rice Oil shares worth the paper they were printed on, for he never paid the engraver.[83]

Rice's finances crashed in late 1918, no doubt partly due to legal pressures. He was forced into bankruptcy when he could not repay a small loan of $2,970.[84]

More Trouble with the Law—and with His Wife

Rice was brought to trial in the Rice Oil case in January 1920. He arrived at the courthouse wearing a $10,000 fur coat. At the same time, Rice was also being sued for divorce by his third wife. In his fraud trial he charged that he had been framed and that the government was pressing the case against him because various district attorneys and a former New York governor had been bribed to leave the real stock crooks alone. It appeared to be more of Rice's grandstanding, but Rice knew what he was doing. The former governor and district attorneys all took Rice's bait and testified to refute the bribery allegations. The jury convicted him of the charges in connection with Rice Oil, and in January 1920 the court sentenced him to three years in prison. Rice appealed, however, that the testimony of the former governor and district attorneys had prejudiced the jury against him in the fraud matter, and he eventually succeeded in having the conviction overturned.[85]

Out on bail Rice plunged right back into mining stocks and underwrote the offering of shares in the Broken Hills Mining Company, which owned a silver prospect in Nevada. He mailed prospective suckers his usual wildly optimistic prospectus, along with his personal endorsement. For those not swayed by Rice's fame, the prospectus included a long and glowing endorsement by both the current governor and the state treasurer of Nevada. Rice followed up with display ads in San Francisco newspapers.[86]

When he sold shares in California, however, Rice neglected to obtain a broker's certificate from the California Corporation Commission, in violation of that state's blue sky law. After an engineer sent by the Corporation Commission to investigate the property returned with an unfavorable report, the Corporation Commission denied a permit to sell shares to the public in California. The San Francisco Mining Stock Exchange, having no such ethical standards, listed the stock and allowed it to be traded. Share priced zoomed up in wash sales and created a sensation on the exchange. When the California Corporation Commission tried to stop sales, it discovered that its jurisdiction did not extend to sales between brokers, and it was unable to stop sales on the exchange.[87]

The adverse publicity destroyed the demand for Broken Hills, and the stock price plummeted from $0.50 to $0.04. Some suckers swore out a warrant for Rice's arrest for fraud. When the matter came to trial, however, the buyers chose not to press charges, and the court dismissed the complaint. New management took over Broken Hills, reincorporated as the Reorganized Broken Hills Silver Corporation, and, without Rice, began the business of honest exploration. They didn't find much of value, however, and the site was given back to the desert.[88]

It was at about this time that the future film director Frank Capra—who was himself in the business of selling worthless mining shares—approached George Graham Rice in the lobby of a Reno hotel. Capra walked up to Rice and began a hard sell of his latest mining swindle. "My son," Rice told him, "stocks were made to sell, not buy."[89]

Bingham-Galena Mining Company

"Mr. Rice asks us to say that he believes that an investment made in Bingham-Galena stock at this time will, in all probability, make up any loss which you may have incurred in Broken Hills stock."

—Letter to investors from Child, Barclay & Company, stockbrokers

In 1921 the organizers of the Bingham-Galena Mining Company of Utah chose George Graham Rice to promote the initial stock offering. Rice was still appealing his latest criminal conviction and was newly arrived in Salt Lake City from his Broken Hills fiasco in California.

Despite protests that Rice's participation would give Utah mining a black eye, the Utah Securities Commission decided that it had no grounds to interfere. Rice joined with the brokerage of Child, Barclay & Company, which sent letters to Rice's list of Broken Hills victims in California telling them that the famous George Graham Rice was now recommending Bingham-Galena stock. The California Corporation Commission charged Rice and his brokers with violating California's blue sky law by selling securities without a permit.[90]

California courts issued warrants for Rice's arrest and tried to extradite him from Utah. The Utah Corporation Commission took Rice's side, arguing that California was trying to control business activity in Utah. Rice traveled to California, surrendered himself, and posted bond. The court eventually dismissed the charges for lack of evidence.[91]

After Bingham-Galena shares fell from $0.38 to $0.06, three directors resigned, and John Barclay, Rice's former partner in the promotion, claimed that Rice had forced him out in order to sell himself company treasury shares at below-market prices. The company was unable to raise money by selling more shares, because its permit from the Utah Corporation Commission allowed no sales for less than $0.15 per share. The company discontinued its association with Rice and eventually recovered $21,000 due from Rice on a promissory note. Bingham-Galena changed its name to the Park-Bingham Mining Company to distance itself from the notoriety its brief flirtation with George Graham Rice had won it.[92]

Idaho Gold Corporation

Rice did not let his failures slow him down. Some Utah men owned the Vishnu mine in Idaho and thought that Rice's money-raising skills would be useful. In December 1921 Rice—still out on bond pending appeal—was in Rocky Bar, Idaho, inspecting his next mining venture.[93]

The owners sold a controlling interest to the newly incorporated Idaho Gold Corporation for forty percent of its shares. Rice's Utah-Idaho Finance Company promised to sell the remaining sixty percent of the shares under a formula that would allow it to pocket most of the proceeds. The California State Corporation Department investigated Idaho Gold and denied its request to be listed on the San Francisco Stock Exchange.[94]

Rice looked for ways to draw public confidence. Two mining engineers reported very high values in the Vishnu Mine vein. In November 1923 Rice convinced the Salt Lake City mining magnate Samuel Newhouse to become the company's president.[95]

Newhouse had been the president for only a few weeks when Idaho State Mine Inspector Stewart Campbell warned the public that the Idaho Gold publicity was misleading and that the company had not filed the information required by Idaho's blue sky law. Campbell had inspected the properties in the company of an assayer hired by an unnamed private client. The assayer's samples contained less than a tenth of the values reported by the previous engineers and in the company prospectus. Campbell's verdict: "The vein does not have the appearance of developing a commercial tonnage."[96]

The story grew stranger when it was revealed that the assayer's confidential client was none other than George Graham Rice. Rice, whose wife owned sixty percent of the stock in Idaho Gold, demanded Newhouse's resignation. Newhouse resigned on January 3, 1924, and George Graham Rice was elected president. Idaho Gold was discredited, however, and Rice abandoned the promotion.[97]

Rice controlled a block of stock in Fortuna Consolidated Mines and arranged to pay John Hogan, the editor of *Facts and Fakes on Wall Street*, a commission on the stock's sales. Hogan obligingly printed glowing articles on Fortuna Consolidated Mining Company, articles written by Rice himself. A federal grand jury indicted Rice and Hogan for mail fraud in January 1926.[98]

Rice charged that the indictment was all a plot by his enemies, which he said included the New York Stock Exchange, the Better Business Bureau, postal investigators, and federal prosecutors. The Fortuna indictment languished for years while the authorities were busy prosecuting Rice for various other swindles.[99]

Emeralds under the Jungle

The ancient Chivor emerald mine in Colombia had been lost for more than a century. Indians had discovered and mined the gems until Spaniards took over in 1538. The mine produced fabulously for years, but exhausted its shallow veins and fell idle. Jungle covered the diggings, concealing its location. A Colombian obsessed with the lost mine discovered a clue in an ancient manuscript in Ecuador and rediscovered the mine in 1904.[100]

But mining in the remote mountainous jungle was difficult, and those few emeralds the miners turned over to the company were of very low quality. Frederick Lewisohn bought the bankrupt mine for $7,800 and organized the Colombia Emerald Development Corporation. In 1926 he turned to George Graham Rice to raise money. Rice bought an option on five hundred thousand shares of Columbia Emerald at $0.75 per share. Rice basked in his association with the Lewisohn name (Frederick Lewisohn was the nephew of the mining magnate Adolph Lewisohn) and arranged to have the company listed on the Boston curb market.

Rice's latest newspaper, the *Wall Street Iconoclast,* extravagantly praised the new company and predicted that the price would rise to $75 per share. The mine was trumpeted as the ancient Aztecs and Incas' font of incredible wealth and was said to be able to corner the world's emerald market and clear $5 million in annual profits. Rice's lies in the *Iconoclast* and his wash sales drove the price up to $17 per share, and Rice sold his shares at huge profits.

The New York attorney general gained an injunction against Rice in January 1927. The state supreme court found that he had engaged in misrepresentation, but the only punishment Rice received was a court order that in the future he had to disclose to prospective stock purchasers the true financial status of Colombia Emerald.[101]

Rice continued to try to buy respectability by buying respectable people. He purchased a pedigreed legal team that included former New York governor Charles Whitman and former federal district attorney Henry Wise. Wise now defended Rice, the very man whom he had sent to prison in 1912 in the Ely Central case. When Rice, in his *Iconoclast,* trumpeted Wise's record of fighting stock fraud (but without mentioning that he had prosecuted Rice himself), he implied that the presence of Wise on his legal staff was an endorsement and proof of Rice's honesty.[102]

A grand jury indicted Rice in September 1927 for grand larceny in selling Colombia Emerald shares, but he escaped prosecution. The Colombia Emerald Development Corporation survived its association with Rice, and mined gems—some of excellent quality—sporadically from the Chivor mine until the company went into receivership in 1951.[103]

Idaho Copper

The Idaho Copper Corporation was owned by a group of Boise businessmen who were unable to raise capital to develop the property. Rice agreed to raise capital in exchange for the right to name four of the seven directors. Using his brother-in-law Frank Silva as a front, Rice took options to buy 1,411,500 shares.

The revived Idaho Copper Corporation owned the South Peacock, a copper prospect in the rugged Hell's Canyon area near the Oregon border. The property had been worked sporadically over the past twenty-four years. To salvage something of the Idaho Gold fiasco, Rice had Idaho Copper buy the Idaho Gold Company. The Boston Curb Exchange listed Idaho Copper at the end of 1924.

Rice resurrected a ridiculously exaggerated twenty-year-old report by the disreputable mining engineer W. Bertram Hancock. Hancock had only recently been convicted of fraud for his role in another mining swindle, but Rice saw no reason not to use his old report on the South Peacock. In 1925 Rice's *Wall Street Iconoclast* newsletter ran advertisements for the company and published glowing excerpts from Hancock's report, and accounts of supposed new discoveries. At the same time, the *Iconoclast* staged wash sales to establish a higher price.[104]

A new law in Idaho required mining promoters to file copies of reports and prospectuses with the inspector of mines. As he had with Idaho Gold, State Mine Inspector Stewart Campbell investigated Idaho Copper and found it far less than claimed in its prospectus. On March 24, 1925, Campbell telegraphed the Boston Curb Exchange and asked that trading in Idaho Copper be halted because the firm's publicity was "grossly misleading and greatly divergent from the facts."[105]

The Boise businessmen who were the original owners of Idaho Copper were appalled by Rice's methods. He was diverting funds from needed development work and plunging the company into debt. All three company directors not appointed by Rice resigned and denounced Rice. The Boston curb market responded by suspending trade in Idaho Copper shares while it investigated the company.[106]

Rice was determined that Campbell not be allowed to stop Idaho Copper as he had done the previous year to Idaho Gold. In May 1925,

Idaho Copper lawyers hit Campbell with a $100,000 libel suit. By the time the trial started a year later, Idaho Copper had raised the damage claim to $500,000. However, a Boise jury acquitted State Mine Inspector Campbell, and Rice lost all his appeals for a new trial.[107]

Rice managed to have company shares put back on the Boston Curb Exchange in July 1925, where prices continued their irrational exuberance. Back in Idaho the property did not show similar activity. Stewart Campbell was unintimidated by the lawsuit pending against him and re-visited the South Peacock in September 1925. He announced that despite Idaho Copper's market value of more than $10 million, the entrances to the mine were still all caved in and the site lacked the necessary surface facilities and mining equipment.[108]

Campbell's statements did not sober the market in Idaho Copper Corporation shares, whose price rose from $0.40 to $6.25 on the strength of wash sales on the Boston curb market and wildly optimistic pronouncements in the *Iconoclast*.[109]

Rice looked for a name to lend respectability to his machinations. The geologist Walter H. Weed edited the *Mines Handbook,* an authoritative listing of mining companies, in which he had previously denounced Rice and his Idaho Gold Corporation. Weed was such a nationally known authority that the swindler Joseph "Yellow Kid" Weil sometimes impersonated him to draw suckers into mining swindles. Rice offered Weed $100,000 per year to work for him. Despite his low opinion of Rice and Rice's companies, Weed accepted in September 1925 and became vice president of Idaho Copper Corporation and also of the Idaho Gold Corporation, which he had previously denounced.[110]

Weed arranged to buy the Irondyke mine, an inactive former copper producer in eastern Oregon. A few months later Idaho Copper Corporation merged with the similarly named Idaho Copper Company Ltd., which owned the Red Ledge prospect in Idaho. On June 1, 1926, the Idaho Copper *Company* was listed on the Boston curb market, which had earlier ejected the Idaho Copper *Corporation*. Rice ran up the share price through wash sales and sold 1,102,051 of his shares for $2,117,518, a profit of more than $1.7 million.[111]

New York authorities obtained a restraining order to stop some of Rice's most egregious practices in Idaho Copper, and in March 1927 a grand jury indicted Rice for illegal dealings in the stock. An appellate judge dismissed the indictment four days later, but the New York Supreme Court reinstated it, declaring, "Clearly Rice was engaged in a fraud." A federal grand jury also indicted Rice for mail fraud, but Rice successfully fought off state and federal actions and continued his manipulations.[112]

The Idaho Copper Company ran out of money in November 1927 and laid off its workers; the court appointed a receiver. Rice still had plenty of Idaho Copper stock to unload and feared that the receiver would try to stop him from cashing in. Rice's brother-in-law Frank Silva claimed that Idaho Copper owed him $915,000 and convinced a New Jersey court to allow share sales to continue. However, the receiver managed to freeze share sales and eventually recovered large blocks of shares from Rice, although Rice denied knowledge of another 219,297 shares that company records showed to be in his possession.[113]

The Idaho Copper Trial

When Idaho Copper went into receivership and its shares became worthless, stockholders clamored for a criminal investigation. A federal grand jury in New York indicted Rice for mail fraud once more on January 17, 1928, along with Walter Weed and the Boise, Idaho, promoter Walter Yorkston. Brother-in-law Frank Silva was indicted the following month.[114]

As he had before, Rice tried to offset his bad reputation by ostentatiously assembling the most respectable legal talent that money could buy. In addition to the former New York governor and the former federal district attorney, Rice hired a former senator, a former justice of the New York Supreme Court, a former New York appellate judge, and a former federal assistant district attorney.[115]

A procession of mining engineers contradicted the glorious lies published in the *Iconoclast*. They testified that the South Peacock property had no known value and that the Irondyke mine could produce copper only at a loss. Witnesses documented the lies about Idaho Copper printed in the *Iconoclast*. Records of a Boston broker showed that Rice had staged wash sales to boost Idaho Copper share prices.[116]

As usual Rice claimed that he was not only innocent of fraud but that the trial was a plot by Wall Street interests. He accused the prosecutor of being in the pay of bucket shop operators.[117]

For the defense summation Rice shrewdly passed over his high-powered New York legal team and chose the attorney James Nichols of Portland, Oregon. Nichols delivered a passionate defense, describing how the people in small towns near the Idaho Copper properties desperately needed Rice to develop the mineral wealth of the region. He described their privations because of lack of development and told how a guilty verdict would plunge the local people into poverty again. Tears filled the eyes of some jurors.

When the jurors went into deliberation, however, they quickly dried their tears and found George Graham Rice and Walter Yorkston guilty, and Walter Weed innocent. The judge sentenced Rice to four years in the federal penitentiary in Atlanta. Yorkston drew a nine-month sentence. The judge admonished Weed to keep out of trouble, saying that the jury had obviously spared him because of his age.[118]

George Graham Rice and Al Capone

While Rice was on bail pending appeal of his Idaho Copper conviction, the government investigated him for tax fraud, a new strategy being tested against a number of criminals, including the gangster Al Capone. In March 1929 Rice was indicted for not paying income taxes on profits from the estimated $17 million he had collected from sales of Idaho Copper Company stock. While the income tax case moved forward, Rice lost his last Idaho Copper appeal and left for the federal penitentiary in Atlanta. He continued to control stock-selling operations from his jail cell. Although he left a string of unpaid debts, he always had enough money to buy big-name legal talent, and his Atlanta attorney was an ex-governor of Georgia.[119]

Rice's first wife, the former Theramutis Ivey, took advantage of his fixed address to serve him another suit for alimony, thirty-five years after the happy event. Mrs. Herzig had last sued Rice for alimony in 1903. She evidently believed that after twenty-seven years this new lawsuit would be more successful.[120]

Rice was brought back New York in 1931 to stand trial for income tax fraud. He declared himself penniless and unable to afford a lawyer. The key prosecution witness was Rice's brother-in-law and recent co-defendant, Frank Silva. As his own attorney, Rice cross-examined Silva. When angry words flew between them on cross-examination, a court attendant rushed in to prevent a fistfight.[121]

Rice was able to show that the tremendous sum of money was received not by him personally but by various corporations including Idaho Copper Corporation, Idaho Gold Corporation, Wall Street Iconoclast Inc., and the George Graham Rice Corporation. The jury acquitted Rice, who returned to the Atlanta federal penitentiary. The government was more successful in its case against Al Capone, who soon joined the inmates at Atlanta. Rice said that he befriended Capone in prison and listened to the gangster complain of the injustice of serving prison time while bankers went unprosecuted. "That Al Capone is a fine fellow," Rice said, adding, "I'd take Al's word quicker than anybody's on the stock exchange."[122]

The Cabaret Owner

Rice left prison in 1933 and returned immediately to mine promotions, touting the Halifax Mining Company in California, and the Buckskin National Company in Nevada. By January 1934 he was publishing *Rice's Financial Watchtower*, a newsletter that promoted his companies, including the International Silver and Gold Corporation and the Texan Oil and Land Company. He called himself "the only honest financial writer in town." Authorities issued a warrant for his arrest in November 1936 because dealing in securities violated his parole. The government froze his assets.[123]

Rice didn't relish the thought of five years in prison, and disappeared. For all his legal problems, Rice had never before fled. He quit mine promotion not because he had run out of gullible investors after thirty years without a single profitable mine to his credit—he still had plenty of suckers ready to give him money—but because lying and fraud in the mining business had become legally unfashionable. Rice was never brought to justice. In 1939 George Graham Rice was reported running his own nightclub in Milwaukee, "working hard and happy as a lark."[124]

Death Valley Scotty: The Showman as Mine Swindler

"Here was an indigent desert rat who gave his name to one of the most famous private residences in America;...a self-confessed con man who became the best-known prospector in the history of the world—and he did it all without ever discovering a damned thing!"

—Hank Johnston, *Death Valley Scotty: "The Fastest Con in the West"*

In April 1902 a young couple sat in their New York City room wondering how they would support themselves. Walter Scott had just quit his job as a trick rider for Buffalo Bill's Wild West Show in an argument with Buffalo Bill Cody himself. It was the start of the season, and Scott had been expected to take a prominent part in the parade down Fifth Avenue to publicize the opening at Madison Square Garden. Scott missed the parade—according to his wife, he "got sidetracked uptown"—and as Cody led the parade he spotted his missing employee "a bit worse for wear" cheering wildly from the sidewalk. Knowing Walter Scott's love for liquor, his wife's description suggests that he had been drunk.

Cody angrily docked Scott a week's pay, and Scott equally angrily quit the show. Cody relented and rescinded the fine, but Scott was too stubborn to return to his job. That left him with the question of how to make a living. He was twenty-nine years old and had spent his entire adult life in Buffalo Bill's show.

Scott had met his wife just two years before, when the exuberant cowboy walked into a candy store on Broadway, tossed his hat over the counter, and asked a young lady behind the counter, the twenty-year-old widow Ella Millius, to be his guest while he performed at a matinee show. Ella afterward corresponded daily with Walter Scott as he traveled with Buffalo Bill. She visited him in Boston that summer, and when the season ended she traveled to Cincinnati, where they married.

Walter Scott had planned to spend the next seven months traveling with the show, as he had every year for the past ten. He and his wife had just returned from Colorado, where he had spent the off-season as a gold miner at Cripple Creek. Walter asked his wife to hand him two souvenir pieces of gold ore, rich specimens given to her by a mine superintendent, that she had brought from Cripple Creek to show her friends at the candy shop. Walter took the rocks and left the hotel room without explanation.

The Invention of "Death Valley Scotty"

Walter Scott headed for the Knickerbocker Trust Company to speak with the assistant secretary-treasurer, Julian Gerard. Scott had met Gerard five years earlier at a party where socialites had mingled with Buffalo Bill's performers.

Walter Scott showed the Colorado gold ore to Gerard and spun a tale of his great mineral discovery in Death Valley, California, for which he needed financial backing. Gerard had the samples assayed and found that Scott's secret vein was worth a fabulous $2,000 per ton. Gerard and Scott signed a "grubstake" agreement, by which Scott gave Gerard a half interest in his discovery in exchange for monetary support for six months.[1]

Scott described the location of his discovery in vague terms. He had worked at a borax plant in Death Valley as a teenager and remembered enough to sound plausible to Gerard. Gerard gave Scott money for a summer of prospecting, and Scott took the next train to Death Valley. Ella Scott moved to Los Angeles in June 1902, where Scott settled her in an apartment then left. She was to seldom see her husband, as he cared little for married life.

Scott returned to the desert and created the fictional character he was to play for the next half-century. The mountainous desert region around Death Valley was still unburdened by roads or buildings, but it was already richly populated with legends of lost mines, such as the Lost Breyfogle and the Lost Gunsight. Some legends had a basis in fact, but it hardly mattered. They were all wonderful stories, and Walter Scott added another: Death Valley Scotty's secret gold mine.

In that summer of 1902, Scott's visits for supplies and recreation made him known in the desert towns of Barstow and Daggett, California. The locals were unimpressed with Scott as a prospector, for he never bought mining tools or explosives. He never showed rock specimens or gold dust, although he sometimes ostentatiously carried fifty pounds of rocks hidden inside a carpet bag. What impressed the locals was Scott's fat roll of paper money, which he spent carelessly. When he first appeared in town, he awed the stool-warmers in a Barstow saloon by displaying $1,500 and bragging that he could get much more.

Walter Scott sent a steady stream of mail to Gerard, his New York sugar daddy. He filled his letters with stories of incredible hardships: rattlesnakes, burros dying of starvation and poison water, the cruel heat of the desert, days without water, and long treks on foot. The letters described the richness of ore on his claim and ended with appeals for more cash.

Scott never shipped his promised ore samples to Gerard. He straggled into the county courthouse at Independence, California, on September 24, 1902, to file a single mining claim, named the Knickerbocker (after Gerard's company), but so vaguely described as to be impossible for anyone but Scott to locate.

In the fall of 1902 the six-month grubstake contract expired, as did Gerard's patience with Scott's oversupply of desert adventure tales, lack of tangible results, and demands for more money. Gerard took up Scott's invitation to visit the desert claim himself and met Scott in Needles, California. Scott procrastinated, took overly long to round up some burros, then told Gerard that the trip to the mine would be too hard for the New Yorker, and refused to lead him to the Knickerbocker claim. Nothing Gerard could say could budge Scott, and Gerard returned to New York.

Stringing the Sucker Along

Although Gerard was mostly convinced that he was the victim of a fraud, Scott was so persuasive that Gerard was soon sending more money. In return he received more of Scott's colorful desert letters and title to the elusive Knickerbocker claim. By the end of 1903 he had lost $5,000 in grubstake money to Scott. Between supply and boozing trips to desert towns and occasional visits to his wife, Scott explored the Death Valley area. He made his home in a natural shelter formed by a rock overhang in a secluded canyon.

In August 1903 Gerard sent the mining engineer E. M. Wilkinson to inspect Scott's mine. As he had with Gerard the previous year, Scott told Wilkinson that the trip would be too arduous. Wilkinson wrote back to Gerard that the visit never took place because Scott refused to take along even tents.

Gerard and Scott exchanged long-distance recriminations. Scott claimed that Wilkinson had insisted on traveling in high style with an entourage of servants and that the mining engineer had a bad reputation. Scott claimed to have piled up fourteen tons of rich ore, enough to pay Gerard back all the money. Once again Scott's bluff worked and Gerard signed another grubstake contract.

Showmanship Worthy of Buffalo Bill

In June 1904 Scott was in Riverside, California, showing a sample of gold amalgam and carrying a heavy satchel that he said was full of the same. He took a train east, supposedly to show his backer the long-promised results, but the gold amalgam never reached New York. In Philadelphia, Scott reported that the $12,000 in gold amalgam had been stolen from his Pullman berth somewhere between Pittsburgh and Harrisburg. The sensational report was carried by newspapers nationwide. Scott alone seemed unperturbed by the theft. "There's plenty more where that came from," he told reporters, adding that he had been headed to Jersey City to get married.[2]

Scott had instantly transformed himself from a local barroom blowhard to a nationally known man of mystery. He was good copy for reporters, and they printed exaggerated stories of his spending. Scott adopted a uniform of blue flannel shirt, bright red tie, Ulster coat, and black Stetson hat. In Los Angeles, conscious of the scrutiny of reporters, Scott took the best hotel suites, tipped lavishly, and handed out $100 bills for small purchases, refusing the change. Newspapers called him "Death Valley Scotty."

Scott linked his own legend to the most compelling legend of the region: he said that he had found the Lost Breyfogle mine, for which prospectors had searched for many years. Scott found that he could drink for free at the expense of clever people hoping to learn the secret location.[3]

His mysterious wealth also drew suspicions. Scott was arrested as a suspect in a 1904 train robbery, but after a few nights in the San Bernardino, California, jail, he was released for lack of evidence. The swindler George Graham Rice feared that the search for Scotty's mine would draw too much attention away from Rice's own mine promotions at Rawhide, Nevada. To discredit Scott, Rice sent out a phony story that Scott's bankroll came from a Wells Fargo pack-train robbery several years earlier. The two bandits had escaped with the loot into the Death Valley country, and although two bodies assumed to be the bandits were later found in the desert, the money was never recovered. Newspapers across the country picked up Rice's lie that Scott had stumbled on the Wells Fargo loot.[4]

Gerard hired mining engineers three more times to examine the Knickerbocker mining claim, which Gerard owned, but whose location was known only to Scott. Each time, Scott's letters assured Gerard that he would lead the engineer to the claim, but each time he invented an excuse. Once he claimed that his donkeys had been run off by Indians; on another occasion he said he was recovering from the bite of a rabid skunk. By May 1905 Scott had taken $10,000, three years, and all of Gerard's pa-

Death Valley Scotty claimed to have a secret mine in the desert.
Courtesy Library of Congress.

tience. Scott tried to entice Gerard with more stories, but Gerard refused to answer Scott's letters and telegrams.

By the time Julian Gerard wised up, Walter Scott was already bilking another sucker. In Chicago in October 1904 Scott convinced the businessman Obadiah Sands to back him. Sands was enthusiastic but without money, so he introduced Scott to two others, the businessman E. A. Shedd, and Albert Johnson, the president of the National Life Insurance Company and a director of the Continental-Illinois Bank. Shedd and Johnson provided $2,500 for two-thirds of any discoveries Scott made in the next thirteen months. Sands was to receive one-sixth as the promoter, and Scott was left with only one-sixth. Scott and Sands left together for Death Valley in late November 1904. Stopping in Goldfield, Nevada, for supplies, Scott pocketed the $1,850 down payment and headed to Los Angeles for a spending spree. Deserted in Goldfield, Sands realized that he had been swindled, and he returned to Chicago. Sands and Shedd would have nothing more to do with Scott, but Johnson was still hooked.

The Coyote Special

Scott decided that the press was growing inattentive, so in Barstow, California, he hired a private train to speed him to Los Angeles. The train, one engine and a single coach with Scott as the only passenger, sped nonstop to Los Angeles. The train cost Scott $200 plus tips, but the stunt received only minimal publicity. Scott drank to his disappointment and went to bed in his Los Angeles hotel thinking his publicity grab a failure.

The next morning, however, a hungover Scott met E. Burdon Gaylord, who was waiting for him in the hotel lobby, with a plan to repeat the publicity stunt on a national scale. Gaylord, who may have wanted to draw attention to mining properties he owned in the Mojave Desert, offered to pay for a train to carry Scott nonstop to Chicago. Scott would get all the publicity, and Gaylord would pay the bills.

Following their plan, Scott announced his intention to hire a special train to set a speed record from Los Angeles to Chicago—then fifty-two hours and forty-nine minutes. The prospect of a train hurtling from Los Angeles to Chicago in just over two days fired the imagination of a nation in love with speeding locomotives, and intrigued the public with the eccentric extravagance of the trip. Scott squeezed a great deal of newsprint out of his announcement then went to the Santa Fe Railroad to arrange the trip. Santa Fe officials had read his plans in the papers, couldn't have been more delighted at their publicity windfall, and began issuing press releases of their own, promising to deliver Scott to Chicago in even less than two days. Scott put down $4,500 in cash, and the railroad stocked the dining car with the best food and liquor. After a week of joint publicity, Scott paid the balance of $1,500 by peeling $100 bills from a large bankroll while reporters watched.[5]

The Coyote Special left Los Angeles at 1 p.m. sharp on July 9, 1905. Scott and his wife were greeted by an enthusiastic crowd of more than a thousand in Chicago, forty-four hours and fifty-six minutes later, cutting nearly eight hours from the previous speed record. The publicity impressed the Chicagoan Albert Johnson, whom Scott had swindled the previous November. Johnson thought that the free-spending train trip proved that Scott must have a secret bonanza, and he met with Scott.[6]

Newspapers fed the frenzy by printing a lot of nonsense about Scott. *The New York Times* reported that Scott's mine was guarded by a crew armed with high-powered rifles and led by a sixteen-year-old boy. Scott's miners had supposedly made three ore shipments by mule train, for which Scott had received $170,000.[7]

After two days in the bridal suite of the Great Northern Hotel, Scott and his wife quarreled. He left her in Chicago and boarded a regular passenger train for New York. She followed the next day. Scott's business in New York was to woo back his recalcitrant sucker, Gerard. The plan partially worked, for Gerard reasoned that Scott could not have paid for the cross-country train unless he had a secret mine. This time, however, Gerard knew better than to give large sums of cash. He paid Scott $100 for a deed giving Gerard a one-half interest in all mines Scott was ever to find, past or future. The sum was small for such a sweeping deed, but Scott no doubt figured he could string Gerard along for more money, and Scott probably needed the cash badly.[8]

Unknown to Gerard, Scott was broadening his clientele among New Yorkers eager to take advantage of the unsophisticated prospector. Scott took a regular passenger train back to Los Angeles but stopped off in Cincinnati to offer $40,000 to a stockbroker for some mining shares. The broker turned down the offer, but he and his mining shares enjoyed the national spotlight that followed Scott.[9]

Tracking Down Scotty's Mine

Many prospectors tried to find Scott's mine, but he allowed none to follow him. His field glasses watched for pursuers, and his Mauser rifle with telescopic sight warned away those who came too close. He did nothing to discourage rumors that he had killed fortune seekers following him. Scott enjoyed the center of attention in saloons for days as he wavered on the edge of revealing the location—then he would leave town in the middle of the night. He would be tracked from one sparsely peopled desert outpost to another, but when he approached the Funeral Range east of Death Valley, he would disappear.[10]

One of his favorite ways to lose someone shadowing him was to avoid the scarce watering holes all afternoon, then water his mules at a spring just before dark. His pursuer, maintaining a respectful distance because of Scott's rifle marksmanship, had no choice but to water his thirsty mount while Scott disappeared into the darkness. Prospectors accustomed to noticing the slightest track of man in the desert speculated that Scott wrapped the hooves of his mules to offer no trace. A prospector who believed that he had an affinity with mules even tried to coax the secret out of Scott's lead mount, but Slim proved no more communicative than his owner.[11]

In August 1905 a prospector arrived in Goldfield, Nevada, and announced that he had discovered Scott's secret Camp Hideout in the Funeral Range, starting a rush by fortune hunters expecting that Scott's secret mine lay nearby. Those who searched the remote canyon found only a nearly inaccessible rock shelter filled with the accumulations of years of bachelor life, a bathtub, canned goods, after-dinner mints, and rock walls covered with pictures of show-business personalities, but no mine. The veteran prospector Shorty Harris told reporters that prospectors had already known the location of Scott's camp and had determined that there was no mine there.[12]

Another rumor inaccurately held that Scott's source was a secret mine near Bullfrog, Nevada, which Scott worked secretly because it was on someone else's property.[13]

Prospectors continued to search for Scott's secret mine, and every so often someone would claim to have found it. Scott led them on by leaving his diary behind at one of his regular camps. The diary detailed millions of dollars in gold ore supposedly mined, but gave no clue to the location. Scott filed twelve mining claims at the Inyo County courthouse, many in the names of friends and relatives, but described them only as "due east of Bennett's Wells." Prospectors could find no trace of the required boundary monuments.[14]

A more formidable threat to Walter Scott's myth-making came from Gerard's latest agent hired to smoke out Scott's mine: Antonio Apache. He was an Apache Indian who had been captured as a boy during the Indian wars and educated in New York and Boston. Newspapers wrote that he held a Harvard degree, but Harvard has no record of his attendance. Gerard threatened Scott with legal action until Scott agreed to reveal the location of the mine to Apache. Antonio Apache went to Scott's Los Angeles hotel in September 1905 to arrange the trip to the mine, but Scott refused. Scott told the newspapers that he didn't deal with Indians and that he was prepared to return all the money advanced by Gerard.[15]

Apache spent the next three weeks doing some urban tracking. He hired two detectives to follow Scott, while he himself interviewed all Scott's acquaintances he could find in Los Angeles and the desert towns. He discovered no one who had seen Scott with a load of ore but found three other New Yorkers who had come west to see Scott, two of whom admitted that they had lent money to the prospector for a share in his mine. Apache advised his client that Scott was a habitual liar and that there was almost certainly no mine.

After Apache returned to New York, Scott peppered Gerard with letters and telegrams trying to get him back on the hook. Gerard, who had already given Scott at least $10,000, refused to answer.

Imaginary Gunfights

One headline not planned by Scott was about an October 1905 auto accident that threw him from a car as he tried to set a speed record, on a bet. Scott landed in the hospital and lost the $350 wager, but he never paid.[16]

Headlines announced Walter Scott's death in December 1905, when his brother Bill Scott reported that Walter's mule had returned to camp without its owner but with a saddle punctured by a bullet hole and stained with blood. Newspapers reported on the search for Scott's body, but the searchers found him in perfect health twenty days later. Scott told a thrilling tale of how he and his faithful Indian companion Shoshone Jack—whom no one but Scott had seen—had been ambushed by outlaws. Shoshone Jack had suffered a flesh wound riding the mule, hence the blood and bullet hole, but Jack and Scotty had barricaded themselves in one of Scott's hideouts until the outlaws withdrew. Scott also showed off the remains of a human foot, which he said he found at Wingate Pass.[17]

Bill Scott was miffed enough after weeks of worry to tell reporters the version his brother had related in private: the blood on the saddle belonged to a mountain sheep that Scotty had shot; he was hoisting the sheep onto his mule when the rifle accidentally fired, putting a bullet hole in the saddle and spooking the animal. A prospector revealed that while the world had searched for Scott's body, the supposed corpse was waiting in camp, to make his reappearance more dramatic.[18]

Scott returned to the desert but within a week rode into Barstow with a bad bullet wound in his leg and another thrilling tale. As he and two imaginary companions, Sam Jackson and Shoshone Jack, rode through Wingate Pass, three desperados hired by Antonio Apache ambushed Scott to steal the directions to his mine, which Scott carried on his person. The desperados caught Scott in a crossfire and hit him in the leg, but Scott and friends returned a rapid fire. He refused to say if he killed any of the ambushers but hinted darkly that he had given as well as he got. His two friends evaporated back into his imagination, and alone he evaded assassins and rode his mule back into Barstow.[19]

This time no one could doubt that the blood was Scott's, and the bullet remained in his leg the rest of his life. After treatment in Barstow, Scott was put on the train to Los Angeles, where friends took him to his hotel to recuperate between dramatic recitals to reporters.[20]

Again a different version surfaced. The doctor who treated Scott's wound told reporters that powder burns on Scott's clothing showed that the shot had been fired point-blank. Two prospectors said that they had met Scott out on the desert, where Scott had accused them of following him. Scott was drawing his rifle from its saddle holster when it discharged, wounding himself in the leg. Although wounded, Scott was still armed with a rifle, so the two prospectors quickly left.

The "Battle of Wingate Pass": Wild West Show from Hell

Scott's tale of ambush on Wingate Pass turned into a rehearsal in his imagination, which he would shortly produce as a Wild West show with real actors and real bullets.

Walter Scott had a mythical mine, and A. Y. Pearl knew wealthy Easterners susceptible to tall tales, so they joined talents to take money from the credulous. Pearl got a group of Bostonians headed by E. W. Quint interested in Scott's mine. Pearl, Scott, and their latest suckers met in New York in January 1906 and arranged to sell part of the mystery mine for $60,000. The Quint syndicate insisted on sending the Boston mining expert Daniel Owen to see the mine before they would pay. Scott foresaw the biggest coup in his career, but first he had to fool Owen.[21]

Scott's expedition assembled at Daggett, California. Walter Scott brought his brothers Bill and Warner, although Bill Scott knew better than to trust his brother Walter and refused to go until one of the potential investors agreed to pay his wages. The Chicago insurance and banking executive Albert Johnson was there, still intrigued by the secret mine even after being swindled by Scott the previous year. Johnson, who shunned publicity, pretended that he was Dr. Jones, a physician.

Scott disappeared from the group in Daggett. When he reappeared after a few hours, he claimed that he had been kidnapped and had escaped but would not discuss it further. He hired the Daggett sheriff to be his temporary bodyguard.

Bill Key was a prospector who had discovered the Desert Hound mine, which he worked on a small scale. At Daggett, he agreed to show the Desert Hound as Scott's secret mine and split the proceeds of the sale. Scott had no sooner made the agreement than he decided it wouldn't work: he had promised Owen a bigger mineral deposit than the Desert Hound. He had to stop the expedition, but it was too late for his usual excuses.

On February 23, 1906, the nine-man expedition rode out of Barstow toward Death Valley. The group took the little-used old forty-mule-team wagon road. Scott kept Owen well supplied with whiskey, which the sociable Owen did not refuse. The rest of the party joined in, except teetotaler Johnson, and the group remained drunk for the next several days as they moved toward Death Valley.

At Willow Spring, Scott told his brother Bill to stay at camp and ordered Bill Key and Bob Belt to scout ahead for danger. Key and Belt were not seen again on the trip. The rest of the group, now six, moved over Wingate Pass and started the long grade down into Death Valley. That afternoon the group wanted to make camp at a dry lake, but Scott insisted that they continue.

The group heard rifle shots ahead, and a rider soon hurried by from that direction, warning them that he had just been shot at from ambush. He was unhurt, but his pack animals had scattered, and one of the bullets had put a hole in his hat. Scott drew his rifle and exclaimed that he could "whip any outlaws singlehanded." They soon reached the ambush point, where a small escarpment overlooked the wagon road. Key and Belt had established themselves behind low rock barriers and had been passing the time and a jug of liquor. Scott, riding ahead, drew his rifle and fired two shots in the air. The horses drawing the wagon behind him bucked, tumbling the tipsy mining engineer Owen into the back of the wagon. A shot came from the top of the escarpment, and Warner Scott, riding with Owen, cried in pain.22

Warner's cries quickly brought Scotty and A. Y. Pearl riding up. Scotty rode toward the assailants on the bluff. "Stop the shooting!" he shouted, "Stop the shooting. You've hit Warner." No more shots were fired by either side.23

The second wagon caught up, and the group camped there for the night, in clear rifle shot of the ambush point. The rifle bullet from the bluff had hit Warner Scott in the groin, and he was in agony and bleeding heavily. Johnson's doctor's satchel came out, and Owen and Pearl put thirty-seven stitches into Warner Scott and saved his life. At dawn the party started back. They reached Dagget on March 1 and put Warner Scott on the train to Los Angeles.

Albert Johnson wanted no publicity from a shooting and quickly left town. Scott uncharacteristically left town without talking to the press, but not before posting two friends at the door to Warner Scott's room, to prevent anyone from interviewing the invalid.24

A. Y. Pearl filled reporters' notebooks with a thrilling account of the ambush and his own bravery under fire. He counted at least four assassins shooting from both sides of the wash, including a bushwhacker who popped up out of the brush right next to him and fired. Pearl told reporters that the bullet fired point-blank came within an eighth of an inch of killing him but merely put a bullet hole in his hat, which he proudly showed to reporters. "My friend Owen is not a fighting man," Pearl said, "and he was not in it. Scott's Chicago friend, the banker[,] was game, but he was not used to such episodes, and the fighting was up to Walter and I. Well, I suppose we fired back and forth ten, or maybe twenty minutes—it seemed a century, but time goes slowly in such moments." He claimed that he and Scott killed or wounded one or two of the attackers before the desperados retreated into the dark. The papers dubbed it the "Battle of Wingate Pass," and it seemed destined to rank in western lore with the gunfight at the OK Corral. Some Barstow, California, merchants who outfitted prospectors spoke of hiring a posse to drive desperados from the area.[25]

Daniel Owen told the sheriff of San Bernardino County that Scott tried to have him murdered but that he was saved when the horses bucked the wagon. He said that when hit, Warner had shouted, "That shot was meant for Owen." Owen also accused Pearl of putting the bullet hole in his hat himself, since he said that only three shots had been fired during the episode, two by Scotty, and the one that hit Warner Scott. Another excursionist, DeLyle St. Clair, filed a criminal complaint against Walter Scott, Bill Key, Bob Belt, and Shorty Smith, a prospector he identified as one of the gunmen.[26]

San Bernardino County sheriff John Ralphs was wary of Walter Scott's publicity stunts, especially one timed just as Scott was about to start an acting career. Ralphs asked the Los Angeles police to verify that Warner Scott had indeed been wounded. When the Los Angeles police were turned away by A. Y. Pearl, Ralphs sent his undersheriff to Los Angeles. The no-nonsense deputy shoved Pearl aside, strode to Warner's bed, and pulled back the sheets to view the wounded groin. The mute testimony of the bullet wound was all he received, however, as Warner Scott's lawyer was there to advise his client to remain silent.[27]

When Warner Scott broke his silence, it was to sue five people, including his brothers Bill and Walter, for $152,000. Warner claimed that the ambush had meant to shoot Daniel Owen to get a code book used for secret messages to his eastern clients.[28]

Sheriff Ralphs, responsible for law enforcement in sprawling San Bernardino County, California, an area larger than the combined states of New Hampshire and Vermont, knew that bringing in the culprits was no easy task. He obtained arrest warrants, and with a deputy and a Los Angeles reporter rode out to Death Valley to inspect the crime scene 120 miles from the county seat and to apprehend the suspects. Considering the vastness of the Death Valley country, the pursuit may seem odd; they knew, however, that the suspects would be seen by someone at one of the handful of desert mining camps. They crossed paths with the wanted men several times, but A. Y. Pearl had warned the fugitives, and the posse returned empty-handed after eleven days. Shorty Smith turned himself in but was released by a judge who found his alibi solid.[29]

The attorney Wallace Wideman, representing Walter Scott, offered the witness DeLyle St. Clair $100 cash and a railroad ticket to Mexico, with a larger sum to be waiting for him there, if he would remain south of the border and out of reach of subpoenas. St. Clair invited the lawyer to a meeting to discuss terms and also invited the police and newspaper reporters, who listened to Wideman's proposal through thin hotel walls. Wideman was promptly arrested, and not for the first time, for attempted bribery. He protested that it was all a misunderstanding.

The Con Man as Actor

While the battle of Wingate Pass was being refought in the press, Walter Scott was out of town starring in the play *Scotty, King of the Desert Mine*. Buffalo Bill, the consummate huckster, had started in show business a generation earlier in a similar manner, playing himself in the wildly successful *Scouts of the Plains. Scotty, King of the Desert Mine,* opened in Seattle, then moved south.[30]

When the play arrived in San Francisco, the audience watched the real-life drama of Scott being arrested on stage. The playwright and a friend of Scott put up the $500 bond so that the play could continue. Scott was rearrested while the play was in Los Angeles, and the judge increased bail to $1,000. The play planned to travel out of state to avoid having its star entangled in the lawsuit by brother Warner, but Walter Scott was rearrested and taken to the San Bernardino jail, and his backers once more posted bond.[31]

A deputy sheriff arrested Bill Key in the saloon at Ballarat, California (now a ghost town). When Key arrived in San Bernardino, A. Y. Pearl was there to try to stop Sheriff Ralphs from questioning the prisoner, so Ralphs arrested Pearl as well, for disturbing the peace. Bill Scott, who at the time of the shooting was miles away at the previous night's camp, was also arrested.[32]

The authorities knew that without a confession, the case was weak—and all the accused refused to admit guilt. In April 1906 the San Bernardino County district attorney dropped all charges after he learned that the ambush had occurred outside his jurisdiction, just over the line in neighboring Inyo County. Bill Scott, Bill Key, and A. Y. Pearl were released from the San Bernardino County jail, and because Inyo County authorities declined to pursue the matter, the case was over.[33]

Scott later said that the ambush was the work of the Butch Cassidy gang. Scott also bragged that he himself had moved the county boundary marker six miles to fool the surveyor. He admitted privately, however, that the ambush was staged to scare off their companions. Scott said that until his brother was shot, "I hadn't had so much fun since Buffalo Bill's Wild West Show." Many years later, Bill Key also admitted that he and Bob Belt had been the ambushers on the bluff, and revealed that Belt had fired the near-fatal shot after having too much to drink.[34]

Whatever the merits of the lawsuit by Warner Scott, he knew that none of the defendants had even a fraction of the $152,000 damages claimed. To settle with his brother, Walter Scott assumed the cost for his brother's medical treatment. It was a typically empty gesture: Scotty never paid the bill.

The play and Scott's legitimate acting career both folded the following month in Saint Louis, Missouri. Scott traveled on to Chicago to meet with his next sucker.

Albert Johnson: Sucker Number Two

Scott's next sucker was Albert Johnson, the insurance executive who had already been swindled once in November 1904, when Scott absconded with his money and left Obadiah Sands in Goldfield, Nevada. Scott now agreed to give half of his mine to Johnson in exchange for financial support, even though it was well known that he had already signed the other half over to Julian Gerard. To clear title, Johnson agreed to a three-way split between himself, Gerard, and Scott.[35]

It is odd that both of Scott's principal suckers, Gerard and Johnson, were successful in the most conservative of businesses: banking and insurance. Although Johnson was an astute businessman, he had a blind spot for get-rich-quick schemes. He appears to have lost money on all of them, from investing in an ill-advised mining scheme to searching Mexico for the fabled lost treasure of Pancho Villa.

In January 1907 Johnson filed the Sheephead mining claim from a location described by Scott. Johnson did not get much for his money, as the claim was located in the middle of alkaline wasteland and contained no ore.[36]

Scott continued to frequent the saloons of the desert country of California and southern Nevada. He told his tall tales and exhibited a bullet he claimed to have dug out from the shoulder of his imaginary employee, Sam Jackson, after another desperate gunfight; he gave newspapermen updates as wounded Jackson hovered between life and death. Scott talked up another special train trip, but it never materialized. Again and again Scott agreed to sell his mine but would never reveal the location. He promised to show his mine to a pair of Goldfield reporters but stuck his auto into a sand dune, forcing the reporters to limp back into town. Again, he promised to show his mine to investors but left them stranded in the middle of Death Valley in the summer heat for a week before returning with their mules to lead them back to civilization.[37]

By the end of 1907 Johnson had given Scott more than $22,000 in cash, plus equipment and supplies. Impatient with Scott's stalling, he sent young Alfred MacArthur to Death Valley to befriend Scott and learn the location of the mine. MacArthur arrived in Rhyolite, Nevada, in January 1908, where he waited until Scott arrived two weeks late. MacArthur spent the next three months traveling on horseback with Scott and Bill Key across the deserts and mountains. He was treated to another gunfight when unknown persons shot at their tents in the middle of the night and ran off with the horses. MacArthur was not scared off, however, and continued wandering with Scott and Key. Scott did not reveal the location of the mine but proposed that he and MacArthur join forces to swindle Albert Johnson.[38]

They reached Barstow, California, in April 1908, and MacArthur accused Scott of being a fraud. When MacArthur took the horses and equipment that their employer Johnson had paid for, Scott did not resist, but told his creditors. Barstow businessmen attached the horses for Scott's unpaid bills. MacArthur objected, but his protests succeeded only in getting himself arrested. Bill Key intervened and secured MacArthur's release after two nights in the Barstow jail. Scott had meanwhile disappeared.[39]

MacArthur returned to Chicago and reported to Albert Johnson that Scott was a habitual liar and that his mine was a ruse. Johnson refused to believe his employee and angrily told him that his job had been to find the mine, not to determine whether or not it existed: "I don't care what happened out there, Alfred, I tell you where there's smoke there's fire!"

The following year Johnson himself went to Death Valley to coax the mine's location from Scott. Although he was thin and in poor health, Johnson spent almost a month in the desert with Scott, while Scott tried to get rid of the tenderfoot by roughing it. Johnson failed to learn the mine's location but survived and even thrived in the outdoors. Again in 1910 Johnson left the city to spend time in the desert with Scott.[40]

Johnson made a habit of joining Scott in his travels for a few weeks each winter, still convinced that Scott would tell him the location of the mine once he gained the prospector's trust. Despite back problems and a limp from a train wreck as a young man, the insurance executive hiked, camped, and became a skilled horseman. He bought hand-tooled boots and chaps and a pair of pearl-handled pistols.

Fencing High-Graded Ore

Remissions from his eastern suckers were not reliable enough for Scott, so he supplemented his income by dealing in stolen gold ore. Across the state line in Goldfield, Nevada, miners did a thriving business stealing ore from the mines that employed them, a practice euphemized as "high-grading." More than fifty assayers set up shop in Goldfield, most of them fronts for buying stolen ore. Scotty's supposed secret gold mine gave him a reason to have rich gold ore, so he bought stolen Goldfield ore, which he tried to sell to legitimate ore buyers. Previously he had flashed wads of bills but no ore, but beginning in the summer of 1905, he had very rich ore to show—closely resembling the rock from certain mines in Goldfield.[41]

Scott showed up in San Francisco in July 1908 with a trunk full of ore that he put in a bank deposit box, but ore buyers would not purchase his ore because they suspected that it was stolen. Scott left town after several days, without selling his ore and without paying his bill at the Saint Francis Hotel. One afternoon, a month later, a liquored-up Scott walked in to a San Francisco police station to charge that someone at the bank had tampered with the lock on his trunk and stolen his gold. The police found no evidence that the lock had been jimmied, but after they examined the rich ore in Scott's trunk, they put Scott himself in jail, on suspicion of receiving stolen goods. Scott stuck to his story of a secret mine, and the police released him.[42]

Scott tried to keep himself in the newspapers, but the papers tired of him. He no longer threw money around, but when a reporter met the disheveled prospector in a Goldfield, Nevada, barroom and accused him of being broke, Scott opened his flannel shirt, pulled a pack of bills from his money belt, and told the reporter to count it: the prospector was carrying $11,000 in cash—yet was arrested a few days later for failing to repay a $100 debt. Scott invited the novelist Elinor Glyn to visit him in Death Valley. Glyn had just made a much-publicized visit to Rawhide, Nevada, but she declined Scott's invitation. He threatened to emigrate to South Africa and showed up at a Marine Corps recruiting station in Chicago to take a physical. Once again, Scott's mule Slim was found wandering about without its owner; some feared the worst, but Scott showed up as healthy as ever. Scott turned everything into grist for the Death Valley Scotty legend. Once he explained a gash in his arm as an attack by a rattlesnake, but witnesses said that it was from a barroom brawl in Los Angeles.[43]

Scott hired his fame out to some Lovelock, Nevada, promoters trying to market their mining claims. The Death Valley recluse said that he had bought an option on the property and had found incredibly rich gold ore, which he was taking on another special train to Denver, New Orleans, and New York to sell the property to wealthy capitalists. The announcement made a splash in the newspapers, but Scott never followed up and who, if anyone, bought the property was not revealed.[44]

The Death Valley Scotty Gold Mining & Development Company

For all his gift for publicity, Walter Scott never tried to sell stock in his secret mine. Exploiting Scott's notoriety to sell worthless mining stock was left to a trio named Goodin, Watt, and Sharp. On June 1, 1912, they advertised the Death Valley Scotty Gold Mining & Development Company in Los Angeles newspapers and announced that they had agreed to buy Scott's mine for $1 million and had already given Scott $25,000 as down payment.[45]

However, Scott still owed more than $1,000 to the doctor who had treated his brother's wound after the battle of Wingate Pass, and the doctor sued to recover his money. Scott maintained in court that he was penniless and didn't know where the $25,000 had gone; he was jailed for contempt of court.[46]

While Scotty was suffering at the hands of his creditor, the Los Angeles Chamber of Mines and Oil, a group trying to keep shady promoters from giving the mining and oil industries a bad name, investigated the Death Valley Scotty Gold Mining & Development Company. The chamber discovered that the new company owned no assets, and they convinced the district attorney to send a detective to investigate. Scott had never hesitated to swindle wealthy individuals, but the detective appealed to Scott's conscience, and after some persuasion Scott decided that he would not allow his name to be used to defraud investors of humble means in a stock scam. Scott had sold no mine to the Death Valley Scotty Gold Mining & Development Company, and in fact, his supposed partner Goodin had exclaimed that he "wanted the fakir kept as far away as possible."

Scott detailed his involvement to a grand jury. He admitted that he had no rich mine to sell and said that he had bought stolen ore to support his pretense of a secret mine, although he denied having sold stolen ore. Scott admitted that Goodin, Watt, and Sharp had paid him only $200 for playing along with the scheme. Of the money, he had given $50 to his long-suffering wife, given $50 to a prospector named Rattlesnake Ike, bought some new clothes, and spent the rest on booze. The grand jury indicted Goodin and Sharp for perjury.[47]

Scott made a last attempt at show business when he starred in the silent motion picture "Death Valley Scotty's Mine." However, an industry periodical panned the production as "hastily thrown together" and a "crude melodrama." The film flopped, and Scott returned to the desert.[48]

Albert Johnson supported Death Valley Scotty even after it became clear that Scott had no mine. When Johnson built a lavish retirement estate in the desert, he let Scott pretend to be the owner of what became known as "Scotty's Castle." Photo by Dan Plazak.

"He Repays Me in Laughs"

It took a lot of lost time and money to convince Albert Johnson that Walter Scott was a fraud, but sometime between 1912 and 1914 Johnson realized that he had been swindled—but he didn't care. Johnson loved outdoor life in the hot dry climate and enjoyed Scott's stories and engaging personality. Johnson kept supporting Scott, and also—knowing that Scott was careless in the matter—sent a regular allowance to Scott's wife. Scott and Johnson joined in promoting Scott's reputation, and Johnson became an enthusiastic purveyor of Scotty tall tales. Johnson, a nondrinker, would go into a saloon in Goldfield or Tonopah, pretend to be broke, and unsuccessfully try to mooch a free drink. Scott would then enter and make a show of handing Johnson a $1,000 bill. The two would play their performance in various saloons and let the news spread that Scott was handing out $1,000 bills to strangers. Asked years later about the large sums he had spent on Scott, Johnson answered, "He repays me in laughs."[49]

Scott and Johnson kept the pretense of Scott's secret mine. Johnson told reporters that he was Scott's banker. In the 1920s Johnson decided to build a retirement home in the desert. He rejected a design by Frank Lloyd Wright and built a more conventional Spanish-revival complex at the northern end of Death Valley. Johnson let Scott live there and even pretend that it was his own. "Scotty's Castle" revived the legend and some speculated that Walter Scott was building the compound directly above his secret mine.[50]

Scott's hard living wore heavily on him. In 1916 he was said to have aged twenty-five years in the past five. He began to stay close to camp and tend his garden. Success begot imitation, and an old prospector in Oregon took to calling himself "Death Valley Scotty," enjoying the attention and free drinks. Scott, the real faker, deeply resented the old man's merely posing as a faker.[51]

The Con Man Confesses: His Wife, the IRS, and an Old Victim Do Not Believe Him

In a moment of rare candor, Scott announced to reporters in 1930: "Thar ain't no gold mine, and thar never was." He admitted that his money came from his benefactor, Johnson. Scott later retracted his confession and said it was a joke, but the truth was out. The legend he had created was so powerful that many would not believe that Scott did not have a secret gold mine. Among those who doubted his confession were his own wife, the Internal Revenue Service, and his old victim Julian Gerard.[52]

Walter Scott was a singularly inattentive husband who was absent for many months at a time. His wife, Ella "Jack" Scott, lived first in Los Angeles in 1902, then in Stockton, California, then again in Los Angeles, then in Goldfield and Reno after the birth of their son in 1914. When little Walter grew up and joined the navy, she moved to an apartment in Long Beach, California. Walter visited his wife and son infrequently, even when he was in town generating newspaper copy at fine hotels. He once even convinced her that he was working on business deals of such a peculiar nature that they would fall through if it became publicly known that he were married.

Scotty won headlines once more when he happily announced that he had lost $6 million in the 1929 stock market crash. "Oh, well, that's not so bad," he told reporters, "I'll earn it back." However, the Depression put Albert Johnson, although still wealthy, in reduced circumstances. Johnson cut his payments to both Walter and Jack Scott. Jack Scott suffered her husband's long absences and never moved into the Castle in Death Valley, but when her allowance was cut to $50 per month, she sued Walter Scott for spousal maintenance, asking for $1,000 per month, $25,000 legal costs, half of Scotty's Castle, and half of his imaginary gold properties. Scott demonstrated that he had no income or possessions to share with his wife. Johnson met with her lawyer and agreed to give her $75 each month and pay what her attorney described as "certain other expenses"—presumably his fee.[53]

The Internal Revenue Service was also determined to get its share of Scott's imaginary wealth. They could not find Scott's mine, and Scott finessed the matter by claiming that before the advent of the income tax law he had found a great deal of gold and hidden it. He later also claimed to have hidden $100,000 in gold certificates, also before the income tax. The IRS dropped the investigation.[54]

Albert Johnson let everyone believe that Walter Scott owned the luxurious estate in Death Valley everyone now called "Scotty's Castle." Scott even told reporters in 1939 that he was selling his castle because Death Valley had grown too civilized for him.[55]

Julian Gerard read the newspaper stories about Scotty's Castle and decided that Scott must have built it with proceeds from his secret mine. He threatened to sue Scott. Since Scott made himself unavailable, Gerard dealt with him through Albert Johnson. Johnson kept up the fiction and pretended to negotiate with Scott on Gerard's behalf. After having been fooled by Scott himself, he enjoyed fooling Gerard the same way. Throughout the 1930s Gerard made occasional attempts to get his twenty-two-and-a-half-percent share of the gold, as Johnson stalled.

In 1940 Gerard finally sued Scott for a share of the secret mine and Scotty's Castle. Albert Johnson had anticipated the suit, however, and two weeks earlier had secured an uncontested judgment against Scott in the Inyo County court for $243,291, the amount that Scott had acknowledged he owed Johnson. Now even if Gerard tried to legally seize what few belongings Scott actually owned, he would have to stand in line behind Johnson.[56]

Scott showed up at the courthouse in Los Angeles with a black eye that he told reporters he received from one of his mules. Other sources, however, said that he got the black eye the night before in a Los Angeles bar fracas. Under oath, Scott then gave his own legend a black eye. He testified that he had never had a mine and that his only income was the largesse of Albert Johnson. The judge denounced Scott but found nothing valuable of Scott's that could serve as compensation. Gerard gained title only to seventeen of Walter Scott's worthless mining claims.[57]

Scott later told people that his testimony was just a smoke screen so that he and Johnson would not have to share the gold—but few believed him.

Scotty's Legend Outlives Him

Albert Johnson died in 1948. The foundation administering the Castle allowed Walter Scott to stay on; his principal recreation was telling tall tales to tourists. Scott insisted on the reality of his mine and said that Johnson pretended to support him to enable him to pay no income tax—for which he gave Johnson half the gold.[58]

When Scott began internal hemorrhaging on January 5, 1954, his doctor put him into a car and started for the nearest hospital 185 miles away in Las Vegas. They had not traveled far when the doctor realized that Scott would never see the hospital. They stopped at the tiny desert outpost of Slim's Corners, Nevada, where the town proprietor and namesake Slim Riggs helped carry Walter Scott into the saloon; they laid him on the pool table, where he died.

Hundreds drove out to the desert to watch Walter Scott's burial on the hill above Scotty's Castle. The Castle became the property of the National Park Service, and for years after Scotty's death guides at the Castle in Death Valley National Monument continued to tell tourists of Death Valley Scotty and his secret mine whose location died with him.[59]

20

The Literary Life: Julian Hawthorne, Mining Swindler

Swindlers often secure famous names to serve as directors of their worthless companies. Such guinea pig directors rarely do more than cash their stipend checks, but Julian Hawthorne was different. Son of the literary icon Nathaniel Hawthorne, and a novelist like his father, this guinea pig director took an active hand in mining swindles. Who better to compose imaginative lies than a professional fiction writer?

Julian Hawthorne grew up in the most lionized literary circles of nineteenth-century America. He entered Harvard in 1863, where he threw himself into the camaraderie and sports but not into routine appearance at classes. Harvard expelled Hawthorne for poor attendance, and after months of idleness young Hawthorne embraced an odd career choice for an indifferent student poor in mathematics: engineering. He at first thought to return to Harvard but then enrolled at the Polytechnic Realschule in Dresden, Germany, in 1869.

Although he experienced some fits of enthusiasm for engineering, short-lived bursts of study could not carry him through to a degree. More frequent than his passing fancies for engineering were his swoons for various young ladies. The athletic and cultured Hawthorne turned the heads of many women, and his romantic temperament often returned the admiration.[1]

Temporary attachments to ladies and engineering ended when he met Minnie Amelung, an American girl in Dresden. He stopped attending classes and could think of little but starting married life together. The only dilemma was how to support himself and his Minnie. He dropped out of the Dresden Realschule in 1870, married Miss Amelung, and took a bureaucratic position with the New York City Department of Docks.

The Literary Life

His position with the Department of Docks gave him time to write, and the Hawthorne name gave him instant entrée with publishers. *Harper's Magazine* sent him $50 for his short story "Love and Counter-Love," a sum that astonished Hawthorne and convinced him that writing was an easy way to make money.[2]

When his position with the Docks Department expired two years later, he was offered a job supervising canal construction in Central America. He chose instead to remain in New York and support himself by writing, even though his father warned him not to try to make a living as a writer. Julian completed his first novel, *Bressant*, in just three weeks, although he had to rewrite it, because, he later noted, the first draft was "too immoral to publish, except in French." He followed up his successful first effort with *Idolatry*, a gothic tale about a Dr. Heiro Glyphic.

Julian Hawthorne's output was prodigious, and he authored sixty-one books. But he noted that in 1887 he "had already ceased to take pleasure in writing for its own sake," but financial pressures demanded more. His son later recalled, "I have indelible memories of tradesmen wearing out our 'Welcome' mats at our various residences, looking for payment for groceries, butchers and bakers, etc."[3]

Both Hawthorne and the public discovered that he possessed only an echo of his father's gift, and he turned increasingly to nonfiction and literary criticism. By 1895 the name Julian Hawthorne had fallen so low in literary circles that when he entered *A Fool of Nature* in the New York *Herald*'s competition for the best American novel of the year, he used the pen name Judith Holinshead. The story, written in twenty-one days, won the *Herald*'s $10,000 prize before its true authorship was discovered. Characteristically and unfortunately, Hawthorne followed his success with a literary blunder, *Love Is a Spirit*, which became his last novel.

Hawthorne turned to journalism and lectures. He covered a famine in India, and presidential elections, and went to Cuba to write up the Spanish-American War in the company of such other literary journalists as Stephen Crane, Richard Harding Davis, and Frank Norris. He spent several years as sports editor of the New York *American*. Hawthorne's family life deteriorated along with his career. He became estranged from his wife and became the steady companion of the painter Edith Garrigues.

Hawthorne was strongly ambiguous about money. He was a utopian socialist, and his artistic side spurned wealth as a base distraction to spiritual fulfillment; but he once confided to his journal: "And yet my whole endeavor as long as I live will be to get rich."

Get Rich Quick

"The time has now arrived when I want more money than the sort of literature I have produced can provide."

—Julian Hawthorne, letter to potential investors, 1908

In July 1908 Hawthorne's old Harvard friend, the surgeon William J. Morton, asked him to join a business venture that, he promised Hawthorne "will cost you nothing." The discovery of silver at Cobalt, Ontario, in 1904 had sparked a rush to the north woods, and a frenzy of Canadian mining company promotions —a goodly share of which were swindles.[4]

William Morton had visited some prospecting camps in northern Ontario and was so excited by the silver boom that he paid $14,000 for some mining claims near Temagami. After his return home, however, geologists told him that his mining claims had little or no value. His properties were spurned by mining men, but his faith in the claims was unshaken, so he determined to form his own mining company.

Morton knew that the Hawthorne name would give legitimacy to the new company, the Temagami-Cobalt Company, Ltd. Hawthorne would receive stock for the use of his name on the board of directors, and Morton promised him that he would have to do nothing but cash his dividend checks. From Bear Lake, Ontario, Morton wrote Hawthorne a letter giddy with optimism:

> "Here is the Diabase Peninsula—that bare rocky chunk of land—good for minks and chipmunks up to this time now erected into a $20 000 000 fairy tale. Yet who shall say that there is not $20 000 000 worth of silver in it from here down to China. I believe there is.[5]

Hawthorne had been increasingly dissatisfied with his writing career. His prolific pen was scratching a living in journalism, but journalism seemed an embarrassing step down the social scale for the son of Nathaniel Hawthorne. In his essay *Journalism the Destroyer of Literature,* he decried the materialistic rush of the modern age and wrote that newspaper and magazine editors were rejecting the best literature (presumably meaning his own) because it would make the rest of their publications look drab. Julian may have seen mining wealth as a more acceptable source of income.[6]

Hawthorne and Morton were joined in the promotion by the distinguished former US assistant secretary of state and mayor of Boston, Josiah Quincey, and the less-than-distinguished stock promoter Albert Freeman, who was experienced in promoting worthless mining companies. The four incorporated the Temagami-Cobalt Mines and began selling stock.

Freeman arranged to pocket most of the proceeds. Of the $400,000 paid by investors into Temagami-Cobalt, $310,000 disappeared into the treasury of the Continental Syndicate, a firm owned by Freeman. Morton received stock and $14,000 cash for his property, leaving only $76,000 out of the $400,000 to develop the Temagami-Cobalt Company.[7]

Literary License

Julian had traded on the Hawthorne name all his life and knew its commercial value. Hawthorne enthusiastically made his name central to the promotion, and people across the country were soon flattered with letters bearing the honored Hawthorne signature—ably forged by the clerical staff.

Hawthorne claimed in his scarlet promotional letters that the prominent mining investment firm of Ricketts and Banks were avidly trying to buy the mining properties, but Ricketts and Banks quickly denied it. The firm wrote to Hawthorne that his literature "misrepresent[ed] the facts." As Ricketts and Banks demanded, Hawthorne dropped their name from future letters, but he retained the false story, only changing the would-be buyer to an unnamed "prominent firm."

Hawthorne wrote that a London mining engineer was sent to buy the property for the giant South African concern Gold Fields Ltd., but the engineer called Hawthorne's account "moonshine." He said that he had had casual contact with Hawthorne once or twice but had never examined the property or tried to buy it.[8]

A former professor of mining engineering at Yale was similarly misquoted in Hawthorne's letters as being awestruck by the richness of the ore bodies of the Temagami-Cobalt Company. The professor angrily wrote that his remarks had described the Cobalt silver mines and not Hawthorne's properties at Temagami, which he had never seen.[9]

The *Engineering & Mining Journal* at first criticized the venture only as a risky investment run by amateurs. But after it became obvious that Hawthorne was acting in bad faith, the *Journal* called his promotional literature "willful deception" and a "miserable attempt to secure money."[10]

Miserable deception though it was, it was highly successful. Encouraged by Hawthorne's ability to lure investors, the group quickly created two more Ontario mining companies: the Montreal-James Mines, Ltd., and the Elk Lake-Cobalt Mines, Ltd. Each incorporated as a holding company in Maine—then known for its lax corporate laws—and also as a separate company of the same name in Canada.

The novelist and journalist Julian Hawthorne, son of Nathaniel Hawthorne, found that he could make more money writing imaginative fiction for mining company prospectuses. Courtesy Library of Congress.

The companies issued press releases of rich ore discoveries. According to one statement, the promoters claimed assays as high as forty thousand ounces of silver per ton of ore, an extraordinary result since a ton weighs only about thirty-two thousand troy ounces to begin with.[11]

Newspapers for Hire

"The jackals of the press are of a lower class than the prostitutes."

—*Engineering & Mining Journal,* December 14, 1912, on newspapers promoting fraudulent Canadian mining companies

Newspapers specializing in investment news offered to sell favorable publicity. Freeman and Morton supplied two laudatory articles to the *United States Investor,* which published them for $128. The *Mercantile and Financial Times* published a glowing article based on the company literature, and Freeman paid them $150. The *Cobalt News Bureau,* which despite its name had no staff outside New York City, received $200 for publishing a favorable article on the Hawthorne companies.[12]

The Hawthorne promoters also found Canadian newspapermen whose ethics unstiffened upon applications of cash. The Toronto *World* printed so many advertisements from fraudulent mining companies that the *Canadian Mining Journal* named the *World* the single most pernicious influence on Canadian mining. For the Hawthorne companies, the *World* printed a highly favorable article, which the promoters then reprinted in their stock-peddling literature.[13]

Hawthorne Silver and Iron Mines, Limited

Although his companies had yet to ship any ore, Hawthorne's endorsement continued to bring in investor dollars. The group exploited the Hawthorne name even further when it incorporated the Hawthorne Silver & Iron Mines, Ltd. Hawthorne wrote potential stock buyers that investments in Hawthorne Silver & Iron would certainly pay thirty percent annually and could easily yield as much as one hundred percent per year. The Hawthorne company found prominent Canadians willing to rent their names to the company, including a member of the Canadian parliament, and an ex-director of the Canadian Geological Survey, who acted as consultant to the company.[14]

The centerpiece of the new company was the Wilbur iron mine, a property that had been worked off and on over the years, as each successive company failed to make a profit from its small pockets of iron ore. The Hawthorne company made a great show of building a pier and other facilities to load the ore onto ships on Lake Ontario, but never shipped any ore.[15]

No one was more ringing in his criticism than the editor of the *Canadian Mining Journal,* who fulminated against Hawthorne and his companies.[16]

> "Julian Hawthorne, unworthy son of worthy Nathaniel Hawthorne, is a magazine writer. He is also a sublimated ass."
>
> —*Canadian Mining Journal,* April 15, 1909

"Except as inspiration for Hawthorne's pellucid prose, shares in Hawthorne's silver and iron mines are worth about as much as Confederate paper money."

—*Canadian Mining Journal,* September 1, 1910

Although denounced by the principal mining papers of New York and Toronto, Hawthorne Silver & Iron received uncritical treatment in London from the *Mining Journal* and from the *Times.* The *Mining Journal* reported his extravagant claims at face value in its news columns, while running a full-page advertisement for Hawthorne Silver & Iron Mines, Ltd. The *Times* announced the stock issue in its news columns on the same day that it printed an advertisement for Hawthorne Silver & Iron Mines that ran the full length of a page. Neither publication alerted readers to the disrepute of the promoters and their properties. London's *Financial Times* also printed a Hawthorne advertisement but was astute and honest enough to warn its readers away from the stock.[17]

The *Canadian Mining Journal* lamented that Hawthorne's advertising attracted English investors to worthless properties, while honest Canadian prospectors could not raise capital for genuinely promising prospects.

Literature on Trial

"Hundreds of men of good material, gentle breeding, and high intelligence are now going to jail for offenses which have been in a manner created by the newborn scruples of lawmakers and the subtle distinctions of public prosecutors."

—Julian Hawthorne, 1913, *Washington Post,* June 8, 1918, p. 4

A US senator asked the Justice Department to investigate the Hawthorne companies after some of the senator's friends complained of losing money. Hawthorne's reaction was typically patrician: he disdained to respond to Justice Department questions and instead gave a list of famous names who could vouch for his fine character. Despite Hawthorne's gentle breeding and impressive social references, a federal grand jury returned indictments against Hawthorne and his partners.[18]

Hawthorne, Morton, Quincey, and Freeman went on trial in New York in late 1912 for mail fraud. Prosecutors charged that investors had lost $4 million in the Hawthorne mining companies. The Hawthorne companies first tried to hide their finances by spiriting the ledgers off to their Canadian subsidiaries, but the judge tolerated no cross-border shell games, and company officers produced the accounts to avoid being jailed for contempt.

On March 14, 1913, the jury found Morton and Hawthorne guilty on seventeen counts, Freeman guilty on twenty-three counts, and Quincey not guilty. The judge sentenced Hawthorne, Morton, and Freeman to prison. Hawthorne and Morton were shocked by the verdicts, because they believed that their family connections entitled them to go free. Hawthorne later wrote that he had thought that no jury would convict a Hawthorne. William J. Morton blamed his prison sentence on the ingratitude of the American public to the contribution his father, William T. G. Morton, had made in pioneering the use of ether as an anesthetic. Hawthorne, Morton, and Freeman were led away in handcuffs to the Tombs jail.[19]

Hawthorne and Morton, the Harvard chums of a half-century before, declined appeals and entered the federal penitentiary in Atlanta. They finished their terms before the end of the year. Albert Freeman, sentenced to five years' imprisonment, appealed the verdict, which was eventually overturned because two separate judges had presided over different parts of the trial. A second trial deadlocked the jury. Faced with a third trial in 1916, Freeman plea-bargained. Freeman pled guilty and was fined $3,000 by an unhappy federal judge who told him that he deserved much harder treatment. Freeman stepped forward, peeled three $1,000 notes from a roll of bills, and walked free.[20]

After release from prison, Hawthorne moved to Pasadena, California, far from his New England home, wife, family, and friends. His first post-prison book, *The Subterranean Brotherhood,* was an account of his own imprisonment and a call to free all criminals immediately, whatever their crimes. His return to books was met with indifference, however, and the Victorian literary figure tried Hollywood screenplays. A movie executive turned down a Hawthorne script in 1920: "As your story stands it is impossible pictures material." His wife Minnie died in Connecticut in 1925; the next day, in California, Hawthorne married his long-time companion Edith Garrigues.[21]

His schoolmate and jailmate William Morton continued to chase elusive fortune. In a letter to Hawthorne, he drew a pot of gold at the end of a rainbow. Morton was still bewitched and impoverished by the strike-it-rich mining schemes of Albert Freeman. Four years out of prison, Morton visited Freeman's latest unfortunate enterprise, a copper mine in Cuba. "As to the copper mine," he wrote Hawthorne, "I believe it is a really big thing and the sooner he sends you your stock the better." After months of disappointment, however, he admitted, "I heartily wish I had my money back."[22]

Morton's wife wrote Hawthorne in 1920: "Freeman has been talking very big of his enterprise to Will and also to me when we saw him last August. Whereupon Will withdrew all his other investments and put the money in with Freeman, all of it." After Morton died, she wrote, "Will had great losses of late and tried to make it up with some investments with Mr. Freeman, and was waiting daily to hear that they should turn out well, and the anxiety of it shortened his life." Years later, she hired an attorney to recover money from Freeman but was not optimistic. "I am here as a governess earning a living, for Mr. Freeman took Will's every penny, and it was a fortune at that....The hardest thing about it is to think the thing caused Will's death and broke his heart."[23]

In addition to mining schemes, in the 1920s Albert Freeman promoted companies offering quack medicines. Despite his many phony promotions, he avoided jail and even got his hands on one mine of great value in the McIntyre Porcupine Mines Ltd., an enormously productive property in Ontario.[24]

Hawthorne died in 1934, maintaining his innocence to the end. Many swindlers stubbornly proclaim their innocence in the face of clear evidence to the contrary, but perhaps Hawthorne was truly convinced of the value of the mines. That he wrote lies to sell stock is evident, but he may have rationalized his blatant falsehoods as mere literary license. The romantic novelist, the utopian socialist, the visionary who believed that society should immediately abolish all prisons, may have believed that the properties' mineral wealth would turn his lies into truths.

21

Selling the Pure Blue Sky

The two decades leading up to World War I were difficult for mine swindlers. Muckraking journalists revealed the methods of financial manipulators, and the public pressed elected officials to curb the abuses. Stock markets were seen by many as no better than casinos. A Colorado judge in 1901 enforced an anti-gambling law by shutting down the Denver Mining Stock Exchange. A New York district attorney, defending his proposed anti-gambling legislation, declared in 1904, "Perhaps these big gambling houses are no more crooked than some of the financial institutions we have heard about recently, but they are just as crooked, and that is saying a whole lot."[1]

A few exchanges, particularly the New York Stock Exchange and the Boston Stock Exchange, required some financial disclosures, but those exchanges were for established corporations, not start-ups. Every large city had an exchange dedicated to penny stocks, and the small exchanges were notoriously lax, or sometimes just bucket shops run by the swindlers themselves.

In 1906 and again in 1910, the post office stepped up prosecutions of stock swindles through the mail. However, it could not stop the placing of false newspaper advertisements or the peddling of worthless shares door to door. When the California state commissioner of corporations suggested that newspapers refuse to accept advertisements for blatantly fraudulent stock offerings, the *Mining & Scientific Press* scoffed, noting the "utter lack of conscience and entire absence of decency on the part of the publishers of our San Francisco papers."[2]

Cleaning Up Mining Fraud

Some states passed laws specifically to curb mining fraud. Massachusetts passed a law in 1903 requiring all mining companies selling shares to file a report with the state describing its location and physical equipment— information previously not available for many companies.[3]

In 1905 the California legislature prohibited any employee or agent of a mining company from giving false information to affect share prices. The California state mineralogist pronounced at the end of 1905 that "since the passage of the bill the fraudulent and exaggerated prospectus has nearly disappeared from California." Mining men held up the California law as a model. Washington State passed a similar law later in 1905, and Arizona in 1907. Nevada mandated that mining corporations file reports twice yearly giving the finances and physical state of the property.[4]

Laws to stop mining fraud helped, but mining swindles were only a small part of an epidemic of fraudulent stock offerings, in everything from oil wells to rubber plantations. Laws against fraud were of little use when the fast-moving cheats were nowhere around by the time their customers discovered the fraud.

Blue Sky Laws

Kansas bankers hated stock swindlers. They hated them because the bankers watched their customers withdraw their savings to buy shares of companies no one had ever heard of, and they watched the foolishly invested money leave on the next train out of town in the valise of some snake-oil stock peddler. Kansas's bank commissioner, J. N. Dooley, estimated that five hundred stock salesmen made a living in Kansas selling from $3 million to $5 million in shares each year and that at least ninety-five percent of the money was wasted on worthless stocks.

In 1911 the Kansas legislature passed comprehensive controls on selling stocks. Any company selling shares in Kansas had to first apply to the bank commission and supply detailed financial statements sworn to by the officers. Individual stock salesmen also needed licenses from the bank commission. Every small-town and big-city banker in Kansas became an enforcement agent. "If you hear of anyone offering any stock for sale," Dooley wrote in a circular to banks, "find out whether he has a state license. If he hasn't[,] wire me and I will send an officer after him on the next train."

Proponents of state regulation likened sales of worthless shares to selling empty space in the air—selling the blue sky. The phrase—and the law—swept through statehouses, and within two years blue-sky laws were in force in twenty-three states and one Canadian province. Even some cities, such as Denver, passed their own blue-sky regulations. The American Mining Congress supported the laws, as did the most influential mining periodicals, the *Engineering & Mining Journal* and the *Mining & Scientific Press*.[5]

The burden was suddenly no longer on the state to prove that a company was a fraud but on the company to prove that it wasn't. Entrepreneurs chafed under the requirement to prove the worth of their mining property in each state, often to someone with no knowledge of mining. A corporation approved in one state might be denied in another. A mining property approved for a capitalization of $1 million in Ohio was approved for a cap of only $250,000 in Michigan. A mining promoter complained that the new laws had "helped destroy the spirit of venture that must be preserved if mining development is to be carried on." One critic charged that blue-sky laws were an eastern conspiracy to stop the flow of capital to the western states. The US Supreme Court turned back a constitutional challenge to blue-sky laws in 1917, and by 1923 all but five states had blue-sky laws.[6]

Of course, not all reform laws were what they appeared to be. Texas was the birthplace of many oil companies, both good and bad. The Texas legislature passed strict corporate controls but left a loophole by which a corporation could exempt itself from the blue-sky law. In the early 1900s Texas stock promoters could point to the consumer protections in state law, without disclosing that they were exempt from those provisions.[7]

Public agencies began to investigate and prosecute stock fraud. Victims who previously had to bring lawsuits at their own expense now had the power of the state governments on their side.

Swindlers of all types who had previously lied and cheated with impunity were increasingly harassed by the new laws and stricter enforcement of old laws. Between 1910 and 1930, many fraudulent mine promoters such as Richard Flower, John Grant Lyman, Eugene and Sheldon Burr, and George Graham Rice were pursued and imprisoned.

Mining fraud is today much less common than a century ago, but swindlers are still among us, perhaps now more careful and more clever, still stealing people's money and harming legitimate enterprise, including mining.

In the summer of 1997, small mining companies reported sudden and extreme difficulty financing their projects. Mining executives named one cause: the Bre-X swindle, in which the public lost millions in a Canadian company with an imaginary gold deposit in Indonesia. How many investment dollars were withheld from mining stocks? How many worthy mining projects were stalled or abandoned because of post Bre-X investor jitters? There is no way of knowing to what extent fraud is still damaging the mining industry.

Notes

Abbreviations

AMS	*Australian Mining Standard* (Sydney, Melbourne)
AZR	*Arizona Republican* (Phoenix)
BFM	*Bullfrog (NV) Miner*
CuH	*Copper Handbook,* written and published by Horace J. Stevens, Houghton, Michigan
DMR	*Daily Mining Record* (Colorado Springs and Denver)
Drep	*Denver Republican*
Dtr	*Denver Tribune*
Econ	*Economist* (London)
EvM	*Everybody's Magazine*
EMJ	*Engineering and Mining Journal/Engineering and Mining Journal-Press* (NY)
FT	*Financial Times* (London)
GHN	*Gold Hill (NV) News*
GFN	*Goldfield (NV) News*
GFTr	*Goldfield (NV) Tribune*
LAT	*Los Angeles Times*
LH	*Leadville (CO) Herald*
LeW	*Leslie's Weekly*
MH	*Mines Handbook*
Minv	*Mining Investor*
MJ	*Mining Journal* (London)
MSP	*Mining and Scientific Press* (San Francisco)
NMN	*Nevada Mining News*
NYT	*New York Times*
NYTr	*New York Tribune*
RMN	*Daily Rocky Mountain News/Denver News*
SBS	*San Bernardino (CA) Sun*
SFAC	*San Francisco Alta California*
SFB	*San Francisco Bulletin*
SFCh	*San Francisco Chronicle*
SFCa	*San Francisco Call*
SFE	*San Francisco Examiner*
SLTr	*Salt Lake Tribune*
TS	*Tonopah (NV) Daily Sun*
TL	*Times* (London)
USGS	United States Geological Survey
VCC	*Virginia City (NV) Chronicle*
VTE	*Virginia City (NV) Territorial Enterprise*
WSJ	*Wall Street Journal*

Chapter 1. Introduction to Mining Fraud

1. Wolfgang Paul, *Mining Lore* (Portland, Ore.: Morris Printing, 1970), 330.

2. Georgius Agricola, *De Re Metallica,* transl. H. C. and L. H. Hoover (1556; repr., N.Y.: Dover, 1950), 20-22.

3. E. F. Roots, "Anniversaries of Arctic Investigation: Some Background and Consequences," Transactions of the Royal Society of Canada, ser. 4, no. 20 (1982): 378-81. William McFee, *The Life of Sir Martin Frobisher* (N.Y.: Harper, 1928).

4. G. B. Parks, "Notes: Frobisher's Third Voyage, 1578," *Huntington Library Bulletin* 7 (1935): 181-90.

5. T. A. Rickard, *The Romance of Mining* (Toronto: MacMillan, 1944), 102-19.

6. D. D. Jackson, "Hot on the Cold Trail Left by Sir Martin Frobisher," *Smithsonian* 23, no. 10 (January 1993): 119-30.

7. Walter Kenyon, "'All Is Not Gold That Shineth,'" *The Beaver* 301, no. 1 (1975): 40–46. W. W. Fitzhugh and J. S. Olin eds., *Archaeology of the Frobisher Voyages* (Washington: Smithsonian, 1993).

8. Vilhjalmur Stefansson, *The Three Voyages of Martin Frobisher* (London: Argonaut, 1938), cxvi.

9. Heather Pringle, "A Tale of Fraud and Frobisher's Gold," *New Scientist* (December 25, 1993): 4.

10. Mark Twain, *Roughing It* (1871; repr., N.Y.: New American Library, 1962), 195.

11. MSP, July 4, 1896.

12. FT, November 5, 1888, p. 3, c.1; January 19, 1889, p. 6, c.3; February 5, 1889, p. 4, c.3.

13. MSP, July 30, 1898, p. 102.

14. EMJ, July 30, 1892, p. 97; September 17, 1892, p. 265.

15. *Mining World* (Las Vegas, N.M.), (September 1880): 4. MSP, September 25, 1880, p. 200.

16. *Collier's,* August 10, 1907, pp. 9-11; August 31, 1907, pp. 12-13.

17. DMR, June 26, 1909, p. 8, c.1. NYT, June 6, 1923, p. 25, c.8; July 14, 1923, p. 15, c.3.

18. Aylmer Vallance, *Very Private Enterprise* (London: Thames and Hudson, 1955), 78.

19. EMJ, February 16, 1907, p. 341.

Chapter 2. The Imaginary Mine of Doctor Gardiner

1. *Congressional Globe,* 32nd Cong., 1st sess., 1853, 27, app., pp. 64-67.

2. U.S. Congress, House, Report of the Special Committee of the House of Representatives in the Gardiner Investigation, 32nd Cong., 2d sess., 1852, H. Rep. 1, p. 122.

3. *Washington Star,* March 8, 1854, p. 2.

4. A. R. de la Cova, "The Taylor Administration versus Mississippi Sovereignty: The Round Island Expedition of 1849," *Journal of Mississippi History* (Winter 2000): 1-33.

5. *Congressional Globe,* 32nd Cong., 2d sess., 1853, 27, app., p. 110.

6. Abner Doubleday, *My Life in the Old Army,* ed. J. E. Chance (Fort Worth: Texas Christian University Press, 1998), 157-74.

7. *Washington Post,* June 21, 1896, p. 11.

8. NYT, September 19, 1851, p. 4; September 30, 1851, p. 4; January 27, 1852, p. 2.

9. NYT, July 8, 1852, p. 4.

10. U.S. Senate, Report of the Secretary of State in Answer to a Resolution Calling for Information in Relation to Gardiner Claim, 32nd Cong., 1st sess., 1852, Exec. Doc. 83. NYT, August 24, 1852, p. 1.

11. *Washington Star,* March 4, 1854, p. 2.

12. *Baltimore Sun,* September 11, 1852, p. 11. NYT, September 13, 1852, p. 2.

13. NYT, October 9, 1852, p. 8.

14. *Congressional Globe,* 32nd Cong., 1st sess., 1852, 24, 3:app. pp. 2302-3. 32nd Cong., 1st sess., 1852, 25, app.: p. 1031. 32nd Cong., 2d sess., 1853, 27, app.: pp. 113-17. NYT, September 30, 1852, p. 1.

15. NYT, September 23, 1852, p. 6; September 28, 1852, p. 6; March 7, 1854, p. 1. *Washington Star,* March 6, 1854, p. 2.

16. *Washington Star,* March 11, 1853, p. 3; May 21, 1853, p. 3; May 27, 1853, p. 3; May 28, 1853, p. 3.

17. *Washington Star,* December 13, 1853, p. 3.

18. *Washington Star,* December 29, 1853, p. 3; January 9, 1854, p. 3; January 11, 1854, p. 3; January 13, 1854, p. 3; February 1, 1854, p. 3. U.S. Congress, House Committee on the Judiciary, "Gardiner and Meers," 33d Cong., 1st sess., 1854, House Rept. 369, pp. 63-64.

19. *Washington Star,* March 3, 1854, p. 3; March 5, 1854, p. 3. *Daily National Intelligencer* (DC), March 4, 1854, p. 3. *Baltimore Sun,* March 6, 1854, p. 11. NYT, March 6, 1854, p. 5.

20. *Baltimore Sun,* March 4, 1854, pp. 1, 2. *Washington Star,* March 4, 1854, p. 3. NYT, March 6, 1854, p. 5.

21. *Washington Star,* March 6, 1854, p. 3; March 10, 1854, p. 3.

Chapter 3. The Comstock: Mother Lode of American Mining Swindles

1. SFB, May 5, 1860, p. 3.

2. Maureen Jung, "The Comstocks and the California Mining Economy, 1848-1900" (PhD diss., University of California, Santa Barbara, 1988).

3. SFB, May 2, 1860, p. 3. *Mining Magazine,* January 1861, p. 1123. VTE, March 11, 1877, p. 2. M. V. Sears, *Mining Stock Exchanges 1860-1930* (Bozeman: University of Montana Press, 1973), 20. R. W. Paul, *California Gold* (Lincoln: University of Nebraska Press, 1947), 182.

4. VTE, December 7, 1877, p. 2. Drep, June 25, 1881, p. 6.

5. SFCh, March 17, 1873, p. 3.

6. SFAC, October 20, 1862, p. 1.

7. MSP, March 15, 1873, p. 170; June 21, 1873, p. 390. SFCh, February 23, 1877, pp. 2, 4. J. L. King, *History of the San Francisco Stock and Exchange Board* (San Francisco: J. L. King, 1910), 91-94.

8. Grant H. Smith, "History of the Comstock Lode," *University of Nevada Bulletin,* Geological and Mining Series 37, 1943, pp. 208-9.

9. Mark Twain, *Roughing It* (1871; repr., N.Y.: New American Library, 1962), 231-32.

10. Mark Twain, *Mark Twain's Letters,* ed. E. M. Branch et al. (Berkeley and Los Angeles: University of California Press, 1988), 245, 247, 253, 260. R. A. Dwyer and R. E. Lingenfelter, *Dan De Quille, the Washoe Giant* (Reno: University of Nevada Press, 1990), 4, 30. Larry Cebula, "For Want of the Actual Necessities of Life," *Journal of the West* 36, no. 4 (1997): 32.

11. NYTr, October 14, 1869, p. 5.

12. MSP, March 16, 1863, p. 5. SFCh, May 7, 1870, p. 3; December 25, 1872, p. 2. *San Francisco Daily Stock Report,* December 22, 1879, p. 11. G. T. Ingram, *Digging Gold among the Rockies* (Philadelphia: Hubbard Bros., 1880), 451-53.

13. SFB, October 2, 1860, p. 3.

14. Elliot Lord, *Comstock Mines and Miners,* US Geological Survey Monograph 4, 1883, pp. 132, 177. Maureen Jung, The Comstocks and the California Mining Economy, 1848-1900 (PhD diss., University of California, Santa Barbara, 1988).

15. SFAC, November 26, 1865, p. 1. Bruce Alverson, "The Limits of Power: Comstock Litigation, 1859-1864," *Nevada Historical Society Quarterly* 43, no. 1 (2000): 74-99.

16. *Virginia City (NV) Union,* December 10, 1864, p. 2. Smith, p. 69.

17. Lord, p. 173. W. M. Stewart, *Reminisces of Senator William M. Stewart of Nevada* (N.Y.: Neal, 1908), 151.

18. SFCa, September 10, 1863, p. 12.

19. Smith, p. 69.

20. Stewart, *Reminisces,* 154-59.

21. SFAC, June 23, 1862, p. 1; August 16, 1863, p. 1.

22. SFAC, September 4, 1863, p. 1.

23. GHN, July 23, 1864, p. 2. *Virginia City (NV) Bulletin,* January 11, 1864, p. 2; January 12, 1864, p. 3 c. 3; January 18, 1864, p. 3 c. 2. *Virginia City (NV) Union,* January 17, 1864, p. 2 c. 2; January 19, 1864, p. 2 c. 2; January 24, 1864, p. 2 c. 1, p. 3 c. 3; January 27, 1864, p. 2 c. 2.

24. SFAC, July 9, 1865, p. 11. Smith, pp. 81-82.

25. GHN, August 5, 1864, p. 2.

26. *Virginia City (NV) Bulletin,* January 25, 1864, p. 2. GHN, August 1, 1864, p. 2; July 29, 1864, p. 2.

27. GHN, May 11, 1864, p. 2; May 27, 1864, p. 2; August 4, 1864, p. 2. SFAC, May 15, 1864, pp. 1-2.

28. GHN, August 11, 1864, p. 2.

29. GHN, July 30, 1864, p. 2.

30. *Virginia City (NV) Union,* August 21, 1864, p. 2; August 24, 1864, p. 2; August 25, 1864, p. 2. SFAC, August 24, 1864, p. 1. Stewart, *Reminisces,* 160-61.

31. SFAC, August 24, 1864, p.^1. Lord, pp. 161-63.

32. R. R. Elliot, *Servant of Power, A Political Biography of Senator William M. Stewart* (Reno: University of Nevada Press, 1983), 44-45.

33. *Washoe (NV) Weekly Times,* October 28, 1864, p. 2.

34. SFAC, April 30, 1865, p. 2.

35. *Sacramento Union,* August 12, 1864, p. 2.

36. MSP, August 13, 1864, p. 104. *Sacramento Union,* August 13, 1864, p. 2.

37. King, 243-44.

38. The Gould & Curry mine is said to have used the scheme in 1863 (Lord, p. 289. G. A. Vent and Cynthia Birk, "Insider Trading and Accounting Reform: The Comstock Case," *Accounting Historians Journal* 20, no.2 [1993]: 72), but the evidence is doubtful. The cited reference of both sources mentions only that the Gould & Curry refused visitors in 1863 but says nothing of shutting in miners (VTE, January 19, 1871, p. 2). When Hale & Norcross miners were sequestered in 1868, the *Virginia City (NV) Enterprise* called the move "novel" (January 11, 1868), indicating the first use of the tactic.

39. VTE, January 11, 1868, p. 3. GHN, January 11, 1868, p. 3. SFAC, January 15, 1868, p. 2.

40. VTE, January 12, 1868, p. 3; February 14, 1868, p. 3.

41. GHN, February 15, 1869, p. 3; February 18, 1869, p. 3; February 19, 1869, p. 3. VTE, February 20, 1869, p. 3.

42. SFAC, February 8, 1872, p. 1; February 9, 1872, p. 1.

43. SFCh, February 4, 1872, p. 1; February 6, 1872, p. 3; February 9, 1872, p. 3; February 13, 1872, p. 3. GHN, February 5, 1872, p. 3; February 6, 1872, p. 3; February 8, 1872, pp. 2, 3. SFAC, February 11, 1872, p. 1; February 19, 1872, p. 1.

44. GHN, February 27, 1872, p. 3; February 28, 1872, p. 3; February 29, 1872, p. 3; March 1, 1872, p. 3; March 5, 1872, p. 3.

45. GHN, October 12, 1872, p. 3. SFCh, October 13, 1872, p. 1.

46. VTE, May 25, 1878, p. 3.

47. SFCh, December 8, 1877, p. 4.

48. SFB, December 8, 1880, p. 3.

49. SFB, May 21, 1862, p. 3.

50. J. B. Farish, Farish Collection, #1052, box 1, pp. 89-91, American Heritage Center, University of Wyoming, Laramie.

51. The mine and sums vary with the teller. C. C. Goodwin, *As I Remember Them* (Salt Lake City: Salt Lake Commercial Club, 1913), 183-84. Dan DeQuille, *The Big Bonanza* (1876; reprint, Las Vegas: Nevada Publishing, Las Vegas, 1974), 313-15.

52. VTE, April 27, 1877, p. 3.

53. VTE, May 4, 1877, p. 3. MSP, May 12, 1877, p. 296.

54. SFAC, November 15, 1876, p. 1; December 12, 1876, p. 1; December 14, 1876, p. 2. SFCh, November 30, 1876, p. 1; December 20, 1876, p. 5.

55. SFCh, December 6, 1876, p. 5; December 13, 1876, p. 8. SFAC, December 20, 1876, p. 2.

56. SFCh, December 29, 1877, p. 3; April 8, 1877, p. 8; September 16, 1877, p. 4; September 30, 1877, p. 8; January 9, 1878, pp. 2-3. GHN, January 16, 1877, p. 4; January 29, 1877, p. 3. VTE, February 27, 1877, p. 2. VCC, December 21, 1877, p. 3.

57. VCC, January 10, 1880, p. 3; January 15, 1880, p. 3; January 16, 1880, p. 3. EMJ, February 7, 1880, p. 105.

58. VCC, January 21, 1880, p. 3; January 22, 1880, p. 3. EMJ, January 31, 1880, p. 90; February 7, 1880, p. 105.

59. VCC, January 23, 1880, p. 3; January 26, 1880, p. 3; January 27, 1880, p. 3; January 28, 1880, p. 3. Wells Drury, *An Editor on the Comstock Lode* (1936; repr., Reno: University of Nevada Press, 1984), 234-435.

60. GHN, February 27, 1869, p. 3; March 11, 1869, p. 3. MSP, March 13, 1869, p. 165.

61. MSP, January 27, 1877, p. 56; May 19, 1877, p. 312.

62. MSP, November 7, 1908, p. 629. Smith, pp. 194-95.

63. G. F. Redmond, *Stock Market Operators* (1924; repr., London: Prentice-Hall, 1999), 5-6.

64. SFCh, April 25, 1876, p. 3. *Eureka (NV) Sentinel,* April 25, 1876, p. 2.

65. EMJ, May 13, 1876, p. 474.

66. SFCh, August 2, 1876, p. 5; September 30, 1876, p. 3. *San Francisco Daily Stock Report,* April 25, 1876, p. 1; August 2, 1876, p. 1.

67. SFCh, March 11, 1877. VTE, March 13, 1877, p. 3.

68. SFAC, February 4, 1877, p. 2. SFCh, April 19, 1877, pp. 2, 4.

69. GHN, November 13, 1878, p. 3. EMJ, November 16, 1878, pp. 343, 352; November 23, 1878, p. 361.

70. SFCh, January 12, 1877, pp. 2-3. VTE, January 12, 1877, p. 2.

71. SFCh, January 18, 1878, p. 1; December 8, 1878, pp. 1, 4. VTE, May 26, 1878, p. 1; May 28, 1878, p. 2. GHN, May 27, 1878, p. 2. NYTr, May 26, 1878, p. 1. S. P. Dewey, "The Bonanza Mines of Nevada, Gross Frauds in the Management Exposed" (1878; repr. in *Speculation in Gold and Silver Mining Stocks* [N.Y.: Arno, 1974].)

72. SFCh, January 18, 1878, p. 1. NYT, May 20, 1878, p. 5; October 13, 1878, p. 1; October 16, 1878, p. 2; October 23, 1878, p. 1; December 12, 1878, p. 2. GHN, May 21, 1878, p. 3. VTE, May 21, 1878, p. 3. EMJ, December 14, 1878, p. 430; August 23, 1879, p. 131; November 22, 1879, p. 378.

73. NYT, August 8, 1880, p. 1; November 12, 1880, p. 2; December 19, 188, p. 10. SFCh, December 8, 1878, p. 1; September 9, 1879, p. 2; September 16, 1879, p. 2. SFAC, December 10, 1880, p. 1; December 14, 1880, p. 1; December 15, 1880, p. 1; December 17, 1880, p. 1. SFB, December 15, 1880, p. 1. EMJ, December 18, 1880, p. 395.

74. SFB, March 30, 1881, p. 4. NYTr, March 31, 1881, p. 1. SFCh, March 31, 1881, p. 4.

75. Smith, pp. 221-22. J. H. Burke, "The Bonanza Suits of 1877," in *History of the Bench and Bar of California* (Los Angeles: Commercial Printing House, 1901).

76. GHN, August 5, 1879, p. 3; August 6, 1879, p. 3. NYT, August 13, 1879, p. 5; August 17, 1879, p. 5. EMJ, August 16, 1879, p. 108.

77. SFCh, December 21, 1877, p. 6; February 10, 1878, p. 5. VCC, December 21, 1877, p. 3.

78. GHN, May 10, 1877, p. 3. MSP, May 12, 1877, p. 300; May 26, 1877, p. 329. SFCh, May 15, 1877, p. 4. EMJ, May 19, 1877, p. 346.

79. SFCh, September 5, 1877, p. 4; September 9, 1877, p. 1.

80. SFB, December 6, 1877, p. 2. VCC, December 7, 1877, p. 3. VTE, December 8, 1877, p. 2. SFCh, December 10, 1877, p. 2; January 16, 1878, p. 2; January 20, 1878, p. 1; March 7, 1878, p. 3.

81. GHN, July 20, 1877, p. 3; July 26, 1877, p. 3; August 1, 1877, p. 3; August 2, 1877, p. 2; August 3, 1877, p. 3; August 9, 1877, p. 2. VTE, December 2, 1877, p. 3. VCC, December 3, 1877, p. 3; December 4, 1877, p. 3; December 5, 1877, p. 3. SFCh, November 29, 1877, p. 2; December 1, 1877, p. 2; December 5, 1877, p. 2.

82. VCC, December 12, 1877, p. 3; December 13, 1877, p. 2. SFB, December 13, 1877, p. 1. SFCh, December 10, 1877; December 13, 1877, p. 3. GHN, December 13, 1877, p. 2. VTE, December 13, 1877, p. 2. *Esmeralda (NV) Herald,* December 15, 1877, p. 2. EMJ, December 22, 1877, p. 464. MSP, December 22, 1877, p. 396.

83. GHN, December 18, 1877, p. 3. SFCh, December 14, 1877, p. 8; December 15, 1877, p. 4; December 19, 1877, p. 6; December 23, 1877, p. 5. VTE, December 19, 1877, p. 3. VCC, December 21, 1877, p. 3; December 24, 1877, p. 3.

84. VCC, December 11, 1877, p. 1.

85. GHN, December 12, 1877, p. 2; December 14, 1877, p. 3. VCC, December 29, 1877, p. 3; December 12, 1877, p. 3. SFCh, December 26, 1877, p. 3; December 27, 1877, p. 3; April 8, 1878, p. 2; April 19, 1878, p. 1.

86. VTE, November 9, 1886, p. 1.

87. MSP, October 9, 1886, p. 240; November 6, 1886, p. 304; November 20, 1886, p. 336. EMJ, October 23, 1886, p. 289; October 30, 1886, p. 309. VTE, November 13, 1886, p. 1.

88. NYT, November 28, 1886, p. 7.

89. NYT, November 30, 1886, p. 1; December 5, 1886, p. 8.

90. SFCh, December 3, 1886, p. 2. NYT, December 3, 1886, p. 1; December 4, 1886, p. 1.

91. SFCh, December 4, 1886, p. 2; December 5, 1886, p. 8; December 7, 1886, p. 2; December 12, 1886, p. 5; December 15, 1886, pp. 4, 5. NYT, December 7, 1886, p. 1. King, 300-3.

92. SFCh, December 8, 1886, p. 5. NYT, December 8, 1886, p. 1; December 10, 1886, p. 5. Smith, pp. 282-84.

93. EMJ, January 15, 1886, p. 54.

94. SFCh, February 17, 1872, p. 3; February 18, 1872, p. 1. Lord, pp. 245-49. EMJ, July 2, 1892, p. 3.

95. SFCh, March 31, 1877, p. 4. EMJ, July 2, 1892, p. 3.

96. MSP, July 17, 1886, p. 35; October 23, 1886, p. 265.

97. EMJ, September 12, 1891, p. 317.

98. EMJ, November 8, 1890, p. 542.

99. J. R. Browne, "Mineral Resources of the States and Territories West of the Rocky Mountains," (House Exec. Doc. 202, 40th Cong., 2d sess., 1868), 366. R. W. Raymond, *Mines of the West* (N.Y.: Ford, 1869), 54. SFAC, July 2, 1862, p. 1; August 20, 1865, p. 1. MSP, June 1, 1889, p. 400; June 14, 1890, p. 400; August 16, 1890, p. 110; October 11, 1890, p. 238. EMJ, March 21, 1891, p. 346.

100. EMJ, May 23, 1891, p. 603; December 26, 1891, pp. 719, 722-23.

101. MSP, November 14, 1891, p. 316; October 24, 1908, p. 570.

102. EMJ, January 25, 1890, p. 117.

103. EMJ, January 29, 1890, p. 367.

104. NYT, December 3, 1879, p. 2; December 12, 1879, p. 5. EMJ, March 19, 1887, p. 209; December 13, 1890, p. 683. MSP, March 12, 1892, p. 192.

105. *Mining Exchange Journal,* August 29, 1890, p. 2. EMJ, September 13, 1890,
p. 304; October 4, 1890, p. 397; November 14, 1891, p. 575. SFCh, November 19, 1891,
p. 12; December 16, 1891, p. 5. NYT, November 26, 1891, p. 1.

106. EMJ, November 7, 1891, p. 539. SFCh, December 3, 1891, p. 3; December 9,
1891, p. 10.

107. SFE, November 20, 1891, p. 6. SFCh, December 2, 1891, p. 5; December 8, 1891,
p. 12. EMJ, November 28, 1891, p. 624; December 5, 1891, p. 652.

108. VTE, November 24, 1891, p. 2. SFCh, November 26, 1891, p. 12; December 3,
1891, p. 5. MSP, November 28, 1891, p. 350. EMJ, December 5, 1891, p. 649; February 6,
1892, p. 189.

109. EMJ, December 12, 1891, pp. 685-86, December 19, 1891, pp. 709-10; February
13, 1892, p. 213; February 20, 1892, pp. 237-38. VTE, February 9, 1892, p. 2.

110. EMJ, October 18, 1890, p. 447; October 25, 1890, p. 476; November 8, 1890,
p. 540; April 15, 1893, p. 338; May 27, 1893, p. 481. SFE, November 26, 1891, p. 3. MSP,
December 12, 1891, p. 384.

111. MSP, June 1, 1888, p. 400; December 5, 1891, pp. 368-69.

112. NYT, December 4, 1891, p. 2. SFCh, December 4, 1891, p. 7. MSP, December 19,
1891, pp. 396-97.

113. VTE, March 2, 1892, p. 2; March 11, 1892, p. 2; April 1, 1892, p. 2; May 27,
1892, p. 2. EMJ, June 4, 1892, pp. 592-93.

114. EMJ, May 28, 1892, p. 563; August 13, 1892, p. 145. MSP, June 4, 1892, p. 410;
June 25, 1892, p. 464.

115. VTE, February 26, 1892, p. 3. NYT, June 6, 1892, p. 4; June 11, 1892, p. 4; June
14, 1892, p. 4. EMJ, July 2, 1892, p. 1. MSP, September 17, 1892, p. 186.

116. *Fox v. Hale & Norcross Silver Min. Co.,* 41 Pac 308, 328 (1895); 44 Pac 1022,
1023 (1896); 53 Pac 32, 169 (1898); 54 Pac 731 (1898). MSP, October 15, 1892, p. 250;
February 25, 1893, pp. 113, 118; May 16, 1896, p. 394; August 14, 1897, p. 143; October
8, 1898, p. 351; October 22, 1892, p. 399; October 29, 1892, p. 423; November 12, 1898,
p. 479. EMJ, September 3, 1892, p. 232; October 1, 1892, p. 327; November 5, 1892,
p. 447; November 12, 1892, p. 472; November 26, 1892, pp. 505, 519, 522; December 3,
1892, p. 544; December 17, 1892, p. 591; March 4, 1893, p. 193; July 21, 1894, p. 63;
August 10, 1895, p. 135; August 17, 1895, p. 159; November 2, 1895, p. 426. SFCh,
November 12, 1895, p. 7; December 10, 1895, p. 8; December 11, 1895, p. 16.

117. MSP, April 30, 1892, p. 322; June 11, 1892, p. 430. EMJ, June 11, 1892, p. 613.
Mining Industry & Tradesman, February 25, 1892, p. 94.

118. EMJ, September 5, 1892, p. 266; October 31, 1891, p. 498. SFCh, January 12,
1892, p. 12. MSP, March 19, 1892, p. 212.

119. SFCh, December 6, 1891, p. 24.

120. SFCh, December 25, 1891, p. 9. VTE, February 24, 1892, p. 2.

121. EMJ, February 27, 1892, p. 262; July 2, 1892, pp. 15-16; August 13, 1892, p. 148.
VTE, March 10, 1892, p. 3; March 11, 1892, p. 3; March 13, 1892, p. 3. MSP, July 30,
1892, p. 74.

122. EMJ, December 19, 1896, p. 597; March 27, 1897, p. 313; June 5, 1897, p. 589;
June 12, 1897, p. 618; June 19, 1897, p. 649; June 4, 1898, p. 689; January 7, 1899, p. 37.

123. EMJ, July 30, 1892, p. 111; August 13, 1892, pp. 160-61; September 3, 1892,
pp. 217, 232.

124. EMJ, August 22, 1891, p. 211; November 7, 1891, p. 525; October 29, 1892,
p. 423. NYT, December 6, 1891, p. 8. SFCh, July 20, 1894, p. 10. MSP, July 28, 1894, p. 50.

125. SFCh, August 23, 1895, p. 9. NYT, August 24, 1895, p. 1.

126. NYT, May 28, 1888, p. 1.

127. VTE, February 12, 1892, p. 2.

128. EMJ, March 5, 1892, pp. 286-87.

129. SFCh, August 27, 1895, p. 16; August 28, 1895, p. 12; December 10, 1895, p. 8. VCC, August 29, 1895, p. 3. EMJ, December 21, 1895, p. 593. Fox v. Mackay et al., 56 Pac. 434, 435 (1899); Fox v. Mackay et al., 57 Pac. 670, 672 (1899).

130. MSP, January 25, 1896, p. 63.

131. J. H. Culver and J. C. Syer, *Power and Politics in California,* 2nd ed. (N.Y.: Macmillan, 1986), 39.

132. *Mining Exchange Journal,* October 1, 1890, p. 2. MSP, February 17, 1894, p. 98. EMJ, July 30, 1892, p. 114. C. G. Yale, in *California Mines and Minerals* (San Francisco: California Mining Association, 1899), 6-7.

133. VTE, June 7, 1891, p. 3. EMJ, July 4, 1891, p. 16; August 27, 1892, p. 210.

134. EMJ, December 12, 1896, p. 552; January 2, 1897, p. 33.

135. EMJ, May 13, 1893, p. 433. MSP, October 24, 1908, p. 572.

136. MSP, October 24, 1908, p. 573.

137. MSP, October 24, 1908, p. 573.

138. MSP, October 24, 1908, p. 574.

139. NMN, October 22, 1908, p. 7. MSP, October 24, 1908, p. 575.

140. MSP, September 30, 1893, p. 210. NMN, March 12, 1908, p. 1. EMJ, May 23, 1914, pp. 1065, 1068.

141. EMJ, January 7, 1899, p. 36. MSP, October 24, 1908, p. 574.

142. NMN, March 12, 1908, p. 8; March 19, 1908, p. 7.

143. MSP, February 10, 1912, p. 232. 151 Cal., pp. 427-31. 153 Pac., 710-12.

144. EMJ, May 14, 1910, p. 1038; September 24, 1910, p. 625.

145. *Mining & Engineering World,* January 10, 1914, p. 59.

146. W. S. Palmer, "Present Status of Comstock Mining," *Pacific Mining News of Engineering & Mining Journal-Press* 2, no. 6 (June 1923): 159-61. EMJ, September 24, 1927, p. 501.

147. Smith, p. 63. Smith defended the Comstock operators on the basis that they met the ethical standards of their time. But Smith violated his own rule to show Mackay in a better light. Mackay and his partners arranged milling contracts with themselves, contrary to the law; but Smith noted in justification of Mackay that the law was changed in 1933 to legalize what had been unlawful in Mackay's own time (pp. 221, 255).

148. Lord, pp. 247-49.

149. Lord, pp. 287-88.

150. SFCh, January 10, 1875, p. 4.

151. Smith, p. 262. EMJ, April 23, 1891, p. 491; October 31, 1891, p. 513; November 7, 1891, p. 525; September 24, 1892, pp. 289-90; October 29, 1892, pp. 410-11. MSP, September 17, 1892, p. 186; October 22, 1892, p. 266; April 29, 1893, p. 258.

152. Smith, pp. 63, 254.

153. SFB, May 25, 1878, p. 2. EMJ, January 28, 1893, p. 73. Smith, p. 219. J. P. Young, *Journalism in California* (San Francisco: Chronicle, 1915), 145. John Bruce, *Gaudy Century* (N.Y.: Random House, 1948), 232. R. E. Lingenfelter and K. R. Gash, *The Newspapers of Nevada* (Reno: University of Nevada Press, 1984), 97, 254. Lucius Beebe, *Comstock Commotion* (Stanford: Stanford University Press, 1954), 96-97.

154. MSP, February 24, 1877, p. 114.

155. M. V. Scars, *Mining Stock Exchanges 1860-1930* (Bozeman: University of Montana Press, 1973), 179. WSJ, March 7, 1962, p. 28; December 12, 1962, p. 1; August 16, 1967, p. 6.

Chapter 4. Snow Job at the Emma Mine

1. SLTr, November 16, 1871, p. 1.

2. U.S. Congress, House Committee on Foreign Affairs, "Emma Mine Investigation," 44th Cong., 1st sess., 1876, H. Rept. 579, pp. 45.

3. EMJ, May 13, 1876, p. 467; August 8, 1914, pp. 248-51.

4. MSP, September 24, 1870, p. 218; October 8, 1870, p. 253. *Daily Central City (CO) Register,* January 7, 1871, p. 1. *Nebraska City News,* February 25, 1871, p. 2.

5. U.S. Congress, House Committee on Foreign Affairs, "Emma Mine Investigation," 44th Cong., 1st sess., 1876, H. Rept. 579, pp. 443-45.

6. NYT, April 13, 1871, p. 5. SFCh, March 3, 1876, p. 3.

7. V. C. Heikes, "Little Cottonwood District," in *The Ore Deposits of Utah* (USGS Prof. Paper 111, 1920), 255.

8. EMJ, September 4, 1875, p. 234. NYT, April 19, 1876, p. 5.

9. SLTr, March 26, 1876, p. 1. U.S. Congress, House Committee on Foreign Affairs, "Emma Mine Investigation," 44th Cong., 1st sess., 1876, H. Rept. 579, pp. 57, 753.

10. SFAC, February 1, 1875, p. 1. U.S. Congress, House Committee on Foreign Affairs, "Emma Mine Investigation," 44th Cong., 1st sess., 1876, H. Rept. 579, p. 78.

11. SFAC, January 31, 1863, p. 1. W. T. Jackson, "The Infamous Emma Mine: A British Interest in the Little Cottonwood District, Utah Territory," *Utah Historical Quarterly* 23 (1955): 339-62. U.S. Congress, House Committee on Foreign Affairs, "Emma Mine Investigation," 44th Cong., 1st sess., 1876, H. Rept. 579, p. 752.

12. NYT, August 31, 1899, p. 7.

13. NYT, June 14, 1874, p. 4.

14. NYT, February 14, 1874, p. 2; March 14, 1874, p. 3; July 17, 1874, p. 4.

15. EMJ, November 27, 1875, p. 523.

16. J. F. Fulton and E. H. Thompson, *Benjamin Silliman* (N.Y.: Schuman, 1947), 240.

17. EMJ, December 12, 1871, p. 377; December 19, 1871, p. 393. SLTr, September 2, 1871, p. 2; December 13, 1871, p. 3.

18. NYT, November 14, 1876, p. 2.

19. NYTr, November 24, 1875, p. 4; November 30, 1875, p. 2.

20. Econ, November 11, 1871, p. 1365.

21. U.S. Congress, House Committee on Foreign Affairs, "Emma Mine Investigation," 44th Cong., 1st sess., 1876, H. Rept. 579, pp. xii, 69.

22. U.S. Congress, House Committee on Foreign Affairs, "Emma Mine Investigation," 44th Cong., 1st sess., 1876, H. Rept. 579, pp. 320-21.

23. EMJ, March 25, 1876, p. 293.

24. MJ, August 12, 1871, p. 710.

25. MJ, November 11, 1871, p. 987; December 9, 1871, p. 1102. NYT, December 15, 1876, p. 2. NYTr, December 15, 1876, p. 2.

26. MJ, November 25, 1871, p. 1052.

27. U.S. Congress, House Committee on Foreign Affairs, "Emma Mine Investigation," 44th Cong., 1st sess., 1876, H. Rept. 579, p. 404. NYT, March 13, 1876, p. 1; March 17, 1876, p. 1, c. 4; March 18, 1876, p. 1, c. 3. NYTr, March 17, 1876, p. 1.

28. U.S. Congress, House Committee on Foreign Affairs, "Emma Mine Investigation," 44th Cong., 1st sess., 1876, H. Rept. 579, p. 135.

29. Econ, June 8, 1872, p. 709.

30. MJ, April 13, 1872, p. 338; April 20, 1872, p. 363. U.S. Congress, House Committee on Foreign Affairs, "Emma Mine Investigation," 44th Cong., 1st sess., 1876, H. Rept. 579, pp. 852-54.

31. MJ, June 8, 1872, p. 532.

32. SLTr, December 13, 1871, p. 3. NYT, December 19, 1876, p. 3. NYTr, December 22, 1876, p. 2. MSP, April 21, 1883, pp. 272-73. MJ, May 11, 1872, p. 443. K. M. Johnson, "Two Mines: Two Books," *Journal of the West* 17, no. 1 (January 1978): pp. 35-38. U.S. Congress, House Committee on Foreign Affairs, "Emma Mine Investigation," 44th Cong., 1st sess., 1876, H. Rept. 579, p. 573. G. A. Lawrence, *Silverland* (London: Chapman and Hall, 1873).

33. Econ, May 4, 1872, pp. 544-45.

34. Asbury Harpending, *The Great Diamond Hoax and Other Stirring Incidents in the Life of Asbury Harpending* (1915; repr., Norman: University of Oklahoma Press), 134-37.

35. U.S. Congress, House Committee on Foreign Affairs, "Emma Mine Investigation," 44th Cong., 1st sess., 1876, H. Rept. 579, pp. 517–21, 754.

36. U.S. Congress, House Committee on Foreign Affairs, "Emma Mine Investigation," 44th Cong., 1st sess., 1876, H. Rept. 579, pp. 82, 128.

37. U.S. Congress, House Committee on Foreign Affairs, "Emma Mine Investigation," 44th Cong., 1st sess., 1876, H. Rept. 579, pp. 406-7.

38. MSP, August 3, 1872, p. 72; November 16, 1872, p. 305. MJ, August 10, 1872, p. 748. TL, November 12, 1872, p. 6, c. 1.

39. Econ, June 15, 1872, pp. 738-39. *Salt Lake Herald,* November 5, 1872, p. 2. MJ, June 8, 1872, p. 538; June 15, 1872, p. 548. U.S. Congress, House Committee on Foreign Affairs, "Emma Mine Investigation," 44th Cong., 1st sess., 1876, H. Rept. 579, pp. 397-400.

40. SLTr, July 30, 1872, p. 2. *Salt Lake Herald,* November 28, 1872, p. 2.

41. NYTr, February 15, 1875, p. 2.

42. EMJ, October 3, 1885, p. 232.

43. NYT, May 24, 1876, p. 5.

44. NYTr, December 20, 1876, p. 2.

45. MJ, January 18, 1873, p. 76; March 1, 1873, p. 241; May 17, 1873, p. 544; May 24, 1873, p. 572; May 31, 1873, p. 603. MSP, March 29, 1873, p. 204; April 5, 1873, p. 216.

46. MSP, June 6, 1874, p. 361.

47. SFAC, February 16, 1873, p. 1. MSP, July 19, 1873, p. 42; May 30, 1874, p. 344; June 13, 1874, p. 376.

48. U.S. Congress, House Committee on Foreign Affairs, "Emma Mine Investigation," 44th Cong., 1st sess., 1876, H. Rept. 579, pp. 632-34.

49. EMJ, February 6, 1875, p. 3. MSP, January 10, 1874, pp. 18, 28. It is more likely that Williams "dressed" the mine, leaving a few inches of ore on the walls to create an illusion that the ore extended indefinitely further.

50. NYTr, December 30, 1874, pp. 2, 4; April 5, 1875, p. 7; January 17, 1876, p. 1. GHN, January 20, 1875, p. 2. EMJ, January 2, 1875, p. 3. MSP, July 4, 1874, p. 10; August 15, 1874, p. 108; April 24, 1875, p. 273. SFCh, August 12, 1874, p. 3; January 8, 1875, p. 1.

51. NYTr, January 21, 1876, p. 1 c. 4; February 7, 1876, p. 1 c. 4.

52. NYTr, March 8, 1876, p. 1.

53. U.S. Congress, House Committee on Foreign Affairs, "Emma Mine Investigation," 44th Cong., 1st, sess., 1876, H. Rept. 579, pp. 348-49. C. C. Spence, "Robert C. Schenk and the Emma Mine Affair," *Ohio Historical Quarterly* 68, no. 2 (1959): 141-60.

54. EMJ, March 18, 1876, April 15, 1876, p. 365. U.S. Congress, House Committee on Foreign Affairs, "Emma Mine Investigation," 44th Cong., 1st sess., 1876, H. Rept. 579, pp. 761-62.

55. U.S. Congress, House Committee on Foreign Affairs, "Emma Mine Investigation," 44th Cong., 1st sess., 1876, H. Rept. 579, p. xvi.

56. Econ, November 18, 1876, p. 1342.

57. NYTr, April 16, 1875, p. 2; December 14, 1876, p. 2. EMJ, March 18, 1876, p. 272. NYT, November 16, 1876, p. 8; December 20, 1876, p. 2. VTE, March 9, 1877, p. 1.

58. NYTr, February 4, 1875, p. 9; December 16, 1876, p. 5.

59. NYT, April 21, 1877, p. 3; April 26, 1877, p. 10.

60. *New York Post,* April 19, 1877, p. 4; April 29, 1877, p. 4. NYT, April 28, 1877, p. 3; April 29, 1877, p. 5. EMJ, May 5, 1877, p. 291. VTE, May 8, 1877, p. 2.

61. NYTr, June 13, 1877, p. 4.

62. NYT, May 3, 1877, p. 2; March 3, 1878, p. 3.

63. Econ, January 15, 1876, p. 67. EMJ, September 16, 1876, p. 189. *Pall Mall Gazette,* November 24, 1876, p. 6.

64. MSP, December 4, 1875, p. 364; April 8, 1876, p. 232; January 26, 1878, p. 54. EMJ, January 12, 1878, p. 25; August 16, 1879, p. 114; November 20, 1880, p. 330. *Esmeralda (NV) Herald,* January 26, 1878, p. 2. *Salt Lake Herald,* March 19, 1879, p. 3. SLTr, January 1, 1881, p. 1.

65. NYTr, March 4, 1878, p. 8. NYT, June 12, 1878, p. 3.

66. NYTr, February 1, 1876, p. 1; February 14, 1876, p. 1.

67. NYTr, May 12, 1877, p. 2; October 15, 1877, p. 1; July 30, 1880, p. 1. NYT, June 11, 1876, p. 10; July 30, 1880, p. 2. EMJ, July 31, 1880, p. 73.

68. NYTr, October 22, 1879, p. 2; October 1, 1880, p. 2. NYT, October 22, 1879, p. 3; October 1, 1880, p. 3; June 9, 1882, p. 8. EMJ, October 2, 1880, p. 225.

69. EMJ, December 4, 1880, p. 362.

70. EMJ, May 27, 1882, p. 282; June 10, 1882, p. 306. MSP, April 21, 1883, pp. 272-73.

71. EMJ, November 20, 1880, p. 329; October 15, 1881, p. 257. MJ, May 10, 1884, p. 535.

72. EMJ, August 1, 1885, p. 83; September 5, 1891, p. 277.

73. EMJ, May 8, 1875, pp. 305-6; June 12, 1875, p. 436; March 17, 1888, p. 196; August 6, 1892, p. 136; July 7, 1894, p. 21; December 29, 1894, p. 621. FT December 31, 1888, p. 2.

74. EMJ, September 26, 1908, p. 629.

Chapter 5. A Brief History of Diamond-Mining Frauds in America

1. SFB, November 27, 1872, p. 3. SFCh, November 28, 1872, p. 3.

2. MSP, April 6, 1861, p. 3. RMN, January 19, 1867, p. 3; April 24, 1867, p. 3; May 30, 1867, p. 4; May 25, 1867, p. 4.

3. MSP, March 11, 1871, p. 147.

4. Arizona Citizen, August 31, 1872, p. 1. New Mexico Republican Review, December 3, 1870, p. 2. EMJ, February 21, 1871, p. 120.

5. A. C. Hamlin, "Origin and Properties of the Diamond," Proceedings of the American Association for the Advancement of Science 22 (1874): 104-8.

6. GDR to Harpending, December 1, 1870, MS 950, box 1, folder 2, Harpending Collection, California Historical Society, San Francisco.

7. TL, December 23, 1874, p. 10.

8. TL, August 29, 1872, p. 5.

9. Agreement between Arnold and Harpending, October 31, 1871, MS 950, box 1, folder 2, Harpending Collection, California Historical Society, San Francisco.

10. TL, August 30, 1872, p. 5.

11. Dodge to Harpending, April 13, 1872, MS 950, box 1, folder 5, Harpending Collection, California Historical Society, San Francisco.

12. R. W. Raymond, Statistics of Mines and Mining in the States and Territories West of the Rocky Mountains (Washington, DC: Government Printing Office, 1872): 490-91.

13. Congressional Globe, 41st Cong., 3d sess., 1871, pt. 2: pp. 896-97, 978-79, 985-86, 997, 1014, 1026.

14. Congressional Globe, 42nd Cong., 2d sess., 1872, pt. 1: pp. 395, 532-35, 2456-62, 2897-99.

15. MS 950, box 1, folder 3, Harpending Collection, California Historical Society, San Francisco. MSP, November 9, 1872, p. 300. SFCa, November 16, 1872, p. 3. RMN, November 20, 1872, p. 4.

16. TL, December 2, 1872, p. 7. A. J. Dahl, "British Investment in California Mining, 1870-1890" (PhD diss., University of California, Berkeley, 1961), 86-89.

17. TL, January 15, 1868, p. 1.

18. SFB, December 5, 1872, p. 1. Some later accounts held that the expeditioners were blindfolded between Rawlins and Diamond Butte, but contemporary accounts do not mention blindfolds.

19. Prescott (AZ) Weekly Miner, August 24, 1872, p. 1.

20. SFAC, November 26, 1872, p. 1.

21. SFCh, August 6, 1872, p. 3.

22. SFCh, July 31, 1871, p. 3; August 1, 1872, p. 3; August 2, 1872, p. 3; August 3, 1872, p. 3. SFAC, August 1, 1872, p. 1. MSP, August 3, 1872, p. 72. EMJ, August 20, 1872, pp. 121, 123; September 3, 1872.

23. SFCh, September 12, 1872, p. 2; December 16, 1872, p. 4. RMN, August 13, 1872, p. 1. *Arizona Citizen*, August 17, 1872, p. 2. MSP, September 7, 1872, pp. 152-53.

24. TL, November 20, 1872, p. 6.

25. SFCh, September 4, 1872, p. 3.

26. TL, December 21, 1874, p. 11.

27. SFCh, October 7, 1872, p. 3. SFCa, October 7, 1872, p. 3.

28. RMN, August 28, 1872, p. 2; August 30, 1872, p. 4; September 4, 1872, p. 1.

29. GHN, October 26, 1872, p. 2.

30. SFCh, August 4, 1872, p. 1; August 9, 1872, p. 3; August 10, 1872, p. 3. *Prescott Weekly Miner*, August 24, 1872, p. 2.

31. *Grass Valley (CA) Union*, November 19, 1871, p. 3. SFCa, August 10, 1872, p. 1. SFCh, October 8, 1872, p. 3; October 11, 1872, p. 3.

32. SFCh, August 8, 1872, p. 2; August 14, 1872, p. 3; August 21, 1872, p. 3; August 30, 1872, p. 3; November 14, 1872, p. 1. *Prescott (AZ) Weekly Miner*, October 12, 1872, p. 3; November 16, 1872, p. 2, November 30, 1872, p. 2. MSP, August 10, 1872, p. 89; August 24, 1872, pp. 121, 124; September 7, 1872, pp. 155, 156. SFAC, August 10, 1872, p. 1; August 27, 1872, p. 1. EMJ, September 17, 1872.

33. RMN, September 18, 1872, p. 1. SFCh, September 20, 1872, p. 2; September 30, 1872, p. 1; October 13, 1872, p. 1. *Prescott (AZ) Weekly Miner*, September 21, 1872, p. 4. TL, December 2, 1872, p. 7.

34. SFB, October 28, 1872, p. 1.

35. RMN, September 7, 1872, p. 4; September 21, 1872, p. 4.

36. *Prescott (AZ) Weekly Miner*, September 7, 1872, p. 2; September 14, 1872, p. 2; December 14, 1872, p. 2. SFB, October 29, 1872, p. 1. SFE, December 11, 1872, p. 1; December 27, 1872, p. 3.

37. RMN, November 19, 1872, p. 4.

38. *Laramie (WY) Independent*, August 19, 1872, p. 2; August 20, 1872, p. 2. SFCh, August 20, 1872, p. 1. RMN, August 22, 1872, p. 4.

39. *Laramie (WY) Independent*, September 9, 1872, p. 2. RMN, September 8, 1872, p. 4; September 11, 1872, p. 2. SFCh, September 15, 1872, p. 8.

40. *St. Louis Globe-Democrat*, December 17, 1883, p. 2.

41. RMN, November 14, 1872, p. 4. MSP, November 16, 1872, p. 316. SFCh, October 29, 1872, p. 1; November 20, 1872, p. 1. SFAC, November 26, 1872, p. 2; November 28, 1872, p. 1; November 25, 1872, p. 1; November 27, 1872, p. 1.

42. *Arizona Citizen*, November 2, 1872, p. 2.

43. A. D. Wilson, "The Great California Diamond Mines: A True Story," *Overland Monthly* 4 (1904): 291-96.

44. R. G. Willson, "'A Salted Field' Ends Hopes of World's Big Jewel Market," *Arizona Ways & Days* (July 28, 1957): 30-31.

45. SFCh, November 24, 1872, p. 5. SFB, December 6, 1872, p. 1.

46. SFCh, November 10, 1872, p. 8.

47. RMN, December 4, 1872, p. 4.

48. SFCh, November 21, 1872, p. 3; November 22, 1872, p. 2.

49. RMN, November 23, 1872, p. 4. TL, December 21, 1872, p. 7.

50. MSP, November 30, 1872, p. 344. SFCh, November 25, 1872, p. 3; November 26, 1872, p. 3.

51. SFCh, December 4, 1872, p. 3.

52. SFCh, November 28, 1872, p. 2. MJ, March 8, 1873, p. 266.

53. SFCh, December 24, 1872, p. 1; January 4, 1873, p. 1. *Louisville Courier-Journal*, December 16, 1872, p. 3; August 16, 1874, p. 4. SFB, March 25, 1873, p. 1. MSP, March 8,

1879, p. 145. D. E. McClure Jr., *Two Centuries in Elizabethtown and Hardin County, Kentucky* (Elizabethtown: Hardin County Historical Society, 1979), 354-56.

54. J. L. Considine, "The Great Diamond Swindle," *Sunset* 52 (February 1924): 49-58.

55. SFCh, December 8, 1872, p. 1.

56. SFCh, November 27, 1872, p. 3. *Prescott Weekly Miner,* December 21, 1872, p. 2.

57. E. M. Muir, "The Great Diamond Swindle," *New Mexico* 26, no. 8 (1948): 47-48. Rita Hill and Janaloo Hill, "Alias Shakespeare the Town Nobody Knew," *New Mexico Historical Review* 42 (1967): 219. H. N. Ferguson, "The Great Diamond Hoax of 1872," *Desert* 20 (February 1957): 4-7. E. L. Conrotto, *Lost Gold and Silver Mines of the Southwest* (Mineola, N.Y.: Dover, 1991), 186-88.

58. M. E. McCallum and C. D. Mabarak, "Diamond in Kimberlitic Diatremes of Northern Colorado," *Geology* 4 (1976): 467-69.

59. RMN, April 5, 1910, p. 5.

60. *Virginia City (NV) Union,* August 11, 1864, p. 1. *New York Post,* September 9, 1875, p. 3. NYT, September 9, 1875, p. 4. L. J. Arrington, *History of Idaho* (Boise: University of Idaho Press, 1994), 220-23. H. H. Bancroft, *History of Washington, Idaho, and Montana* (San Francisco: History Company, 1890), 448-67.

61. *Owyhee (ID) Avalanche,* November 18, 1865, p. 1. *Idaho Statesman,* November 9, 1865, p. 2; January 11, 1865, p. 3.

62. *Owyhee (ID) Avalanche,* November 25, 1865, p. 3. MSP, December 16, 1865, p. 369. T. G. McFadden, "We'll All Wear Diamonds," *Idaho Yesterdays* 10, no. 2 (1966): 2-7.

63. SFAC, August 2, 1865, p. 1.

64. SFAC, February 15, 1866, p. 1.

65. *Owyhee (ID) Avalanche,* December 16, 1865, p. 2.; December 23, 1865, p. 2.

66. *Owyhee (ID) Avalanche,* January 13, 1866, p. 3. MSP, February 10, 1866, p. 97.

67. *Owyhee (ID) Avalanche,* January 27, 1866, p. 3; April 7, 1866, p. 3; April 21, 1866, p. 3; April 28, 1866, p. 3; January 7, 1893, p. 1. MSP, March 17, 1866, p. 174.

68. *Owyhee (ID) Avalanche,* March 10, 1866, p. 2. NYT, December 21, 1866, p. 1; December 28, 1866, p. 1. Bancroft, 467.

69. *Idaho Statesman,* June 2, 1866, p. 2.

70. NYT, January 3, 1893.

71. TS, July 21, 1905, p. 6. DMR, August 25, 1905, p. 1.

72. *Tonopah (NV) Miner,* July 22, 1905, p. 4. TS, July 22, 1905, pp. 1-2, 5.

73. TS, July 23, 1905, p. 1. P. I. Earl, "Tonopah's Great Diamond Rush," *California Mining Journal* 52, no. 5 (January 1983): 36-37.

74. TS, July 25, 1905, p. 1; July 26, 1905, pp. 1-2; July 26, 1905, p. 5.

75. TS, July 27, 1905, p. 1; July 28, 1905, p. 1.

76. *Tonopah (NV) Bonanza,* July 29, 1905, p. 8. *Goldfield (NV) Review,* August 10, 1905, p. 6.

77. EMJ, July 17, 1928, pp. 113-14. NYT, December 14, 1928, p. 36.

78. *Milwaukee Sentinel,* May 19, 1884, p. 3; July 9, 1884, p. 3. W. A. Hobbs, "The Diamond Field of the Great Lakes," *Journal of Geology* 7, no. 4 (1899): 375. *Saturday Evening Post,* January 7, 1950, p. 97. A. A. Vierthaler, "There Are Diamonds in Wisconsin," *Lapidary Journal* 15, no. 1 (1958): 21-22. *Milwaukee Journal,* November 9, 1978, Accent section, p. 1. Kevin Krajick, *Barren Lands* (N.Y.: Holt, 2001), 114-15. G. F. Kunz, "Precious Stones," in *Mineral Resources of the United States 1886* (Washington, DC: USGS, 1887), 598.

79. Drep, June 30, 1881, p. 6. SLTr, November 8, 1882, p. 2. NYTr, November 23, 1882, p. 4. SFCh, December 3, 1891, p. 10. *Jerome (AZ) Mining News* June 26, 1899, p. 2. G. F. Kunz, "Precious Stones," in *Mineral Resources of the United States 1901* (Washington, DC: USGS, 1902), 731. *Engineering & Mining World,* August 12, 1911, p. 278. NYT, January 4, 1925, p. 26.

80. *Denver Post,* December 27, 1901; May 8, 1921, magazine section, p. 3. RMN, December 27, 1901; January 6, 1909, p. 1. SFCa, January 1, 1901, p. 1. *Pueblo (CO) Chieftain,* January 6, 1909, p. 1. D. B. Sterrett, "Precious Stones," in *Mineral Resources of the*

Year 1906 (Washington, DC: USGS, 1907), 1219. R. B. Cook, *Minerals of Georgia* (Atlanta: Georgia Geologic and Water Resources Division, 1978), pp. 19-20.

Chapter 6. Tin Men

1. J. T. Lovewell, "Gold in Kansas Shales," Transactions of the Kansas Academy of Science, pp. 129-37. *Topeka State Journal,* June 29, 1903.

2. Henry Schoolcraft, *Historical and Statistical Information Respecting the History, Condition, and Prospects of the Indian Tribes of the United States* 1 (Philadelphia: Lippincott, Grambo, 1851), 157-59.

3. B. F. Mudge, *First Annual Report on the Geology of Kansas* (Lawrence, Kans.: Speer, 1866), 30-31.

4. Other Native Americans turned to swindling for fun and profit. For a ruse by Indians in Minnesota, see DMR, February 18, 1905, p. 14.

5. A. C. Koch, *Journey through a Part of the United States of North America in the Years 1844 to 1846*, transl. and ed. by E. A. Stadler (1846; repr., Carbondale: Southern Illinois University Press, 1972), xvii-xxxv. M. J. O'Brien, *Paradigms of the Past* (Columbia: University of Missouri Press, 1996), 77-87.

6. R. H. Dott Jr. and W. M. Jordan, "Doctor Koch's Horrendous Hydrarchus," *Geotimes* 44, no. 3 (March 1999): 20-24.

7. *St. Louis Democrat,* December 7, 1859, p. 2. "Reported Discovery of Gold Mines in Missouri," *Mining Magazine,* 2nd ser. 1 (March 1860): 406-7.

8. *American Journal of Mining,* April 13, 1867, p. 48; August 31, 1867, p. 138. *Iron County (MO) Register,* July 13, 1867, p. 3. *St. Louis Democrat,* July 17, 1867, p. 4. *Chicago Tribune,* August 7, 1867, p. 2.

9. *St. Louis Democrat,* July 22, 1867, p. 2. John Rowe, *The Hard-Rock Men* (N.Y.: Barnes & Noble, 1974), 163.

10. *American Journal of Mining,* August 3, 1867, p. 69; December 14, 1867, p. 373. Erasmus Haworth, "Annual Report for 1894," *Missouri Geological Survey* 8(1895): 134-35.

11. *St. Louis Democrat,* July 4, 1867, p. 2. *Iron County (MO) Register,* August 8, 1867, p. 2; October 17, 1867, p. 2. *American Journal of Mining,* September 14, 1867, p. 165; October 26, 1867, p. 265; November 2, 1867, p. 277; November 23, 1867, p. 329; December 14, 1867, p. 377.

12. MJ, August 17, 1867, p. 538; October 19, 1867, p. 699; November 28, 1867, p. 731; January 18, 1868, p. 54; March 1, 1873, p. 244. TL, August 23, 1867, p. 5. *American Journal of Mining,* September 28, 1867, p. 201.

13. *St. Louis Democrat,* August 17, 1867, p. 2. *Iron County (MO) Register,* August 22, 1867, p. 2; August 29, 1867, p. 2. House Committee on Agriculture, "Irrigation of Public Lands," House Exec. Doc. 293 (1868), p. 3. *American Journal of Mining,* April 3, 1869, p. 213; May 1, 1869, p. 277. Charles P. Williams, "Tin in Missouri," *Journal of the Franklin Institute* 57, no. 6 (1869): 376. EMJ, May 24, 1870, p. 322.

14. EMJ, February 25, 1904, p. 323. H. C. Thompson, *Our Lead Belt Heritage* (1955; reprint, n.p.: Walsworth, 1992), 141-42.

15. *American Journal of Mining,* June 5, 1869, p. 357; May 24, 1870, p. 322. *St. Louis Democrat,* September 22, 1870, p. 2, c. 3; September 23, 1870, p. 4, c. 4; September 29, 1870, p. 3, c. 4; October 1, 1870, p. 2, c. 2. EMJ, October 10, 1871, p. 228; February 27, 1872, p. 137.

16. *St. Louis Democrat,* September 16, 1870, p. 4, c. 3; September 20, 1870, p. 4, c. 4; September 21, 1870, p. 4, c. 4; October 17, 1870, p. 3, c. 2; October 21, 1870, p. 2, c. 3. EMJ, November 8, 1870, pp. 298-99; December 13, 1870, p. 378.

17. *St. Louis Democrat,* January 25, 1871, p. 1; August 30, 1871, p. 4.

18. *Iron County Register,* March 2, 1871, p. 2. MSP, March 4, 1871, p. 136; April 8, 1871, p. 211. *History of Southeast Missouri* (Chicago: Goodspeed, 1888), 215-16. *Mining Industry and Tradesman* (March 10, 1892): 113.

19. H. M. Strickland, *Silver under the Sea* (Cobalt: Highway Book Shop, 1979), 134-37. Elinor Barr, *Silver Islet: Striking It Rich in Lake Superior* (Toronto: Natural Heritage/Natural History, 1988), 54-55.

20. MSP, August 16, 1873, p. 104. EMJ (January 10, 1874): 17; March 21, 1874, p. 181; July 14, 1877, p. 25.

21. DMR, February 18, 1905, p. 14.

22. EMJ, March 18, 1873, p. 172. MSP, August 16, 1873, p. 104.

23. EMJ, January 10, 1874, p. 17. Walter McDermott, "Mine Reports and Mine Salting," *Transactions of the Institute of Mining & Metallurgy* 3 (1894): 120-21. DMR, February 18, 1905, p. 15. The site of the swindle is now in Pukaskwa National Park.

24. *Ogden (UT) Junction,* September 23, 1871, p. 2; October 11, 1871, p. 3; October 14, 1871, p. 3; October 25, 1871, p. 3.

25. *Ogden (UT) Junction,* November 1, 1871, pp. 1, 3; November 4, 1871, p. 2. MSP, October 28, 1871, p. 257; November 25, 1871, p. 329; December 16, 1871, p. 371; October 25, 1873, p. 258. EMJ (November 28, 1871).

26. MJ, December 30, 1871, p. 1173.

27. EMJ, August 23, 1884, p. 118.

28. "That Tin Fraud," *Mining Industry* 5 (1889): 166. NYT, November 22, 1890, p. 8.

29. EMJ, November 10, 1883, p. 300.

30. EMJ, November 10, 1883, p. 300.

31. EMJ, May 29, 1886, p. 387; November 6, 1886, p. 325. *Financial & Mining Record* (June 19, 1886).

32. EMJ, December 11, 1886, p. 424.

33. EMJ, May 28, 1887, p. 388; June 4, 1887, p. 397; June 25, 1887, p. 451; April 6, 1889, p. 331. MSP, November 12, 1887, p. 309. NYT, March 30, 1889, p. 1; March 31, 1889, p. 8.

34. EMJ, October 29, 1887, p. 307; December 14, 1889, p. 519. NYT, December 6, 1887, p. 9.

35. EMJ, December 14, 1889, p. 519.

36. MSP, May 31, 1913, p. 836.

37. EMJ, September 15, 1888, p. 212; October 20, 1888, p. 323; January 19, 1889, p. 72; October 26, 1889, p. 358.

38. NYT, December 1, 1890, p. 2.

39. *Rapid City (SD) Journal,* January 29, 1888, p. 2. EMJ, June 23, 1888, p. 461; July 28, 1888, p. 70. FT, October 30, 1888, p. 4. NYTr, August 11, 1889, p. 13. W. T. Jackson, "Dakota Tin: British Invest in Harney Peak, 1880-1900," *North Dakota History* 33, no. 1 (1966): 22-63.

40. *Rapid City (SD) Journal,* December 6, 1887, p. 2; December 17, 1887, p. 1; January 3, 1888, p. 2. NYT, December 7, 1887, p. 9; July 19, 1888, p. 5; September 16, 1891, p. 9. *Mining Industry & Tradesman* 10 (1892): 53. EMJ, February 6, 1892, p. 189; August 6, 1892, p. 136.

41. EMJ, February 25, 1888, p. 140; March 31, 1888, p. 230.

42. EMJ, October 20, 1888, p. 331; December 15, 1888, p. 508; January 5, 1889, p. 18; April 13, 1889, p. 352; August 17, 1889, p. 145; January 25, 1890, p. 117; November 29, 1890, p. 634; April 11, 1891, p. 453; September 3, 1892, p. 232. NYT, April 17, 1889, p. 4; November 26, 1890, p. 9.

43. EMJ, February 8, 1890, p. 169; November 10, 1890, p. 1, c. 7.

44. *Mining World* (Chicago), May 15, 1909, p. 926. MSP, July 24, 1909, p. 115. Frank Hebert, *40 Years Prospecting and Mining in the Black Hills of South Dakota* (Rapid City: Rapid City Journal, 1921), 163.

45. *Rapid City (SD) Journal,* July 12, 1887, p. 1; July 13, 1887, p. 2; July 14, 1887, p. 2; July 16, 1887, p. 2; July 19, 1887, p. 2; August 4, 1887, p. 1; August 5, 1887, p. 2; August 13, 1887, p. 2; August 14, 1887, p. 2; August 15, 1887, p. 2; August 19, 1887, p. 2; September 28, 1888, p. 1. NYT, November 11, 1888, p. 1. EMJ, December 1, 1888, p. 455.

46. EMJ, September 7, 1889, p. 201; December 14, 1889, p. 527.

47. EMJ, September 14, 1889, p. 219; October 12, 1889, p. 312.

48. *Rapid City (SD) Journal,* December 25, 1887, p. 2; October 17, 1889, p. 2. Despite the prediction of the *Rapid City Journal,* the *Engineering & Mining Journal* is still publishing.

49. *Rapid City (SD) Journal,* September 14, 1888, p. 1. EMJ, September 29, 1888, p. 255.

50. NYT, February 19, 1888, p. 1; April 23, 1888, p. 2. *Rapid City Journal,* October 18, 1889, p. 2.

51. *Rapid City (SD) Journal,* October 5, 1887, p. 2; October 6, 1887, p. 2. *New York Sun,* August 31, 1889, p. 12. EMJ, October 19, 1889, p. 333.

52. EMJ, July 16, 1892, p. 64; November 19, 1892; November 26, 1892, pp. 512-14, 536. *Engineering News,* November 24, 1892, p. 484-85.

53. EMJ, October 29, 1892, p. 424; December 17, 1892, p. 592. NYT, December 24, 1892, p. 11, c. 4. TL, December 24, 1892, p. 11.

54. NYT, February 3, 1893, p. 1. EMJ, February 4, 1893, p. 97; February 11, 1893, p. 135; September 9, 1893, p. 275; June 30, 1894, p. 604. MSP, February 11, 1893, p. 82. E. D. Gardner, "Tin Deposits of the Black Hills, South Dakota," US Bureau of Mines, Information Circular 7069 (Washington, DC: Government Printing Office: 1939), 6.

55. NYT, June 30, 1894, p. 9; August 21, 1894, p. 8; December 11, 1894, p. 6; November 12, 1895, p. 9. EMJ, June 30, 1894, p. 604; July 7, 1894, pp. 2, 5; July 14, 1894, pp. 25, 40; July 21, 1894, p. 64; August 25, 1894, p. 183. MSP, July 28, 1897, p. 50.

56. NYTr, June 30, 1894, p. 9. EMJ, December 15, 1894, p. 568; November 16, 1895, p. 475.

57. EMJ, November 17, 1894, p. 471; December 7, 1895, p. 546; December 21, 1895, p. 594; May 23, 1896, p. 489; January 8, 1898, p. 53; February 19, 1898, p. 233; May 7, 1898, p. 562; December 16, 1899, p. 721. MSP, May 31, 1913, p. 836; November 21, 1914, p. 790.

58. EMJ, November 18, 1893, p. 516. George Wheeler, *Pierpont Morgan and Friends* (Englewood Cliffs, N.J.: Prentice-Hall), 289-303.

59. *Mining World* (Chicago), June 13, 1908, p. 970; May 8, 1909, p. 910; May 15, 1909, pp. 925-26; July 10, 1909, p. 152; August 28, 1909, p. 457; January 21, 1911, p. 188.

60. L. S. Dean, "The Broken Promise of Tin at Ashland," *Alabama Review* 39 (July 1986): 174-86. EMJ, December 9, 1882, p. 309.

61. EMJ, January 13, 1883, p. 13.

62. L. S. Dean, "The Prospecting Career of William Hugh Smith," *Alabama Review* 40 (April 1987): 95-110.

63. LeW, November 12, 1903, p. 471. AMS, October 18, 1905, p. 367.

64. MSP, April 22, 1876, p. 258; December 15, 1888, p. 394. EMJ, February 14, 1903, p. 272.

65. R. H. Bedford and F. T. Johnson, "Survey of Tin in California," US Bureau of Mines, Report of Investigations 3876 (Washington, DC: Government Printing Office, 1946), 2.

66. H. D. Clark, "Author of 'Lost Mines of the West' Says Cartel Blocked U.S. Tin Mining," *California Mining Journal* 36 (April 1967): 3-4.

Chapter 7. Professor Samuel Aughey: "A First-Class Charlatan"

1. R. N. Manley, "Samuel Aughey: Nebraska's Scientific Promoter," *Journal of the West* 6, no. 1 (1967): 108-18.

2. *Cheyenne Daily Leader,* January 13, 1884, p. 3. In 1919 Aughey's daughter offered to endow a scholarship in his name at the University of Nebraska, but the chancellor declined, citing a "forgotten record that might possibly be brought to light."

3. *Cheyenne Daily Leader,* October 16, 1884, p. 3.

4. *Cheyenne Democratic Leader,* May 19, 1886, p. 3. "A Gigantic Fake," *Wyoming Industrial Journal,* 1, no. 4 (September 1899), pp. 98-99. *Casper Journal,* September 29, 1984, p. 6. *Cody Enterprise,* October 3, 1984, p. B4.

5. *Cheyenne Democratic Leader,* May 21, 1886, p. 3; May 27, 1886, p. 3; June 13, 1886, p. 3; July 7, 1886, p. 3.

6. *Cheyenne Democratic Leader,* August 26, 1886, p. 3; October 10, 1886, p. 3.

7. *Cheyenne Democratic Leader,* January 5, 1887, p. 3.

8. *Cheyenne Democratic Leader,* January 11, 1887, p. 3; April 1, 1887, p. 3. William Bryans, *History of the Geological Survey of Wyoming* (Wyoming Geological Survey Bulletin 65, 1986), 13-18.

9. *Little Rock Weekly Arkansas Democrat,* October 24, 1878, p. 3.

10. *Little Rock Arkansas Gazette,* March 16, 1887, p. 5.

11. Drep, July 23, 1881, p. 6; December 10, 1881, p. 3. RMN, July 23, 1881, p. 8; October 2, 1881, p. 6; February 9, 1882, p. 3; February 11, 1882, p. 4; February 16, 1882, p. 6; January 25, 1883, p. 3. EMJ, July 8, 1882, p. 21; September 2, 1882, p. 124.

12. EMJ, July 18, 1865, p. 47; April 10, 1886, p. 271.

13. *Cheyenne Leader,* October 30, 1887, p. 3. *Mining Industry,* May 25, 1888, p. 6.

14. *Little Rock Arkansas Gazette,* January 24, 1886, p. 1. EMJ, March 27, 1886, p. 236; July 2, 1887, p. 10.

15. EMJ, July 28, 1888 pp. 63-64; August 18, 1888 pp. 123-24.

16. EMJ, April 23, 1887, p. 289.

17. EMJ, August 18, 1888 pp. 128-29.

18. Donald Harington, *Let Us Build Us a City* (N.Y.: Harcourt Brace Jovanovich, 1986), 386-421.

19. EMJ, September 1, 1888, p. 168; September 8, 1888, p. 189; October 6, 1888, p. 278; October 20, 1888, pp. 325-27. President Herbert Hoover and his wife, Lou Henry Hoover, later dedicated their translation of the mining classic *De Re Metallica* to Branner.

20. *Pacific Coast Miner,* May 16, 1903 pp. 362-66. N.Y. *Commercial,* February 7, 1907, p. 7. MSP, February 23, 1907, p. 226; February 25, 1911, p. 286. *Mining & Engineering World,* February 24, 1912, p. 482. *Colfax (WA) Gazette-Commoner,* December 12, 1935.

Chapter 8. Swindle in South Park

1. The Hassayamper tradition is kept alive by Jim Cook, who proclaims himself the "Official State Liar of the State of Arizona and Director of the Wickenburg Institute for Factual Diversity."

2. Dtr, April 23, 1884, p. 2. *Sierra (CO) Journal,* April 24, 1884, p. 2.

3. Walter McDermott, "Mine Reports and Mine Salting," *Transactions of the Institution of Mining and Metallurgy* 3 (1894), p. 135.

4. *Pueblo (CO) Chieftain,* April 15, 1884, p. 4.

5. *Fairplay (CO) Flume,* April 10, 1997, p. 1. Dtr, April 23, 1884, p. 2.

6. *Colorado Springs Gazette,* p. 1. *Pueblo (CO) Chieftain,* April 22, 1884, p. 6.

7. Dtr, April 23, 1884, p. 2. LH, April 19, 1884, p. 3.

8. LH, April 22, 1884, p. 1. Dtr, April 23, 1884, p. 2. The Leadville papers were in fact not above denigrating legitimate rivals. At the time, the *Leadville Herald* was denouncing the rush to Idaho's Coeur D'Alene district as a fraud (April, 2, 1884, p. 2). Despite the *Herald's* low opinion of it, Coeur D'Alene soon surpassed Leadville in silver production.

9. LH, April 18, 1884, p. 3.

10. LH, April 22, 1884, p. 3.

11. Dtr, April 23, 1884, p. 2. LH, April 22, 1884.

12. Dtr, April 22, 1884, p. 1; April 23, 1884, p. 2. The *Colorado Springs Gazette,* in jealously calculating how much money was spent in Canon City—rather than in Colorado Springs—said the camp attracted five thousand people, but eyewitnesses put the figure much lower. If estimates are accurate that the maximum number at any one time was five hundred, it is probable that the total did not exceed two thousand.

13. LH, April 24, 1884, p. 3.

14. T. A. Rickard, "The Cripple Creek Goldfield," *Transactions of the Institution of Mining and Metallurgy* 8 (1900): 51.

15. *Pueblo (CO) Chieftain,* April 22, 1884, p. 6.

16. Dtr, April 23, 1884, p. 1. *Colorado Springs Gazette,* April 24, 1884, p. 1; April 26, 1884, p. 1.

17. Dtr, May 12, 1884, p. 2.

18. Dtr, April 28, 1884, p. 2; May 12, 1884, p. 2.

19. *Denver Times,* December 26, 1901, p. 10.

Chapter 9. Leadville: A Comstock in the Colorado Rockies

1. Duane Smith, *Silver Saga* (Boulder: Pruett, 1974), 63-81. Drep, November 23, 1881, p. 4, c. 2. C. C. Spence, "Colorado's Terrible Mine: A Study in British Investment," *Colorado Magazine* 34 (January 1957): 48-61. The nearby town of Nederland, Colorado, is named in honor of the foolish Caribou investors.

2. EMJ, April 26, 1879, p. 304.

3. EMJ, May 24, 1879, p. 380; June 7, 1879, pp. 420, 422. George Bishop, *Charles H. Dow and the Dow Theory* (N.Y.: Appleton-Crofts, 1960), 249-354.

4. EMJ, June 14, 1879, p. 438; June 21, 1879, p. 456; June 28, 1879, pp. 461, 477. RMN, September 4, 1879, p. 8.

5. EMJ, January 24, 1880, p. 63. *Leadville Democrat,* April 16, 1880, p. 2, c. 1. Horace Tabor soon scandalized Colorado by divorcing his wife to marry young divorcee Elizabeth "Baby" Doe. He was wiped out by a financial panic in 1893, and died in 1899. Baby Doe Tabor returned to Leadville, where she lived in poverty and insanity for years, until she froze to death in her cabin at the Matchless mine. *Rocky Mountain (CO) Globe,* April 20, 1899, p. 1.

6. EMJ, March 20, 1880, pp. 199-200.

7. RMN, March 13, 1880, p. 4. *Leadville Democrat,* March 27, 1880, p. 6; April 13, 1880, p. 2.

8. *Leadville Democrat,* April 1, 1880, p. 6; April 7, 1880, p. 2.

9. *Leadville Democrat,* March 14, 1880, p. 6; March 19, 1880, p. 1; May 5, 1880, p. 2. EMJ, March 27, 1880, p. 224; April 3, 1880 pp. 232-35.

10. Dtr, March 11, 1881, p. 4. EMJ, April 17, 1880, p. 276; May 8, 1880, p. 328. RMN, May 7, 1880, p. 3.

11. NYT, April 23, 1885, p. 8; April 28, 1885, p. 8; April 29, 1885, p. 8; April 30, 1885, p. 8; May 1, 1885, p. 2, c.5. EMJ, April 25, 1885, p. 285; May 2, 1885, p. 303.

12. *Leadville Democrat,* April 4, 1880, p. 6. "Last of the Little Pittsburgh," *Mining Industry* 4 (1889), p. 237. EMJ, August 20, 1881, p. 126; May 21, 1887, p. 375; June 4, 1887, p. 411; June 25, 1887, p. 468; June 14, 1890, p. 690.

13. *Leadville Chronicle,* July 6, 1886, p. 3.

14. NYTr, August 24, 1897, p. 3.

15. LH, July 6, 1886. Tabor later insisted that he had paid little attention to Lovell's salted shaft. "H. A. W. Tabor—by himself," H. A. W. Tabor dictation, n.d., The Bancroft Library, University of California, Berkeley.

16. *Denver Times,* July 6, 1886, p. 4.

17. Virginia McConnell, *Bayou Salado* (Denver: Sage, 1966), 148.

18. EMJ, January 24, 1880, p. 62.

19. W. R. Balch, *Mines, Miners & Mining Interests of the United States, 1882* (Philadelphia: Philadelphia Industrial, 1882), 482.

20. EMJ, February 15, 1879, p. 108.

21. *Colorado Gold and Silver Mines,* 2nd ed. (N.Y.: Crawford, 1880), 429, 451.

22. *Leadville Weekly Democrat,* February 28, 1880, p. 8. RMN, February 11, 1882, p. 4.

23. EMJ, January 24, 1880, p. 62; March 13, 1880, p. 189; March 27, 1880, p. 220.

24. *Eureka (NV) Sentinel,* August 25, 1877. EMJ, March 27, 1880, p. 224.

25. EMJ, April 24, 1880, pp. 283, 292; May 8, 1880, p. 328.

26. EMJ, June 5, 1880, p. 383.

27. EMJ, July 24, 1880; August 14, 1880, p. 105.

28. *Leadville Weekly Herald,* August 21, 1880, p. 3.

29. EMJ, October 9, 1880, p. 233; November 6, 1880, p. 297; August 13, 1881 pp. 101-2.

30. MSP, October 11, 1919, p. 504.

31. NYTr, March 6, 1880, p. 5; March 15, 1880, p. 3. EMJ, July 29, 1882, p. 54.

32. EMJ, August 20, 1881, p. 126; February 4, 1882, p. 65.

33. Drep, December 11, 1881, p. 8. NYTr, December 14, 1881, p. 6; December 21, 1881, p. 2; January 6, 1882, p. 3. RMN, January 10, 1882, p. 3. Stanley Dempsey and J. E. Fell Jr., *Mining the Summit* (Norman: University of Oklahoma Press, 1986), 161-67.

34. NYTr, November 29, 1881, p. 6.

35. NYTr, November 30, 1881, p. 6; December 1, 1881, p. 6; December 5, 1881, p. 6; December 9, 1881, p. 6. EMJ, December 10, 1881, p. 395. Drep, December 25, 1881, p. 10.

36. LH, December 17, 1881, p. 4.

37. RMN, February 17, 1882, p. 7.

38. Drep, December 19, 1881, p. 6. EMJ, December 31, 1881, pp. 435-37; January 7, 1882, p. 9; June 10, 1882, p. 306; June 11, 1887, p. 429; January 4, 1896, p. 28. *Boston Daily Advertiser,* February 2, 1882, p. 5. LH, February 14, 1882, p. 4; April 7, 1882, p. 4; June 2, 1882, p. 4; June 13, 1882, p. 4.

39. VTE, August 24, 1879, p. 3; August 26, 1879, p. 3; August 27, 1879, p. 1; August 28, 1879, p. 2; August 29, 1879, p. 2; September 17, 1879, p. 2; September 23, 1879, p. 2; September 26, 1879, pp. 2, 3. GHN, August 25, 1879, p. 3. *Esmeralda (NV) Herald,* November 24, 1877, p. 3; August 30, 1879, p. 3; September 20, 1879, p. 2; October 4, 1879, p. 2; August 27, 1881, p. 2.

40. EMJ, May 21, 1881, p. 356.

41. EMJ, March 22, 1879, p. 210.

Chapter 10. Tichenor's Gold: Calistoga Natural Springwater

1. *Chicago Tribune,* February 22, 1865, p. 3; November 6, 1878, p. 7.

2. *Chicago Tribune,* October 31, 1878, p. 8.

3. *Chicago Tribune,* November 2, 1878, p. 8. A. E. Sheldon, *Nebraska: the Land and the People* 1 (Chicago: Lewis, 1931).

4. SLTr, May 6, 1873, p. 2.

5. *Salt Lake Herald,* May 4, 1872, p. 2.

6. SLTr, May 3, 1873, p. 2; May 5, 1873, p. 2.

7. *Salt Lake Herald,* May 13, 1873, p. 2. SLTr, May 17, 1873, p. 2.

8. *Little Rock Arkansas Gazette,* August 8, 1877, p. 4.

9. *Chicago Tribune,* October 29, 1878, p. 7; November 8, 1878, p. 8. SFCh, September 12, 1880, p. 8.

10. SFCh, September 16, 1880, p. 4.

11. *History of Napa and Lake Counties, California* (San Francisco: Slocum, Bowen, 1881), 349-50.

12. *Napa (CA) Daily Register,* September 9, 1880, p. 3; September 11, 1880, p. 3; September 13, 1880, p. 2. SFAC, September 11, 1880, p. 1; September 14, 1880, p. 2.

13. SFCa, September 13, 1880, p. 3.

14. *N.Y. Post,* September 16, 1880, p. 4.

15. *Napa Daily Register,* September 16, 1880, p. 1. MSP, September 18, 1880, p. 188.

16. *Polk's Washington City Directory* (Washington, D.C.: Polk, 1890). H. A. Tichenor, *Tichenor Families in America* (Napton, Mo.: 1988), 514.

Chapter 11. Richard Flower: Master Swindler

1. *A History of Edwards County, Illinois* 1 (Albion, Ill.: Edwards Co. Historical Soc., 1980), 11-14.

2. *Combined History of Edwards, Lawrence and Wabash Counties, Illinois* (Philadelphia: McDonough, 1882), 224A-C.

3. *Chicago Times,* February 10, 1884, p. 7. *Cincinnati Enquirer,* April 17, 1886, p. 5.

4. *Louisville Courier-Journal,* March 3, 1886, p. 3.

5. *Rocky Mountain Mining Review,* June 5, 1884, p. 11.

6. *Denver Post,* June 3, 1885, p. 6. *Boston Evening Traveler,* September 2, 1885, p. 3. EMJ, September 25, 1886, p. 226; December 20, 1890.

7. Schaeffer to Tabor, December 15, 1881, First National Bank of Denver Collection: 1860-1974, manuscript collection #929, file folder #44, Colorado Historical Society, Denver.

8. EMJ, February 19, 1887, p. 128.

9. EMJ, May 31, 1879, p. 393; June 7, 1879, p. 420; June 21, 1879, p. 456; June 28, 1879, p. 477; March 27, 1880, p. 224; December 10, 1881, p. 395. *Colorado, Its Gold and Silver Mines,* 2nd ed. (N.Y.: Crawford, 1880), 473.

10. EMJ, February 26, 1887, p. 146; March 5, 1887, p. 163; March 5, 1887, p. 180; March 26, 1887, p. 227; April 2, 1887, p. 252; April 30, 1887, p. 321; May 28, 1887, p. 393; June 4, 1887 pp. 407, 411; June 11, 1887, p. 429; June 18, 1887 pp. 434, 447; June 25, 1887, p. 468; July 2, 1887, p. 15; July 16, 1887, p. 51; July 30, 1887, p. 74. *Boston Globe,* June 7, 1887, p. 4; June 11, 1887, p. 5; June 13, 1887, p. 3; June 14, 1887, p. 5.

11. EMJ, August 6, 1887, p. 105; August 13, 1887, p. 123; September 10, 1887, p. 198.

12. EMJ, September 24, 1887, p. 234; October 22, 1887, p. 306; October 29, 1887 pp. 317, 324; March 24, 1888, p. 228; May 5, 1888, p. 336.

13. EMJ, October 20, 1888, p. 337; November 3, 1888, p. 386; November 24, 188, p. 441; December 8, 1888, p. 491. *Sierra Journal,* October 19, 1882.

14. EMJ, October 5, 1889, p. 298. MSP, November 11, 1899, p. 543. RMN, January 13, 1901, p. 9; March 3, 1901, p. 13; June 23, 1901, p. 9; November 22, 1901, p. 13.

15. *Arizona Bulletin,* March 17, 1899, p. 3.

16. *Deming (NM) Headlight,* October 7, 1893, p. 1, October 23, 1894, p. 4. *Galveston Daily News,* December 1, 1894, p. 2. NYT, December 3, 1894; December 7, 1894.

17. *Galveston Daily News,* December 4, 1894, p. 2.

18. EMJ, April 16, 1898, p. 472; October 15, 1898. AZR, April 11, 1899, p. 7. George H. Smalley, "The Spenazuma Mining Swindle," *Arizona Historical Review* 2, no. 1 (1929): pp. 86-102.

19. *Arizona Bulletin,* July 8, 1898, p. 3; July 15, 1898, p. 3; October 7, 1898. AZR, April 13, 1899, p. 7.

20. EMJ, December 24, 1898, p. 774. *Arizona Citizen,* March 31, 1899, p. 4. AZR, April 26, 1899, p. 7.

21. EMJ, April 15, 1899, p. 434.

22. *Arizona Bulletin,* October 21, 1898, p. 7; January 13, 1899, p. 11; August 18, 1899, p. 4.

23. H. B. Clifford, 1883, *Years of Dishonor, or the Cause of Depression in Mining Stocks, Also the Remedy* (N.Y.: Martin & Brown, 1883; repr. in *Speculation in Gold and Silver Mining Stocks* [N.Y.: Arno, 1974]). AZR, June 24, 1899, p. 1; July 8, 1899, p. 7.

24. *Arizona Bulletin,* March 17, 1899, p. 2.

25. AZR, July 23, 1899, p. 1.

26. *N.Y. Sun,* April 2, 1899, p. 11.

27. EMJ, January 18, 1899, p. 221; April 15, 1899 pp. 434-35; April 29, 1899, p. 495. AZR, April 17, 1899, p. 7. R. G. Wilson, *Arizona Ways & Days,* August 18, 1963, pp. 28-29.

28. AZR, May 17, 1899, p. 1.

29. AZR, May 27, 1899, p. 2; June 6, 1899, p. 1; June 30, 1899, p. 2; July 8, 1899, p. 2. *Arizona Daily Star,* May 26, 1899, p. 2. NYTr, June 22, 1899, p. 3.

30. EMJ, May 27, 1899, p. 616, June 3, 1899, pp. 647-49; July 8, 1899, pp. 32, 40; July 22, 1899, pp. 92-93; August 5, 1899, p. 151; September 3, 1899, pp. 314-15. *Arizona Citizen,* April 14, 1899, p. 4. *Arizona Bulletin,* June 30, 1899, p. 3; July 14, 1899, p. 3; August 4, 1899, p. 1. *Phoenix Gazette,* July 12, 1899, p. 4. *Tucson Star,* July 1, 1899, p. 4.

31. *N.Y. Herald,* July 1, 1899, p. 10. AZR, July 9, 1899, pp. 1-2. *Arizona Citizen,* August 31, 1899, p. 1. MSP, September 2, 1899.

32. *Arizona Bulletin,* July 21, 1899, p. 3.

33. *Arizona Bulletin,* August 18, 1899, p. 3, January 12, 1900, January 19, 1900, p. 2.

34. Smalley places the Lone Pine mine at Celestia, on the west side of the Graham Mountains. In fact, the Lone Pine mine is far away in Yavapai County, Arizona, and I have been unable to confirm the place name "Celestia." It seems that this part of Smalley's story belongs in the Spenazuma Company's camp at Aura, in Graham County. George H. Smalley, "The Spenazuma Mining Swindle," *Arizona Historical Review* 2, no. 1 (1929): 86-102.

35. *Arizona Bulletin,* February 23, 1900, p. 2; November 23, 1900, p. 2; December 14, 1900, p. 8; February 1, 1901, p. 2. CuH 8 (1908): 730-31; CuH 10 (1911): 272, 874, 1595.

36. NYT, July 12, 1899, p. 8; March 7, 1903, p. 3.

37. AZR, June 5, 1899, p. 7. EMJ, July 29, 1899, p. 135. *Phoenix Daily Herald,* September 21, 1899, p. 4. *Arizona Weekly Journal-Miner,* November 1, 1899, p. 1.

38. EMJ, July 22, 1899, p. 105; September 23, 1899, p. 375. *Arizona Weekly Journal-Miner,* September 13, 1899, p. 2; December 20, 1899, p. 1. MSP, December 16, 1899, p. 694. DMR, September 23, 1899, p. 1.

39. *Phoenix Daily Herald,* September 21, 1899, p. 4. *Jerome (AZ) Mining News,* September 25, 1899, p. 1. EMJ, October 7, 1899, p. 435; February 10, 1900, p. 163; March 10, 1900, p. 283; March 17, 1900, p. 313; March 24, 1900, p. 344.

40. *Arizona Daily Journal-Miner,* January 17, 1900, p. 4. EMJ, February 10, 1899, p. 163; March 10, 1899, p. 283; March 17, 1899, p. 313. DMR, January 8, 1901, p. 1. NYT, October 18, 1904, p. 5.

41. EMJ, October 27, 1900, p. 496; November 10, 1900, p. 557; January 12, 1901, pp. 46, 60; February 9, 1901, p. 173. *Arizona Bulletin* January 18, 1901, p. 3. *Arizona Daily Journal-Miner,* November 30, 1901, p. 1. DMR, January 8, 1901, p. 1.

42. NYTr, February 25, 1903, p. 1; February 26, 1903, p. 1. NYT, December 26, 1903, p. 1.

43. NYTr, March 17, 1903, p. 4. Fanny left Delabarre after a few months but never divorced him. After his death in 1910, she sued to break the will because it left little to her. NYT, March 17, 1903, p. 2; December 26, 1903, p. 1; January 16, 1904, p. 10; April 2, 1914, p. 20.

44. NYT, March 11, 1903, p. 3; March 29, 1903, p. 12. NYTr, March 11, 1903, p. 2.

45. CuH 8 (1908): 1100.

46. DMR, November 17, 1906, p. 10; January 5, 1907, p. 3. H. B. Clifford, *Rocks in the Road to Fortune* (N.Y.: Gotham, 1908).

47. NYT, March 12, 1903, p. 2; March 26, 1903, p. 1. NYTr, June 2, 1903, p. 14; June 5, 1903, p. 6.

48. NYT, March 28, 1903, p. 2; June 2, 1903, p. 6; June 5, 1903, p. 6.

49. NYTr, March 13, 1903, p. 1; March 24, 1903, p. 3; March 26, 1903, p. 1; March 27, 1903, p. 5; March 28, 1903, p. 6; March 31, 1903, p. 6. NYT, May 27, 1903, p. 16.

50. NYT, May 15, 1903, p. 3.

51. NYT, April 4, 1903, p. 16.

52. NYT, November 7, 1903, p. 6.

53. NYT, May 24, 1904, p. 16; May 25, 1904, p. 2; June 8, 1904, p. 6.

54. NYT, May 30, 1904, p. 1; October 28, 1904, p. 10; January 31, 1905, p. 5.

55. NYT, January 5, 1907, p. 1; January 6, 1907, p. 12.

56. *N.Y. Commercial,* January 5, 1907, p. 1.

57. NYT, November 23, 1907, p. 1.

58. NYT, September 11, 1908, p. 1; September 12, 1908, p. 2.

59. NYT, September 13, 1908, p. 18; September 15, 1908, p. 18.

60. NYT, September 14, 1908, p. 7.

61. NYT, September 20, 1908, p. 18.

62. NYT, October 22, 1914, p. 18.

63. NYT, November 5, 1914, p. 8; November 14, 1914, p. 8; December 8, 1914, p. 7.

64. GFN October 21, 1916, p. 4. *Mining American,* November 11, 1916, p. 6.

Chapter 12. Thomas Lawson: Champion of the People

1. Frank Fayant, "The Real Lawson," *Success* 10 (October 1907), pp. 663-703.

2. *Boston Directory* (Boston: George Adams, 1887), p. 1676. "The Secret of Lawson's Career," *Current Literature,* March 1908, pp. 261-65.

3. EMJ, August 29, 1891, p. 250. *Public Opinion,* January 26, 1905, pp. 112-13.

4. Frank Fayant, "The Real Lawson," *Success* 10 (November 1907), pp. 719-85. Instead of becoming another Pittsburgh or Birmingham, Grand Rivers today lures tourists eager to get away from Pittsburgh and Birmingham. Deloris Martin and Barbara Heater, "Grand Rivers," in *Livingston County Kentucky* (Smithland, Ky.: Livingston Co. Historical and Genealogical Society, 1989), 24-25.

5. Thomas W. Lawson, "Frenzied Finance," EvM 11 (November 1904), pp. 601-13. *Public Opinion,* February 18, 1905, pp. 229-32.

6. *Philadelphia Inquirer,* January 28, 1904, p. 6. WSJ, March 8, 1904, p. 7; March 18, 1904, p. 5; March 21, 1904, p. 5; March 22, 1904, p. 5; March 26, 1904, p. 5. EvM 11 (October 1904), pp. 455-68; Thomas W. Lawson, "Frenzied Finance," EvM 11 (December 1904), pp. 747-60.

7. Thomas W. Lawson, "Frenzied Finance," EvM 13 (August 1905), pp. 195-207; Thomas W. Lawson, "Frenzied Finance," EvM 14 (January 1906), pp. 73-75; Thomas W. Lawson, "Frenzied Finance," EvM 14 (February 1906), pp. 250-55. WSJ, March 16, 1907, p. 5.

8. Allan Nevins, *John D. Rockefeller* 2 (N.Y: Scribner's, 1940), 431-32. John D., whatever his faults, was a company builder, never a stock manipulator. Rogers's stock manipulations eventually angered John D. Rockefeller such that he bought out Rogers's interest in Standard Oil. Ron Chernow, *Titan* (N.Y.: Random House, 1998), 378-81.

9. *National Cyclopaedia of American Biography* 26 (N.Y.: J. T. White, 1937), 23-24.

10. *Arizona Weekly Republican,* January 16, 1902, p. 1.

11. NYT, March 24, 1899, p. 8; April 28, 1900, p. 11. NYTr, October 5, 1899, p. 4. *Nation,* August 14, 1902, pp. 127-28. EvM, December 1905, pp. 822-31.

12. WSJ, May 10, 1899, p. 2; the variety was named the Mrs. Lawson Pink.

13. WSJ, March 23, 1904, p. 5. EvM, April 1905, pp. 485-95. Frank Fayant, "The Real Lawson," *Success* 11 (February 1908), pp. 71-107.

14. NYT, June 25, 1898, p. 9. EMJ, March 11, 1899, pp. 301, 308; March 18, 1899, p. 337; March 25, 1899, p. 367.

15. *Boston Herald,* August 6, 1899, p. 18. *Houghton (MI) Daily Mining Gazette,* November 13, 1899, p. 6. EMJ, January 27, 1900, p. 119.

16. CuH 2 (1902): 109-14; CuH 8 (1908): 319-20; CuH 10 (1911): 346, 1276. EMJ, June 13, 1903, p. 910; January 6, 1906, pp. 6, 69; February 8, 1908, p. 336; May 30, 1908, p. 1120; April 10, 1909, p. 775. Angus Murdoch, *Boom Copper* (N.Y.: MacMillan, 1943), 184-87. W. B. Gates, *Michigan Copper and Boston Dollars* (Cambridge: Harvard University Press, 1951), 87-89. The mine, near Hancock, Michigan, was for years open as a tourist attraction.

17. In that dullest and most obsequious of official corporate histories, Isaac Marcosson's *Anaconda* (N.Y.: Dodd Mead, 1957), 93, the idea for Amalgamated is taken from Lawson and credited to A. C. Burrage.

18. NYT, January 2, 1899, p. 12; January 17, 1899, p. 11; August 4, 1900, p. 2.

19. NYT, April 28, 1899, p. 1; May 5, 1899, p. 2; May 6, 1899, p. 2. CuH 2 (1902): 292-93. *N.Y. Post,* April 14, 1904, p. 6. EvM, June 1905, pp. 717-29.

20. FT, May 2, 1899, p. 4. EMJ, November 18, 1899, p. 625.

21. WSJ, March 31, 1904, p. 5. LeW, April 21, 1904, p. 380. EvM, July 1905, pp. 28-45; September 1905, pp. 363-71.

22. LeW, August 21, 1902, p. 188. WSJ, December 3, 1907, p. 1; December 11, 1907, p. 3; December 12, 1907, p. 1; December 27, 1907, p. 1.

23. DMR, June 13, 1904, p. 2. CuH 5 (1905): 177; CuH 6 (1906): 182.

24. EMJ, February 11, 1899, p. 183; May 25, 1901, p. 665; January 25, 1902, p. 153; February 22, 1902, p. 291; January 3, 1903, p. 67. WSJ, August 27, 1904, p. 5.

25. C. J. Monette, *Trimountain and Its Copper Mines* (Lake Linden, Mich.: Welden Curtin, 1991), 74-77.

26. WSJ, July 13, 1904, p. 5; July 19, 1904, p. 5; March 13, 1905, p. 5; March 15, 1905, p. 5; March 17, 1905, p. 8; April 13, 1905, p. 8; August 10, 1905, p. 5. DMR, August 12, 1905, p. 4. W. A. Paine to F. W. Denton, March 24, 1905, and n.d., Trimountain Mining Co., files 1905-1906, Michigan Technological University Archives and Copper Country Historical Collections.

27. WSJ, November 21, 1900, p. 4. EMJ, November 24, 1900, p. 625; December 1, 1900, pp. 655-56. LeW, November 15, 1906, p. 475.

28. NYT, December 9, 1901, p. 1. EMJ, February 23, 1901, p. 262.

29. EMJ, March 16, 1901, p. 343; October 17, 1908, pp. 772-73. *Boston Transcript,* November 28, 1904, p. 7. WSJ, December 22, 1904, p. 7. NYT, January 10, 1907, p. 7. *Public Opinion,* February 11, 1905, pp. 189-95.

30. NYT, December 28, 1906, p. 23; December 31, 1906, p. 13; January 3, 1907, p. 14; January 4, 1907, p. 11; January 5, 1907, p. 15; January 7, 1907, p. 13; January 8, 1907, p. 12; January 9, 1907, p. 7; January 11, 1907, p. 12; January 12, 1907, p. 14; January 13, 1907, p. 22; January 14, 1907, p. 8; January 15, 1907, p. 14; January 16, 1907, p. 14; January 17, 1907, p. 16; January 18, 1907, p. 14; January 21, 1907, p. 7; January 21, 1907, p. 7; January 22, 1907, p. 12; January 23, 1907, p. 12; January 24, 1907, p. 12; January 25, 1907, p. 12; January 26, 1907, p. 12; January 28, 1907, p. 13; January 29, 1907, p. 12; January 30, 1907, p. 14; January 31, 1907, p. 15; February 1, 1907, p. 12; February 9, 1907, p. 12; February 11, 1907, p. 12; February 12, 1907, p. 12; February 13, 1907, p. 10; February 14, 1904, p. 7; February 15, 1907, p. 14; February 16, 1907, p. 14; February 21, 1907, p. 14; February 25, 1907, p. 16. WSJ, January 8, 1907, p. 3. LeW, January 14, 1907, p. 92. DMR, January 16, 1907, p. 5. *N.Y. Commercial,* February 1, 1907, p. 5.

31. WSJ, February 28, 1907, p. 6. DMR, October 3, 1907, p. 1; October 26, 1907, p. 4; March 10, 1910, p. 1. CuH 10 (1911): 1701-2.

32. MSP, October 19, 1912, p. 506. EvM, January 1913, pp. 100-1. EMJ, September 4, 1920, p. 494; June 9, 1928, p. 940.

33. CuH 9 (1909): 349-51, 685; CuH 10 (1911): 388-91, 813.

34. *Arizona Daily Star,* June 11, 1899, p. 4. MSP, May 15, 1909, pp. 673-74. EMJ, May 22, 1909, p. 1054; June 19, 1909, pp. 1245-46. *Mining Magazine* 1 (1909): 33. WSJ, August 17, 1910, p. 7. LeW, September 1, 1910, p. 216. John M. Sully, "The Story of the Santa Rita Copper Mine," *Old Santa Fe* 3, no. 10: 133-49. EMJ, November 10, 1923, pp. 803-8.

35. DMR, July 5, 1905, p. 4, c.1; August 3, 1905, p. 1, c.7; April 18, 1906, p. 1, c.7; November 11, 1909, p. 1, c.1; December 18, 1909, p. 3, c.2. EMJ, April 25, 1908, p. 881; May 28, 1910, p. 1135; January 7, 1911, p. 73. MSP, May 9, 1908, p. 621.

36. NYT, September 30, 1903, p. 1.

37. DMR, November 14, 1906, p. 1. *Randsburg (CA) Miner,* December 20, 1906, p. 1. EMJ, December 29, 1906, p. 1231. Minv, December 24, 1917, p. 807.

38. WSJ, February 2, 1904, p. 5. I. F. Marcosson, *Anaconda* (N.Y.: Dodd Mead, 1957), 254.

39. NYT, December 14, 1904, p. 1. *Boston Morning Journal,* December 15, 1904, p. 1. LeW, December 29, 1904, p. 633.

40. WSJ, March 11, 1904, p. 7; March 12, 1904, p. 5; March 17, 1904, p. 5; March 24, 1904, p. 5; March 29, 1904, p. 5; April 5, 1904, p. 6; April 13, 1904, p. 7; July 6, 1904, p. 2.

41. N.Y. *World*, April 10, 1904, p. 1. EvM 11 (July 1904), pp. 1-10; Thomas W. Lawson, "Frenzied Finance," EvM 11 (August 1904), pp. 154-64; "With 'Everybody's' Publishers," EvM 18 (March 1908), pp. 431-32.

42. Thomas W. Lawson, "Frenzied Finance," EvM 11, no. 3 (September 1904), pp. 289-301; Thomas W. Lawson, "Frenzied Finance," EvM 12 (May 1905), pp. 606-7.

43. N.Y. *Post*, April 13, 1904, p. 6. LeW, June 2, 1904, p. 524. WSJ, January 18, 1905, p. 8. EMJ, March 2, 1905, p. 427. *Nation*, February 18, 1925, p. 203. Louis Filler, *Voice of the Democracy* (University Park: Pennsylvania State University Press, 1978), 91-94.

44. DMR, August 21, 1905, p. 1.

45. *Public Opinion*, January 26, 1905, pp. 109-13; February 2, 1905, pp. 149-52; February 29, 1905, pp. 271-74; March 4, 1905, pp. 314-16; March 18, 1905, pp. 404-7; April 1, 1905, pp. 487-505; April 8, 1905, pp. 521-45; April 15, 1905, pp. 571-72. WSJ, February 21, 1905, p. 5.

46. NYT, December 8, 1904, p. 16; December 14, 1904, p. 1; December 17, 1905, pt. 5, p. 1. WSJ, December 7, 1904, p. 6; December 8, 1904, pp. 4, 8; December 9, 1904, p. 2; December 13, 1904, p. 4. *Boston Morning Journal*, December 15, 1904, p. 12; December 17, 1904, p. 1; December 19, 1904, p. 5; January 4, 1905, p. 6; January 21, 1905, p. 7; April 21, 1905, p. 7, c.2. Econ, December 24, 1904, p. 2102, December 17, 1904, p. 2050. *Public Opinion*, January 19, 1905, pp. 69-74.

47. WSJ, December 13, 1905, p. 2. Frank Fayant, "The Real Lawson," *Success* 10 (December 1907), pp. 819-68. NYT, August 23, 1908, pt. 5, p. 3.

48. NYT, July 11, 1905, p. 1; July 14, 1905, p. 1. Hugh O'Neill, "Lawson of Boston," EvM 13 (October 1905), p. 64. *Boston Herald*, February 7, 1906, p. 1.

49. LeW, September 14, 1905, p. 259. WSJ, September 4, 1905, p. 3; October 20, 1905, pp. 5, 7; November 15, 1905, p. 7.

50. NYT, May 18, 1906, p. 2.

51. FT, January 2, 1906, p. 3; January 3, 1906, p. 5. *Boston Herald*, February 3, 1906, pp. 1, 6; February 8, 1906, p. 1. NYT, February 9, 1906, p. 5.

52. WSJ, February 26, 1907, p. 4. NYT, February 27, 1907, p. 14.

53. WSJ, February 28, 1907, p. 4. NYT, February 28, 1907, p. 11; March 1, 1907, p. 11; March 2, 1907, p. 11.

54. WSJ, March 1, 1907, p. 4; March 6, 1907, pp. 4, 7. NYT, March 4, 1907, p. 16; March 7, 1907, p. 14.

55. NYT, March 12, 1907, p. 12; March 13, 1907, p. 12. WSJ, March 13, 1907, p. 7; March 14, 1907, p. 7. LeW, March 28, 1907, p. 304.

56. Thomas W. Lawson, "Friday, the 13th," EvM 15 (December 1906), pp. 821-23; Thomas W. Lawson, "Friday, the 13th," EvM 16 (March 1907), pp. 370-77.

57. NYT, March 18, 1907, p. 9; June 14, 1907, p. 2.

58. EvM 18 (February 1908), pp. 287-88.

59. Frank Fayant, "The Real Lawson," *Success* 11 (January 1908), pp. 23-42; Frank Fayant, "The Real Lawson," *Success* 11 (March 1908), pp. 145-71. NYT, August 23, 1908, p. 3.

60. *Boston Herald*, February 4, 1906, p. 6. "Mr. Lawson and the Personal Equation," *Bookman* 25 (April 1907), pp. 120-24.

61. LeW, February 13, 1902, p. 165; February 27, 1902, p. 211. NYT, May 22, 1903, p. 2. *Oil Investors' Journal*, June 1, 1903, pp. 5-7; June 15, 1903, p. 6; July 15, 1903, p. 6; May 3, 1903, p. 20; June 5, 1907, p. 14. *Boston Morning Journal*, April 11, 1906, p. 10. DMR, April 14, 1906, p. 7.

62. WSJ, August 17, 1908, p. 7. LeW, October 15, 1908, p. 381. CuH 8 (1908): 1086-87. EMJ, May 27, 1911, p. 1043.

63. LeW, April 2, 1908, p. 332. *Oil Investors' Journal,* April 5, 1908, p. 14; January 20, 1909, p. 16. NYT, July 22, 1913, p. 3. Louis Filler, *Appointment at Armageddon* (Westport, Conn.: Greenwood, 1976), 336-39.

64. Gatenby Williams and C. M. Heath, *William Guggenheim* (N.Y.: Lone Voice, 1934), 225-26. E. P. Hoyt, Jr., *The Guggenheims and the American Dream* (N.Y.: Funk & Wagnalls, 1967), 189-91.

65. WSJ, April 12, 1904; April 14, 1904, p. 8; April 21, 1904, p. 5; May 18, 1904, p. 7; March 19, 1907, p. 6; December 13, 1907, p. 8; August 10, 1908, p. 7. DMR, August 12, 1908, p. 4; August 13, 1908, p. 2.

66. WSJ, March 20, 1907; August 15, 1908, p. 7. DMR, August, 14, 1908, p. 2; August 15, 1908, pp. 1, 10; August 22, 1908, p. 2. *Outlook,* September 5, 1908, pp. 9-10. David Graham Evans, "Modern Speculation," *Success* 11 (October 1908), p. 656.

67. WSJ, August 12, 1908, p. 7; August 13, 1908, p. 7; August 14, 1908, pp. 6-7; August 18, 1908, p. 7. MSP, August 22, 1908, pp. 235-36. *New Orleans Daily Picayune,* August 22, 1908, p. 12.

68. "The Innocent Investor and the Mining Boom," *World's Work* 13 (January 1907), pp. 8383-85.

69. *N.Y. Press,* July 4, 1907, p. 3. NYT, May 24, 1912, p. 1; February 26, 1913, p. 3; September 27, 1913, p. 3; October 24, 1918, p. 17.

70. Thomas W. Lawson, "Fools and Their Money," EvM 14 (May 1906), pp. 690-93; Thomas W. Lawson, "The Remedy," EvM 27 (October 1912), p. 472c; Thomas W. Lawson, "The Remedy," EvM 27 (December 1912), pp. 777-92; Thomas W. Lawson, "The Remedy," EvM 28 (January 1913), pp. 89-104; Thomas W. Lawson, "The Remedy," EvM 28 (March 1913), pp. 404-13; Thomas W. Lawson, "The Remedy," EvM 28 (April 1913), pp. 550-59. *Literary Digest,* October 5, 1912, pp. 548-49.

71. WSJ, December 13, 1916, p. 4; December 15, 1916, p. 4; December 20, 1916, p. 4; December 22, 1916, p. 4; December 23, 1916, pp. 1-2.

72. NYT, December 27, 1916, p. 6; December 31, 1916, pt. 1, 15; January 1, 1917, p. 6.

73. NYT, January 2, 1917, p. 10; January 3, 1917, p. 4; January 5, 1917, p. 1; January 7, 1907, p. 3.

74. NYT, January 9, 1917, p. 1; January 10, 1917, p. 4; January 11, 1917, p. 1; January 13, 1917, p. 1; January 16, 1917, p. 1; February 15, 1917, p. 20.

75. NYT, March 26, 1920, p. 17; April 1, 1920, p. 14; April 22, 1920, p. 21. EMJ, April 3, 1920, p. 816.

76. NYT, December 24, 1915, p. 13; October 7, 1922, p. 15; October 14, 1922, p. 14; April 9, 1923, p. 31; May 27, 1923, p. 19; November 22, 1923, p. 2.

77. NYT, December 13, 1923, p. 21; February 8, 1925, p. 1; February 9, 1925, p. 19; February 10, 1925, p. 22; June 25, 1925, p. 13; May 24, 1936, p. 35. C. C. Regier, *The Era of Muckrakers* (Gloucester, Mass.: Peter Smith, 1957), 130-31. A. S. Link, *American Epoch* (N.Y.: Knopf, 1960), 76. Arthur Weinberg and Lila Weinberg, *The Muckrakers* (N.Y.: Putnam, 1964), 261-85. Louis Filler, *The Muckrakers* (University Park: Pennsylvania State University Press, 1976), 171-91.

Chapter 13. Aron Beam and His Gold Process

1. U.S. Bureau of the Census, 1900 Census, Colorado enumeration district 44, sheet 15, line 39. *Biographical and Historical Memoirs of Western Arkansas* (Chicago: Goodspeed, 1891).

2. MSP, August 7, 1897, p. 119. *B.C. Review,* June 18, 1898, pp. 143-44.

3. *Silver Cliff (CO) Rustler,* March 7, 1889, p. 1.

4. EMJ, July 28, 1894, p. 85; December 25, 1897, p. 754.

5. EMJ, September 12, 1896, p. 252; January 23, 1897, p. 96; March 13, 1897, pp. 263-64; July 2, 1898, p. 15; August 20, 1898, p. 225. *Denver Post,* December 26, 1896, p. 6. RMN, December 27, 1896, p. 6. *Mining Record,* July 14, 1898, p. 13.

6. MSP, September 18, 1897, p. 272. EMJ, July 17, 1897, p. 73; November 20, 1897, pp. 613-14; December 11, 1897, p. 694; July 23, 1898 pp. 91-92; May 6, 1899, p. 523; May 27, 1899, p. 616; November 18, 1899, p. 616; October 2, 1897 pp. 391-92; November 6, 1897, p. 542; November 13, 1897, p. 574; November 27, 1897; December 18, 1897, p. 724; December 25, 1897, p. 751; May 29, 1897, p. 547; January 18, 1902, p. 112; February 8, 1902, p. 219; September 27, 1902, p. 418; November 5, 1903, p. 361; December 10, 1903, p. 900; July 15, 1905, p. 76. "The 'Beam' Process and British Columbia Ores," *B.C. Mining Record* 5 (June 1899), pp. 11-13. *LaBelle (NM) Cresset*, September 2, 1897, p. 4; September 9, 1897, p. 1; September 23, 1897, p. 2. U.S. patents 582843, 689946, 691582, 695265, 708615, 737059, 745765, 793816. *Mining World* (Chicago), July 15, 1905, p. 47. *Denver Clerk & Recorder Book* 3917, p. 148.

7. *Denver Post*, December 26, 1896, p. 6. RMN, December 27, 1896, p. 6. "The Beam Process," *Mining & Industrial Reporter* 15, no. 1 (July 1897), pp. 21-22. DMR, September 3, 1897, p. 1; October 23, 1897, p. 1; November 4, 1897, p. 1; November 4, 1897, p. 2; November 27, 1897, p. 1; March 12, 1898, p. 2. *Mining Record*, July 14, 1898, p. 13.

8. Steve Wilson, "Dauntless Gold Seekers of the Wichitas," *Great Plains Journal* 22 (1983): 42-78. *Clarksville (TX) Northern Standard*, May 26, 1849, p. 2. *Rocky Mountain Mining Review*, September 10, 1885, p. 5.

9. D. K. Hale, "Gold in Oklahoma: The Last Great Gold Excitement in the Trans-Mississippi West, 1889-1918," *Chronicles of Oklahoma* 59, no. 3 (1981): 304-19.

10. NYT, November 2, 1890, p. 8. *Ft. Worth Gazette*, April 2, 1891, p. 13. EMJ, December 4, 1880, p. 367; April 11, 1891, p. 453; December 11, 1886, p. 424; April 11, 1896, p. 453. NYTr, November 28, 1897, p. 8. *Mining Industry & Tradesman*, October 16, 1890, p. 192.

11. Charles N. Gould, "Beginnings of the Geological Work in Oklahoma," *Chronicles of Oklahoma* 10, no. 2 (1932): 196-203.

12. *Lawton (OK) News-Republican*, September 24, 1903, p. 1. EMJ, January 28, 1904, p. 148.

13. H. F. Bain, "Reported Gold Deposits of the Wichita Mountains," in *Contributions to Economic Geology, 1903* (USGS Bulletin 225, 1904), 120-22. C. N. Gould, "Gold and Silver," in *Brief Chapters on Oklahoma's Mineral Resources* (Oklahoma Geological Survey Bulletin 6, 1910), 60-62.

14. Pete Hanraty, *Third Annual Report, Oklahoma Chief Inspector of Mines* (1910): 180-87.

15. *Lawton (OK) News-Republican*, September 24, 1903, p. 2. *Mt. Sheridan (OK) Miner*, January 7, 1904, p. 4. The doubtful existence of a platinum lode resurfaced in 1924, in a murder over a secret platinum mine in the Wichitas. *Daily Oklahoman*, June 10, 1924, p. 1.

16. *Mt. Sheridan (OK) Miner*, August 25, 1904, p. 1. *Mineral Kingdom*, January 26, 1905, p. 1.

17. *Oklahoma Geological Survey Bulletin* 1, pp. 75-78. Steve Wilson, "Economic Possibilities of Mining in the Wichita Mountains" (Proceedings of the Oklahoma Academy of Science 43 [1962], pp. 160-63.)

18. *Macon (MO) Republican*, December 12, 1908, p. 1; January 9, 1909, p. 2; January 16, 1909, p. 3; January 23, 1909, p. 6; January 30, 1909, p. 1.

19. *Macon (MO) Republican*, February 6, 1909, p. 6; February 27, 1909, p. 6; March 30, 1909, p. 2; March 27, 1909, p. 1; May 8, 1909, p. 1; June 5, 1909, p. 2; June 19, 1909, p. 6; July 10, 1909, p. 6; August 14, 1909, p. 2; November 13, 1909, p. 3. MSP, April 10, 1909, pp. 498-99.

20. P. L. Russell, *History of Western Oil Shale* (East Bruswick, N.J.: Center for Professional Advancement, 1980), 13-14, 21-24. Colorado Carbon Company: 1918-1925, manuscript collection 961, file folder #4,Colorado Historical Society, Denver. "American Continuous Process Claims Precious Metal Values," *Shale Review* 3, no. 8 (October 1921), pp. 10-11.

21. "Gold Mining in the Shale Fields," *Shale Review* 3, no. 6 (July 1921), p. 11; "Does Shale Carry Gold, Silver, Platinum in Paying Quantities?" *Shale Review* 3, no. 9 (November 1921), p. 12; W. C. Kirkpatrick, "Successful Test at Indiana Shale Plant," *Shale Review* 4, no. 1 (January 1922), pp. 5-6; "Mill Run of American Continuous Retort Is Promised Soon," *Shale Review* 4, no. 2 (February 1922), p. 11; "Plant for Recovering Precious Metals from Shale Is Now Running," *Shale Review* 4, no. 3 (March 1922), p. 14; B. A. Hayden, "Gold, Silver, Platinum and Other Rare Metals Are Found in Oil Shale Deposits," *Shale Review* 4, no. 4 (April 1922), pp. 5-7; "Tests for Metal Recovery Continue at Beam Plant," *Shale Review* 4, no. 5-6 (May-June 1922), pp. 5-6, 7-8; "Does Oil Shale Carry Gold?" *Shale Review* 4, no. 9 (September 1922), p. 9; Thomas Varley, "Bureau of Mines Investigates Gold in Oil Shales and Its Possible Recovery," U.S. Bureau of Mines, Report of Investigation 2413 (1922), p. 9; S. C. Lind, C. W. Davies, and M. W. von Bernewitz, "Platinum Assays and Platinum Promotions," US Bureau of Mines, Report of Investigations 2496. MSP, March 4, 1922, p. 283; April 17, 1920, pp. 557-58. RMN, December 31, 1922, p. 1, c. 3. EMJ, February 10, 1923, pp. 273-76.

22. American Shale and Metals Company (advertisement), *Shale Review* 4, no. 4 April 1922, p. 13; "Placer Machine Recovers Free Gold from Oil Shale Deposits," *Shale Review* 4, no. 9 (September 1922), p. 6. "New Capital for Beam Process," *Shale Review* 4, no. 9 (September 1922), 13. *Denver Clerk & Recorder, Book* 3581, pp. 64-65.

Chapter 14. Golden Sands of the Adirondacks

1. NYTr, December 30, 1892, p. 9. NYT, February 17, 1885, p. 3; July 5, 1896, p. 16.

2. NYT, July 5, 1879, p. 2; July 8, 1879, p. 1; June 2, 1880, p. 8.

3. NYT, June 25, 1880, p. 2; July 27, 1880, p. 2. EMJ, May 1, 1880, p. 304; July 17, 1880, p. 41; August 21, 1880, p. 126. *Boston Daily Advertiser,* August 18, 1880, p. 4.

4. EMJ, May 22, 1909, p. 1065.

5. EMJ, September 3, 1898 pp. 275-76.

6. NYT, October 24, 1897.

7. NYTr, September 14, 1897, p. 3; September 16, 1897, p. 2; March 13, 1898, supp., p. 1. NYT, September 15, 1897, p. 3; September 16, 1897, p. 3; September 21, 1897, p. 5.

8. NYTr, September 10, 1897, p. 2; September 11, 1897, p. 7. NYT, September 17, 1897, p. 5.

9. EMJ, May 19, 1900, p. 582.

10. Harpending Collection, folder 23, box 4, manuscript 950, California Historical Society, San Francisco. *Boston Herald,* August 6, 1899, p. 18, c. 1. EMJ, August 26, 1899, p. 241.

11. DMR, May 22, 1905, p. 3. EMJ, February 24, 1912, p. 392; May 14, 1910, p. 1002.

12. EMJ, October 13, 1900, p. 439.

13. TS, July 27, 1905, p. 1.

14. EMJ, March 19, 1910, p. 620; May 2, 1914, p. 927; March 4, 1916, p. 449; April 25, 1925, p. 674. "A Warning," *Oil & Mining Bulletin* 1 (March 1915), p. 92. MH 16 (1925): 1590.

15. EMJ, September 24, 1898, pp. 363-64; March 2, 1912, pp. 437-38.

Chapter 15. Mining Gold from the Sea

1. The best account of Jernegan is Arthur Railton's "Jared Jernegan's Second Family," and "Charlie Fischer, The Perfect Partner," *Dukes County Intelligencer* 28, no. 2 (1986): 51-91, 92-94.

2. NYT, November 11, 1897, p. 1. NYTr, November 16, 1897, p. 6; July 31, 1898, p. 12. MSP, December 11, 1897, p. 547.

3. EMJ, August 6, 1898, pp. 151, 154.

4. EMJ, January 29, 1910; July 1, 1911. CuH 8 (1908): 853, 1193.

5. NYT, March 18, 1898, p. 12. NYTr, March 23, 1898, p. 9; March 27, 1898, p. 5. EMJ, March 26, 1898, pp. 366-67.

6. EMJ, April 2, 1898, p. 395.

7. EMJ, July 30, 1898, pp. 122-24.

8. MSP, July 2, 1898, p. 6. NYTr, July 29, 1898, p. 14.

9. NYT, July 27, 1898, p. 7; July 29, 1898, p. 12.

10. NYT, July 30, 1898, p. 3. NYTr, July 30, 1898, p. 14.

11. NYTr, August 2, 1898, p. 10.

12. N.Y. Herald, July 31, 1898, p. 7.

13. NYT, August 2, 1898, p. 6.

14. NYTr, August 6, 1898, p. 9.

15. NYTr, August 3, 1898, p. 14; August 11, 1898, p. 9. EMJ, August 13, 1898, p. 66; August 27, 1898, p. 257.

16. NYT, August 17, 1898, p. 9. NYTr, August 17, 1898, p. 9; November 11, 1898, p. 12. EMJ, September 3, 1898; October 15, 1898, p. 467.

17. EMJ, January 21, 1899, p. 93; April 29, 1899, p. 509; May 20, 1899, p. 583.

18. NYT, October 12, 1898, p. 2; February 5, 1899, p. 6. NYTr, December 19, 1898, p. 1; December 20, 1898, p. 7. EMJ, December 24, 1898; January 21, 1899, p. 93; February 11, 1898, p. 183.

19. EMJ, August 15, 1903, p. 225. AMS, March 3, 1904, p. 299.

20. AMS, March 3, 1904, p. 299.

21. Journal of the Society of Chemical Industries, January 31, 1903, p. 97; March 15, 1902, pp. 349-50; December 31, 1902, p. 1536; August 31, 1903, p. 953.

22. EMJ, June 10, 1899, p. 674.

23. EMJ, February 16, 1905, p. 354. AMS, March 29, 1905, pp. 404-5.

24. EMJ, February 9, 1905, p. 282. AMS, May 17, 1904, p. 556; April 12, 1905, p. 447.

25. NYT, April 9, 1905, p. 3. Mining World (Chicago), April 15, 1905, p. 393.

26. NYT, April 9, 1905.

27. AMS, October 18, 1905, p. 371.

28. AMS, March 29, 1905. TL, March 8, 1907, p. 3; March 9, 1907, p. 3.

29. AMS, December 29, 1904, p. 898; February 28, 1906, p. 208.

30. NYT, October 3, 1911, p. 9; October 7, 1911, p. 3. EMJ, October 7, 1911, p. 671; October 14, 1911, p. 735; October 28, 1911, p. 834.

31. J. L. Mero, Mineral Resources of the Sea (N.Y: Elsevier, 1965), 41. Morris Goran, The Story of Fritz Haber (Norman: University of Oklahoma Press, 1967), 91-98.

32. EMJ, October 2, 1920, p. 655; December 25, 1926, p. 1017; May 28, 1927, p. 889. Fritz Haber, "Das gold im meerwasser," Zeitschrift für Angewandte Chemie 40 (1927): 303-14.

33. Mining Record, June 19, 1902, p. 569.

34. EMJ, April 30, 1898, p. 515. NYT, February 1, 1924, p. 1.

35. J. L. Mero, Mineral Resources of the Sea (N.Y.: Elsevier, 1965).

Chapter 16. The Spoilers

1. K. O. Bjork, "Reindeer, Gold, and Scandal," Norwegian-American Studies 30 (1985): 130-95.

2. A. J. Collier et al., "The Gold Placers of Parts of the Seward Peninsula, Alaska," USGS Bulletin 328 (1908): 18, 21.

3. C. M. Naske, "The Shaky Beginnings of Alaska's Judicial System," Western Legal History 1, no. 2 (1988): 163-210.

4. National Cyclopaedia of America Biography 32 (N.Y.: J. T. White, 1945), 92-93. R. P. Wilkins, "Alexander McKenzie and the Politics of Bossism," in The North Dakota Political Tradition (Ames: Iowa State University Press, 1981), 3-39.

5. *Congressional Record,* 56th Cong., 1st sess., April 9, 1900, vol. 33, pt. 5:pp. 3928-34.

6. *Congressional Record,* 56th Cong., 1st sess., April 20, 1900, vol. 33, pt.5:pp. 4471.

7. NYT, May 1, 1900, p. 1.

8. NYT, April 20, 1900, p. 5; April 26, 1900, p. 10; May 2, 1900, p. 5.

9. James Wickersham, *Old Yukon, Tales—Trails—and Trials* (Washington, D.C.: Washington Law Book, 1938), 4.

10. W. E. Lillo, "The Alaska Gold Mining Company and the Cape Nome Conspiracy," PhD diss., University of North Dakota, 1935, p. 172.

11. Sam C. Dunham, "The Poor Swede" (1913) in C. L. Lokke, "From the Klondike to the Kougarok," *Norwegian-American Studies* 16 (1950): 161-89.

12. R. E. Beach, "The Looting of Alaska," *Appleton's Booklovers Magazine* 7, no. 3 (March 1906): 294-408.

13. MSP, February 17, 1900, p. 181.

14. W. W. Morrow, "The Spoilers," *California Law Review* 4, no. 2 (January 1916): 89-113.

15. R. E. Beach, "The Looting of Alaska," *Appleton's Booklovers Magazine* 7, no. 4 (April 1906): 540-47.

16. Lanier McKee, *The Land of Nome* (N.Y.: Grafton, 1902), 114-16.

17. The $150,000 value of the gold was at the then price of about $20 per troy ounce.

18. *Washington Post,* February 19, 1901, pp. 4, 6.

19. *Congressional Record,* 56th Cong., 2d sess., December 17, 1900, 34, pt. 1:p. 355; February 26, 1901, 34, pt. 4:pp. 3055-63.

20. *In re Alexander McKenzie,* 180 U.S. 536 (1901). *Tornanses v. Melsing et al.,* 106 F. 775 (9th Cir. 1901).

21. C. L. Lokke, "From the Klondike to the Kougarok," *Norwegian-American Studies* 16 (1950): 161-89.

22. C. L. Lokke, "A Madison Man at Nome," *Wisconsin Magazine of History* 33 (December 1949): 164-83.

23. Evangeline Atwood, *Frontier Politics: Alaska's James Wickersham* (Portland, Ore.: Binford & Mort, 1979), 83-84.

24. NYT, August 28, 1901, p. 3.

25. NYT, January 8, 1901, p. 2. *Congressional Record,* 56th Cong., 1st sess., February 5, 1902, 35, pt. 2:pp. 1333-39. *Washington Post,* February 9, 1902, p. 2. *Anderson v. Comptois,* 109 F. 971 (9th Cir. 1901). In re Noyes, 121 Fed 209(1903).

26. *Washington Post,* February 3, 1902, p. 1; February 4, 1902, p. 4; February 5, 1902, p. 4.

27. R. E. Beach, "The Looting of Alaska," *Appleton's Booklovers Magazine* 7, no. 5 (May 1906), 606-13. James Wickersham, *Old Yukon* (Washington, D.C.: Washington Law Book, 1938), 369-72.

28. T. M. Cole, "A History of the Nome Gold Rush: The Poor Man's Paradise," PhD diss., University of Washington, 1983, p. 211.

29. F. G. Carpenter, *Alaska: Our Northern Wonderland* (Garden City, N.Y.: Doubleday, Doran, 1928).

Chapter 17. Whitaker Wright's House of Cards

1. Drep, March 21, 1903, p. 5.

2. *Leadville Weekly Herald,* December 20, 1879, p. 3; January 31, 1879, p. 4.

3. LH, October 21, 1880, p. 4; December 3, 1880, p. 4; January 1, 1881, p. 7; March 4, 1881, p. 4; November 25, 1881, p. 4; December 18, 1881, p. 4; December 30, 1881, p. 4.

4. LH, January 10, 1882, p. 4; January 17, 1882, p. 4; January 24, 1882, p. 4; March 5, 1882, p. 4; April 18, 1882, p. 1; April 23, 1882, p. 4; May 2, 1882, p. 4.

5. Drep, May 10, 1881, p. 6. *Las Vegas (NM) Mining World,* July 1881, p. 7; August 1, 1882, p. 278; August 15, 1882, p. 289. NYTr, November 14, 1881, p. 6. RMN, February 18, 1882, p. 3.

6. NYTr, August 27, 1881, p. 12. *Las Vegas (NM) Mining World,* September 15, 1881, p. 25. Bernard MacDonald, "Discussion of 'Genesis of the Lake Valley, New Mexico, Silver Deposits,'" *American Institute of Mining Engineers Bulletin* 26 (1909): 211-16.

7. LH, January 18, 1882, p. 4; January 27, 1882, p. 4. *Las Vegas (NM) Mining World,* August 1, 1882, p. 274; September 15, 1882, p. 26; April 1, 1883, p. 154.

8. LH, June 9, 1882, p. 4. EMJ, July 15, 1882, pp. 36-37, September 9, 1882, p. 141; November 4, 1882, p. 249. *Las Vegas (NM) Mining World,* October 1, 1882, p. 35; November 1, 1882, p. 67. LeW, February 11, 1904, p. 139. *Oil Investor's Journal,* February 15, 1904, p. 9.

9. EMJ, October 28, 1882, p. 231; July 7, 1883, p. 9; July 28, 1883, p. 55. *Las Vegas (NM) Mining World,* April 1, 1883, p. 147.

10. EMJ, September 22, 1883, p. 189; December 1, 1883, p. 346; October 9, 1886, p. 254; May 28, 1887, p. 389; August 6, 1887, p. 102. MSP, September 6, 1884, p. 146.

11. LH, June 26, 1884, p. 3. *Philadelphia Public Ledger,* January 27, 1904, p. 1.

12. EMJ, April 30, 1881, p. 380. *Philadelphia Public Ledger,* January 28, 1904, p. 8.

13. FT September 26, 1894, p. 2, 4; November 6, 1894, p. 2; December 31, 1894, p. 3. EMJ, October 13, 1894, p. 357.

14. FT December 29, 1894, p. 4.

15. FT June 15, 1904, p. 3.

16. TL, January 29, 1901, p. 3. EMJ, August 26, 1899, p. 265. FT, February 9, 1897, p. 2; February 10, 1897, p. 2.

17. *B.C. Review,* October 30, 1897, p. 36. *Spokane Spokesman-Review,* January 27, 1904, p. 1.

18. FT, November 13, 1897, p. 5; December 11, 1897, p. 10.

19. EMJ, March 31, 1900, p. 372.

20. FT, December 1, 1898, p. 5; December 6, 1898, p. 6; December 10, 1898, p. 4. "The London Flotation of the Le Roi," *B.C. Mining Record* 5 (February 1899), pp. 36-37. *B.C. Review,* December 10, 1898, pp. 494-95.

21. *B.C. Mining Record* 4 (October 1898), p. 16. EMJ, March 24, 1900, p. 366. *Spokane Spokesman-Review,* January 29, 1904, p. 8.

22. *B.C. Review,* June 16, 1900, pp. 278-79.

23. EMJ, June 30, 1900, p. 786.

24. "Le Roi No. II," *Canadian Mining Review* 21 (January 1902), p. 16.

25. *B.C. Review,* July 7, 1900, p. 312; February 22, 1902, p. 91.

26. NYT, January 28, 1902, p. 6.

27. Econ, June 11, 1898, p. 869. TL, November 8, 1898, pp. 7, 12.

28. *Pall Mall Gazette,* September 20, 1898, p. 7; September 23, 1898, p. 7; October 5, 1898, p. 7. "The London Flotation of the Le Roi," *B.C. Mining Record* 5 (February 1899), pp. 36-37.

29. TL, January 15, 1902, p. 12; January 23, 1902, p. 9. NYTr, January 29, 1902, p. 14. NYT, January 15, 1904, p. 9. MJ, March 3, 1906.

30. NYTr, March 29, 1903, pt. 2, p. 12.

31. FT, December 16, 1897, p. 2.

32. EMJ, October 20, 1900, p. 476.

33. EMJ, July 14, 1900, p. 56; July 28, 1900, p. 110; October 13, 1900, p. 446; November 17, 1900, p. 596.

34. EMJ, December 30, 1899, p. 806; February 3, 1900, p. 156; February 24, 1900, p. 232. Geoffrey Blainey, *The Rush That Never Ended* (Melbourne: Melbourne University Press, 1963), 203-4.

35. TL, February 26, 1902, p. 15.

36. *Pall Mall Gazette,* December 18, 1900, p. 4; December 29, 1900, p. 4; December 31, 1900, p. 1. TL, December 31, 1900, p. 15.

37. "London and Globe," *B.C. Mining Record* 8 (February 1901), pp. 54-55.

38. TL, August 1, 1901, p. 12; August 8, 1901, p. 2.

39. EMJ, March 23, 1901, p. 377.

40. EMJ, July 6, 1901, p. 25. TL, November 14, 1901, p. 14, c.2.

41. FT, August 30, 1901, p. 3; December 27, 1901, p. 3. NYT, August 31, 1901, p. 6.

42. TL, August 7, 1902, p. 2.

43. TL, January 28, 1902. NYT, August 8, 1902, p. 9.

44. FT, June 3, 1902, p. 3; June 4, 1902, p. 5; June 5, 1902, p. 4; June 6, 1902, p. 5; June 7, 1902, p. 5; June 10, 1902, p. 3; June 11, 1902, p. 3; June 12, 1902, p. 6; June 13, 1902, pp. 4, 7.

45. AMS, March 10, 1904. TL, February 3, 1904, p. 8.

46. TL, March 13, 1903, p. 4.

47. TL, March 16, 1903, p. 6.

48. TL, March 16, 1903, p. 6.

49. NYTr, March 15, 1903, p. 2; March 19, 1903, p. 4; March 21, 1903, p. 4.

50. NYT, August 6, 1903, p. 7.

51. *Manchester Guardian,* January 20, 1899, p. 3.

52. *Manchester Guardian,* January 21, 1904, p. 5; January 22, 1904, p. 3.

53. *Philadelphia Press,* January 28, 1904, p. 2.

54. AMS, February 4, 1904, p. 157. NYT, January 29, 1904, p. 2.

55. *Denver Times,* January 27, 1904, p. 1.

56. AMS, April 4, 1904, p. 146; March 10, 1904, p. 333.

57. FT, May 10, 1904, p. 4; May 28, 1904, p. 4; July 27, 1904, p. 3; November 24, 1904, p. 7; November 26, 1904, p. 5; February 28, 1905, p. 7. Jeremy Mouat, *Roaring Days* (Vancouver: University of British Columbia Press, 1995), 65-66. The Le Roi mine is now open for tourists.

58. FT, May 10, 1904, p. 4; October 29, 1904, p. 2; November 9, 1904, p. 5.

59. WSJ, January 28, 1904, p. 1. *Philadelphia Inquirer,* January 28, 1904, p. 1. *Spokane Spokesman-Review,* January 29, 1904, p. 1.

Chapter 18. George Graham Rice's Adventures with Your Money

1. NYT, November 11, 1894, p. 16; April 20, 1895, p. 7. NYTr, March 8, 1895, p. 2; March 10, 1895, p. 14.

2. *N.Y. Press,* June 28, 1903, pt. 3, p. 1.

3. Minv, May 22, 1911, p. 5.

4. National Archives, Washington, List of post office fraud orders. NYT, June 12, 1903, p. 3, c. 5. G. G. Rice, *My Adventures with Your Money* (Boston: Gorham, 1911), 363.

5. *N.Y. Press,* June 19, 1903, p. 7; June 23, 1903, p. 7.

6. LeW, September 9, 1897, p. 175.

7. GFN, February 24, 1905, p. 7.

8. MSP, October 22, 1910, p. 542; May 29, 1915, p. 830.

9. *Denver Times,* April 25, 1905, p. 1; May 22, 1905, p. 3. *Denver Post,* May 23, 1905, p. 1. TS, May 24, 1905, p. 1; May 27, 1905, p. 6. Some people cannot stay out of trouble. A grand jury in 1909 indicted Sydney Flower, alias Parmeter Kent, for mail fraud. Francis Burton was convicted for malfeasance in his bank failure but appealed and was acquitted on retrial. Two years later, however, after conducting a few more swindles, he took exception to a news item about himself in the Rawhide, Nevada, *Rustler,* and died in a gunfight with that paper's no-nonsense editor. GFN, April 24, 1909, p. 3. GFTr, December 11, 1907, p. 1.

10. GFN, November 24, 1905, p. 4. DMR, December 2, 1905, p. 1. C. B. Glasscock, *Gold in Them Hills* (Indianapolis: Bobbs-Merrill, 1932), 176-82.

11. SFCa, May 5, 1905, p. 2. SFCh, May 6, 1905, p. 1; May 7, 1905, p. 40. *Tonopah (NV) Bonanza*, March 4, 1905, p. 1; April 29, 1905, p. 3. SFE, May 7, 1905, p. 44. GFN, November 10, 1905, p. 4; November 17, 1905, p. 4. Arkell fled San Francisco a year later, to avoid creditors. TDS, April 27, 1905, p. 4; May 5, 1905, p. 6; May 9, 1905, p. 1; May 12, 1905, p. 3; July 1, 1905, p. 9; April 7, 1906, p. 6. *Goldfield (NV) Sun,* April 7, 1906, p. 1.

12. DMR, November 4, 1907, p. 1; November 11, 1906, p. 1; November 18, 1905, p. 1.

13. Minv, October 10, 1910, p. 220. The same *Daily Mining Record* had breathlessly quoted Stanton's highly optimistic predictions (May 20, 1905, p. 5). Stanton took his own life a few years later by turning his pistol on himself (GFN, April 17, 1909, p. 4).

14. DMR, September 15, 1906, p. 1; September 20, 1906, p. 1; September 25, 1906, p. 1; October 6, 1906, p. 1. *Denver Post,* September 20, 1906, p. 2; October 4, 1906, p. 10. *Denver Times,* September 25, 1906, p. 1; October 4, 1906, p. 1. *Denver Republican,* October 5, 1906, p. 5.

15. *Tonopah (NV) Miner,* December 29, 1906. GFTr, October 20, 1910, p. 1. S. S. Zanjani, *Goldfield* (Athens: Ohio University Press, 1992), 161.

16. D. M. Alborn, "Crimping and Shanghaiing on the Columbia River," *Oregon Historical Quarterly* 63 (Fall 1992): 263-91.

17. NMN, July 1907, p. 2.

18. SFE, March 4, 1906, p. 62. WSJ, March 23, 1906, p. 6. *Tonopah (NV) Bonanza,* March 24, 1906, p. 3. *N.Y. Commercial,* January 9, 1907; *Annual Mining Review,* p. 7.

19. BFM, May 25, 1906, p. 2. *Tonopah (NV) Miner,* September 15, 1906, p. 12. S. S. Zanjani, *Goldfield* (Athens: Ohio University Press, 1992), 171. S. S. Zanjani, "Wildcats and Bank Wreckers," *Nevada Historical Society Quarterly* 35 (1992): 105-24.

20. SFCa, September 17, 1906, p. 2.

21. GFN, November 10, 1906, p. 6.

22. EMJ, October 19, 1907, p. 727.

23. *Inyo (CA) Register,* January 17, 1907.

24. BFM, June 16, 1906, p. 1. BFM, November 2, 1906, p. 9.

25. DMR, November 12, 1906, p. 1; November 14, 1906, p. 1; November 20, 1906, p. 2. SFB, January 5, 1907, p. 3. R. R. Elliott, *Nevada's Twentieth-Century Mining Boom* (Reno: University of Nevada Press, 1966), 92.

26. SFE, January 5, 1907, p. 1; January 6, 1907, p. 48; January 8, 1907, p. 1; January 9, 1907, p. 3. *N.Y. Commercial,* January 7, 1907, pp. 1, 7; January 8, 1907, p. 7; January 9, 1907, pp. 1, 17; January 12, 1907, p. 7. SFCa, January 13, 1907, p. 24. NMN, October 23, 1907, p. 1. DMR, October 24, 1907, p. 5. *Tonopah (NV) Bonanza,* January 11, 1912, p. 4.

27. SFCa, June 6, 1907, p. 3; August 6, 1907, p. 4.

28. DMR, January 19, 1907, p. 7. NMN, July 27, 1907, p. 6; July 23, 1908, p. 8; March 27, 1909, p. 7. *Tonopah (NV) Miner,* August 10, 1907, p. 4. *Reno (NV) Gazette,* August 31, 1907, p. 1; September 3, 1907, p. 5.

29. Merrill S. Teague, "Bucket-Shop Sharks," EvM 14, no. 6 (June 1906), pp. 729-31; Merrill S. Teague, "Bucket-Shop Sharks," EvM 14, no. 7 (July 1906), pp. 33-43; Merrill S. Teague, "Bucket-Shop Sharks," EvM 14, no. 8 (August 1906), pp. 245-54. NMN, July 27, 1907, p. 4. *Reno (NV) Evening Gazette,* November 8, 1907, p. 8. TS, November 9, 1907, p. 1. *Tonopah (NV) Bonanza,* November 10, 1907, p. 3. GFT, January 25, 1912, p. 1. *Oakland (CA) Tribune,* August 1, 1948, p. C1.

30. *Goldfield (NV) Chronicle,* June 11, 1907. N. C. Goodwin, *Nat Goodwin's Book* (Boston: Gorham, 1914), 293-301.

31. GFN, April 17, 1909, p. 4.

32. SLTr, August 21, 1907, p. 9. *Rhyolite (NV) Herald,* October 11, 1907, p. 3. DMR, October 12, 1907, p. 1. *Reno (NV) Evening Gazette,* November 5, 1907, p. 5. Dan Edwards was later suspected of salting the Keystone Mine. GFTr, July 29, 1910, p. 3.

33. NMN, October 23, 1907, pp. 5, 6; December 19, 1907, p. 4. MInv 50 (1908): 382.

34. SFCa, June 27, 1908, p. 16.

35. MInv 50 (1908): 140-49, 265-69. NMN, January 2, 1908, p. 2; February 20, 1908, pp. 3, 5; February 27, 1908, pp. 3, 6; March 5, 1908, p. 2; March 12, 1908, p. 1; March 19, 1908, p. 3; April 9, 1908, p. 1, p. 5; May 7, 1908, p. 5. *Mining Financial News,* May 6, 1909, p. 1. GFTr, November 28, 1911, p. 1. Although E. W. King formed the Barnes-King mining company, whose stock was later sold with false promises, the promotion was done by others after King left the company.

36. GFTr, March 24, 1908, p. 1.

37. Elinor Glyn is best known for her essay *It,* the movie adaptation of which propelled Clara Bow to stardom in the 1920s.

38. NMN, December 3, 1908, p. 1; December 10, 1908, p. 1; December 31, 1908, p. 1; January 7, 1909, p. 1; March 11, 1909, p. 1. *N.Y. Sun,* December 11, 1908, p. 1. DMR, April 14, 1909, p. 1; April 21, 1909, p. 1.

39. GFN, December 29, 1908, p. 1.

40. NMN, July 27, 1907, p. 6.

41. Minv, September 5, 1910, p. 68. GFTr, November 11, 1910, p. 1.

42. Joseph Weil and W. T. Brannon, *"Yellow Kid" Weil* (Chicago: Ziff-Davis, 1948), 99.

43. *N.Y. Commercial,* February 1, 1907, p. 7. NYT, February 2, 1907, p. 13. WSJ, February 4, 1907, p. 7. *Ely (NV) Mining Record,* June 13, 1908, p. 1. *Mining Financial News,* May 27, 1908, p. 1.

44. *Ely (NV) Mining Record,* June 4, 1909, p. 1; June 11, 1909, p. 1. DMR, June 14, 1909, p. 1; August 18, 1909, p. 1. GFTr, December 12, 1911, p. 1.

45. MInv 56 (1909): 129-30.

46. Minv, March 28, 1910, p. 149; April 25, 1910, p. 280; May 16, 1910, p. 403; June 6, 1910, p. 69. *Mining World* (Chicago), June 4, 1910, p. 1150; September 24, 1910, p. 584. GFN, July 28, 1910.

47. *White Pine (NV) News,* October 2, 1910, p. 1. EMJ, October 9, 1909, p. 739; October 23, 1909, p. 835; October 30, 1909, p. 883; November 6, 1909, pp. 931-35; November 13, 1909, p. 988; November 27, 1909, p. 1078; February 19, 1910, pp. 399-400. *Denver Post,* November 7, 1909, p. 1. DMR, November 8, 1909, p. 1. MInv (1910): 304.

48. GFN, November 27, 1909, p. 5.

49. Minv, February 7, 1910, p. 272. *Mining World* (Chicago), February 26, 1910, pp. 456-61.

50. MSP, November 27, 1909, pp. 709, 722-23; October 8, 1910, p. 460. Minv (1909): 62. EMJ, February 12, 1910, p. 353; February 26, 1910, pp. 448, 450.

51. *Rawhide (NV) Press-Times,* July 29, 1910, p. 1; August 10, 1910, p. 1. GFN, August 2, 1910, p. 3; August 20, 1910, p. 1.

52. CuH 10 (1911): 1587.

53. GFN, August 20, 1910, p. 1. NYT, August 20, 1910, p. 1.

54. NYT, September 30, 1910, p. 1. WSJ, September 30, 1910, p. 7. EMJ, October 8, 1910, pp. 709-10.

55. GFTr, November 3, 1910, p. 1; November 10, 1910, p. 2; November 13, 1910, p. 1; November 19, 1910, p. 1; November 20, 1910, p. 2. NYT, November 10, 1910, p. 7; January 18, 1911, p. 2. *Mining World* (Chicago) October 8, 1910, p. 674; October 22, 1910, p. 783; November 5, 1910, p. 878; November 26, 1910, p. 1020; December 17, 1910, p. 1161; January 14, 1911, p. 83; January 28, 1911, p. 267; March 11, 1911, p. 567. *White Pine (NV) News,* October 9, 1910, p. 1; November 20, 1910, p. 1. Minv, September 11, 1911, p. 90.

56. DMR, February 18, 1913, p. 1. A. C. Spencer, *The Geology and Ore Deposits of Ely, Nevada* (USGS Professional Paper 96, 1917), 149-51.

57. *Rawhide (NV) Press-Times,* September 30, 1910, p. 1; November 25, 1910, p. 1. GFTr, December 1, 1912, p. 1.

58. GFN, October 30, 1909, p. 4; July 25, 1910, p. 1; August 3, 1910, p. 1; August 6, 1910, p. 1; August 25, 1910, p. 2. *Reno (NV) Evening Gazette,* November 4, 1910, p. 1.

59. GFTr, October 21, 1910, p. 1; October 30, 1910, p. 1; November 5, 1910, p. 1; November 24, 1910, p. 2. *Reno (NV) Evening Gazette,* October 27, 1910, pp. 2, 4, October 29, 1910, p. 1; November 4, 1910, p. 1.

60. *Reno Nevada Weekly,* October 8, 1910, p. 11.

61. *Reno (NV) Evening Gazette,* November 2, 1910, p. 5; November 3, 1910, p. 8; January 31, 1911, p. 2; February 1, 1911, p. 6. GFTr, November 1, 1910, p. 1; November 5, 1910, p. 1. November 6, 1910, p. 4; February 2, 1911, p. 2. NYT, February 1, 1911, p. 8; February 16, 1930 sec. 2, p. 8.

62. GFN, April 10, 1909, p. 4; April 17, 1909, p. 5. GFTr, April 12, 1911, p. 2; April 14, 1911, p. 2; April 17, 1911, p. 2; April 24, 1911 pp. 1-2; April 29, 1911 pp. 3-4; May 4, 1911, p. 1; December 15, 1911, p. 1; January 17, 1912, p. 1. Louis Guenther, "Pirates of Promotion," *World's Work* 37 (November 1918), pp. 29-31. Charles Stoneham later sold his bucket shops and bought the New York Giants baseball team in 1919. By the time he died in 1936, his most infamous misdeed was trading away the hall-of-fame hitter Rogers Hornsby. NYT, January 7, 1936, p. 21.

63. NYT, November 1, 1911, p. 6; November 14, 1911, p. 7; December 9, 1911, p. 6; December 15, 1911, p. 9.

64. NYT, December 2, 1911, p. 6; December 8, 1911, p. 19.

65. NYT, November 23, 1911, p. 1.

66. NYT, December 22, 1911, p. 7; December 23, 1911, p. 11; December 24, 1911, p. 3; December 27, 1911, p. 12; December 28, 1911, p. 5; December 30, 1911, p. 20; January 9, 1912, p. 17. GFTr, December 28, 1911, p. 1. *Tonopah (NV) Bonanza,* January 11, 1912, p. 1. Minv, January 15, 1912, p. 154.

67. *Tonopah (NV) Bonanza,* January 12, 1912, p. 1; January 19, 1912, p. 3.

68. GFTr, January 20, 1912, p. 1; January 23, 1912, p. 1; January 24, 1912, p. 1.

69. NYT, March 3, 1912, pt. 2, p. 13; March 6, 1912, p. 10; March 8, 1912, p. 9. *Wonder Mining News,* March 11, 1912, p. 1. EMJ, March 16, 1912, p. 542. "Not Much, But Some," *Mines and Minerals* 32, no. 10 (May 1912), p. 577.

70. *Mining Science,* October 10, 1912, p. 226. NYT, October 27, 1912, p. 13.

71. Louis Guenther, "Pirates of Promotion," *World's Work* 37 (October 1918), pp. 584-91.

72. EMJ, April 17, 1915, p. 714; April 24, 1915, p. 753; September 4, 1915, p. 405; September 11, 1915, p. 456; October 2, 1915, p. 565; November 20, 1915, pp. 850-51.

73. EMJ, September 26, 1908, p. 629; March 20, 1915, p. 555; September 4, 1915, p. 407; July 16, 1916, p. 160; October 14, 1916, p. 728; November 4, 1916, p. 846. *Salt Lake Mining Review,* March 15, 1915, p. 21. MH 16 (1925): 1697.

74. Minv, September 6, 1915, p. 26. *Salt Lake Mining Review,* October 15, 1916, pp. 36-37; October 30, 1916, pp. 22, 29; November 15, 1916, p. 30; November 30, 1916, p. 30; December 15, 1916, p. 30; December 30, 1916, p. 18.

75. Robert Sobel, *Amex: a History of the American Stock Exchange 1921-1971* (N.Y.: Weybright & Talley, 1972), 8-12. NYT, October 23, 1908, p. 1.

76. NYT, October 5, 1916, p. 19.

77. NYT, October 6, 1916, p. 6; October 21, 1916, p. 16; October 26, 1916, p. 24.

78. NYT, October 7, 1916, p. 6; October 8, 1916, sec. 7, p. 8.

79. *Salt Lake Herald-Republican,* October 8, 1916, p. 22; October 10, 1916, pp. 10-11, October 16, 1916, p. 13. EMJ, February 23, 1918, p. 400. *Salt Lake Mining Review,* January 30, 1917, p. 30; February 28, 1917, p. 26; March 15, 1917, p. 35.

80. MH (1920): 1392-93, 1398-99.

81. NYT, February 22, 1919, p. 5.

82. NYT, July 30, 1918, p. 9; August 1, 1918, p. 16. SFE, July 30, 1918, p. 5. SFCh, August 1, 1918, p. 3. *Goldfield (NV) News & Weekly Tribune,* August 3, 1918, p. 2. *Oil & Gas Journal,* August 9, 1918, p. 2.

83. Louis Guenther, "Pirates of Promotion," *World's Work* 37 (October 1918), pp. 584-91; December 1918, p. 152.

84. NYT, October 26, 1918, p. 13; February 5, 1919, p. 14; June 7, 1919, p. 15. *Salt Lake Mining Review,* November 15, 1918, p. 35.

85. NYT, January 8, 1920, p. 11; January 16, 1920, p. 3; January 30, 1920, p. 10; March 12, 1920, p. 25; March 14, 1920, p. 14; May 1, 1922, p. 1; July 3, 1924, p. 7. SFCh, February 15, 1920, p. 3.

86. MSP, August 6, 1921, p. 205; August 14, 1920, p. 222. SFCh, October 7, 1920, p. 20; October 8, 1920, p. 21.

87. SFCh, October 8, 1920, p. 8. EMJ, November 13, 1920, p. 948.

88. SFE, July 8, 1921, p. 7. EMJ, July 23, 1921, p. 151; February 11, 1922, p. 263. SFCh, January 20, 1922, p. 8; January 27, 1922, p. 3. MH 16 (1925): 1462.

89. *Saturday Evening Post,* May 14, 1938.

90. EMJ, July 2, 1921, pp. 27-28; July 9, 1921, p. 43; July 30, 1921, p. 162. MSP, July 9, 1921, p. 47.

91. SFE, July 12, 1921, p. 10; January 20, 1922, p. 15; January 31, 1922, p. 7. MSP, July 16, 1921, p. 97; August 6, 1921, p. 205; August 20, 1921, pp. 272-73. EMJ, August 30, 1921, p. 313.

92. EMJ, July 29, 1922, p. 210; June 2, 1928, p. 902. SLTr, October 4, 1921, p. 16; October 5, 1921, p. 18; October 6, 1921, p. 20; October 8, 1921, p. 20; October 16, 1921, p. 10; October 26, 1921, p. 4. MSP, October 22, 1921, pp. 557-58. MH 16 (1925): 1738.

93. EMJ, December 10, 1921, p. 935.

94. EMJ, April 1, 1922, p. 546; April 8, 1922, p. 562.

95. *Salt Lake Mining Review,* November 30, 1923, p. 17.

96. *Idaho Daily Statesman,* December 27, 1923, p. 1; December 28, 1923, p. 2. EMJ, January 5, 1924, p. 27.

97. EMJ, March 1, 1924, p. 376; March 29, 1924, pp. 514-15; April 12, 1924, p. 615; May 10, 1924, pp. 753-54.

98. NYT, January 6, 1926, p. 25. SFCh, January 6, 1926, p. 5. EMJ, January 23, 1926, pp. 154-55.

99. NYT, September 22, 1929, p. 19.

100. I. A. Mumme, *The Emerald* (Port Hacking, New South Wales: Mumme, 1982), 20-39.

101. NYT, August 7, 1926, p. 20; August 10, 1926, p. 29; October 12, 1926, p. 40; January 6, 1927, p. 29; January 9, 1927 pt. 2, p. 1; February 20, 1927, p. 8; May 7, 1927, p. 8.

102. EMJ, October 16, 1926, p. 602; January 5, 1927, p. 106; February 26, 1927, p. 372; April 16, 1927, p. 635.

103. NYT, September 21, 1927, p. 31. Two engineers for Colombia Emerald Development wrote books about the experience, but neither had a single good word about the New York management. P. W. Ranier, *Green Fire* (Garden City, N.Y.: Blue Ribbon, 1942). Russ Anderton, *Tic-Polonga* (Garden City, N.Y.: Doubleday, 1953). Over beers in a small desert town, the author once listened to an American mining engineer tell of using the mine as a front to buy black-market emeralds in the 1960s.

104. MH 17 (1926): 181-82, 838. EMJ, January 31, 1925, p. 221.

105. EMJ, March 21, 1925, p. 500; May 16, 1925, pp. 794-95.

106. EMJ, December 17, 1927, pp. 961-62.

107. EMJ, June 6, 1925, p. 936; August 8, 1925, p. 224; August 29, 1925, p. 345; October 17, 1925, p. 625; May 22, 1926, p. 853; August 14, 1926, p. 266; October 23, 1926, p. 663; February 5, 1927, p. 253. *Arizona Mining Journal,* June 15, 1926, p. 50; April 30, 1927, p. 47. *Salt Lake Mining Review,* September 30, 1926, p. 12.

108. EMJ, October 24, 1925, p. 667.

109. EMJ, December 19, 1925, p. 963.

110. MH 16 (1925): 1607. Joseph Weil and W. T. Brannon, *"Yellow Kid" Weil* (Chicago: Ziff-Davis, 1948), 95, 193, 277.

111. EMJ, December 19, 1925, p. 981. *Arizona Mining Journal,* January 15, 1926, p. 37; February 15, 1926, p. 43; March 30, 1926, p. 40; April 15, 1926, p. 42; April 30, 1926, pp. 40, 44; May 15, 1926, p. 40; July 15, 1926, p. 49; August 30, 1926, pp. 38-39, 44; September 30, 1926, p. 42; November 15, 1926, p. 46; December 30, 1926, p. 34; April 15, 1927, p. 34; April 30, 1927, p. 62. NYT, March 8, 1929, p. 13.

112. NYT, February 4, 1926, p. 10; January 29, 1927, p. 7; March 2, 1927, p. 16; March 6, 1927, p. 17; March 19, 1927, p. 19; July 16, 1927, p. 22. EMJ, June 16, 1928, p. 988.

113. NYT, November 27, 1927, p. 12; November 29, 1927, p. 38; December 30, 1927, p. 33; March 30, 1930, p. 22. EMJ, December 10, 1927, p. 923.

114. NYT, January 18, 1928, p. 7; January 24, 1928, p. 4; February 8, 1928, p. 41; February 28, 1928, p. 6; March 2, 1928, p. 18; November 8, 1928, p. 21.

115. NYT, November 14, 1928, p. 16; November 15, 1928, p. 18. EMJ, November 24, 1928, p. 841.

116. NYT, November 17, 1928, p. 21; November 23, 1928, p. 30; November 28, 1928, p. 29; November 29, 1928, p. 19; December 1, 1928, p. 36.

117. NYT, November 14, 1928, p. 16; December 4, 1928, p. 26; December 5, 1928, p. 31; December 22, 1928, p. 7.

118. NYT, December 15, 1928, p. 15; April 11, 1929, p. 16; November 5, 1929, p. 20; November 9, 1929, p. 21; November 10, 1929, p. 25; November 11, 1929, p. 25. EMJ, December 22, 1928, pp. 967, 997.

119. NYT, March 15, 1930, p. 22; July 19, 1931, p. 2; July 22, 1932, p. 35.

120. NYT, February 16, 1930, sec. 2, p. 8.

121. NYT, September 3, 1931, p. 44; September 24, 1931, p. 16, October 21, 1931, p. 48.

122. NYT, October 20, 1931, p. 23; October 21, 1931, p. 48; October 22, 1931, p. 5; October 29, 1931, p. 4; October 30, 1931, p. 2; November 1, 1931, p. 18. *Goldfield (NV) News & Weekly Tribune,* February 9, 1934.

123. *Goldfield (NV) News & Weekly Tribune,* July 21, 1933, p. 2. NYT, November 21, 1936, p. 36. Brian Levine, "George Graham Rice," *Financial History,* no. 40 (1992): 15-18.

124. "Located," *Mining & Metallurgy* 20, no. 393 (September 1939), p. 417. Hugh Shamberger, introduction to *My Adventures with Your Money,* by G. G. Rice (1911; repr., Las Vegas: Nevada Publications, 1986), 21-22.

Chapter 19. Death Valley Scotty: The Showman as Mine Swindler

1. Paul DeLanay, "Death Valley Scotty," *Death Valley Magazine* 1, no. 3 (May 1908), pp. 83-84.

2. SFCa, June 29, 1904, p. 3. SFCh, June 29, 1904, p. 1.

3. NYT, February 9, 1930, sec. 9, p. 11.

4. DMR, April 6, 1906, p. 5. G. G. Rice, *My Adventures with Your Money* (Boston: Gorham, 1911).

5. LAT, July 1, 1905, p. 1. NYT, July 1, 1905, p. 1.

6. NYT, July 10, 1905, p. 7; July 11, 1905, p. 1; July 12, 1905, p. 5.

7. NYT, July 12, 1905, p. 5.

8. NYT, July 14, 1905, p. 1.

9. NYT, July 26, 1905, p. 1.

10. DMR, July 21, 1907, p. 1.

11. *Goldfield (NV) Review,* July 27, 1905, p. 3. BFM, July 20, 1906, p. 5. Bourke Lee, *Death Valley Men* (N.Y.: MacMillan, 1932), 1-67.

12. SFCh, August 27, 1905, p. 32. NYT, August 29, 1905, p. 1. GFN December 29, 1905, p. 5.

13. NYT, September 5, 1905, p. 6.

14. *Goldfield (NV) Review,* July 27, 1905, p. 3. DMR, April 6, 1906, p. 5; July 21, 1906, p. 1; August 8, 1906, p. 4. *Rhyolite (NV) Herald,* August 11, 1905, p. 8; August 25, 1905, p. 1. *Randsburg (CA) Miner,* August 16, 1906, p. 4; September 13, 1906, p. 1. *Las Vegas (NV) Age,* January 19, 1907, p. 1. BFM, October 19, 1907, p. 6.
15. E-mail communication, 2000, Harvard University Archives.
16. LAT, October 31, 1905, p. 1. TS, October 31, 1905, p. 1.
17. NYT, December 21, 1905, p. 1. *Tonopah (NV) Bonanza,* December 23, 1905, p. 4. *Rhyolite (NV) Herald,* January 5, 1906, p. 4.
18. BFM, January 6, 1906, p. 2.
19. O. S. Merrill, *Mysterious Scott* (Chicago: Orin Merrill, 1906).
20. *Tonopah (NV) Bonanza,* December 23, 1905, p. 4. *Rhyolite (NV) Herald,* January 5, 1906, p. 4. SBS, March 22, 1906, p. 3.
21. NYT, January 23, 1906, p. 1. SBS, April 11, 1906, p. 1.
22. L. Burr Belden, "The Battle of Wingate Pass," *Westways* 48, no. 11 (November 1956), pp. 8-9. *Boston Globe,* March 16, 1906, p. 1. SBS, March 20, 1906, p. 1.
23. E. C. Driscoll, "Death Valley Scotty," *Gold!* no. 10 (Spring 1975): 6-17, 48-53.
24. SBS, March 23, 1906, p. 6.
25. SFE, March 2, 1906, p. 3. SBS, March 3, 1906, p. 3.
26. SBS, March 24, 1906, p. 1; April 5, 1906, p. 2.
27. LAT, March 9, 1901, p. 14. SBS, March 9, 1906, p. 5.
28. SBS, April 8, 1906, p. 1.
29. SBS, March 30, 1906, p. 1; March 31, 1906, p. 8; April 1, 1906, p. 4.
30. SFCa, p. 14.
31. LAT, March 21, 1906, p. 4. SBS, March 25, 1906, p. 1; March 27, 1906, p. 1; April 12, 1906, p. 1; April 14, 1906, p. 1.
32. SBS, April 15, 1906, p. 6.
33. LAT, March 21, 1906, p. 7.
34. SBS, April 27, 1906, p. 3; April 28, 1906, p. 6; January 19, 1958, p. B10. BFM, July 27, 1907, p. 6. Eleanor Houston, *Death Valley Scotty Told Me* (Palm Desert: Desert Printers, 1954), 72-73.
35. BFM, August 10, 1907, p. 5.
36. *Goldfield (NV) Review,* January 26, 1907, p. 6. *Randsburg Miner,* January 31, 1907, p. 1. *Rhyolite (NV) Herald,* August 11, 1905, p. 8.
37. *Randsburg (CA) Miner,* March 21, 1907, p. 3, May 16, 1907, p. 4. *Rhyolite (NV) Herald,* April 5, 1907, p. 5. BFM, June 15, 1907, p. 12; June 22, 1907, p. 3; July 6, 1907, p. 3; January 25, 1908, p. 8. LAT, July 19, 1908, p. 10. NMN, August 20, 1908, p. 7.
38. BFM, February 15, 1908, p. 6.
39. BFM, April 11, 1908, p. 5.
40. BFM, June 12, 1909, p. 6.
41. *Kansas City (MO) Journal,* July 30, 1905 pt.1, p. 3. NYT, June 29, 1906, p. 2. DMR, August 1, 1906, p. 3. BFM, June 29, 1907, p. 1. F. A. Crampton, *Deep Enough* (Norman: University of Oklahoma Press, 1956), 32-33.
42. *Tonopah (NV) Miner,* June 29, 1907, p. 4. SFCh, July 24, 1908, p. 7; July 27, 1908, p. 1. SFCa, August 28, 1908, p. 14.
43. BFM, July 31, 1909, p. 4. GFN News August 13, 1910, p. 1. GFTr July 7, 1911, p. 1, July 10, 1911, p. 1. R. G. Lilliard, *Desert Challenge* (N.Y.: Knopf, 1942), 268.
44. GFN August 27, 1910, p. 1. *White Pine (NV) News,* September 25, 1910, p. 3.
45. MSP, September 28, 1912, p. 393.
46. LAT, June 11, 1912, p. 2; June 16, 1912, p. 9.
47. LAT, June 18, 1912, pt. 2, p. 3; June 19, 1912, p. 4; June 20, 1912, p. 8. Scott's reluctance to join in a stock fraud was shared by the better class of con men. "Yellow Kid" Weil said that it was a point of honor among those of his caliber to swindle only the wealthy.
48. *Moving Picture World,* October 12, 1912, p. 118; October 19, 1912, p. 274; November 9, 1912, p. 553.

49. NYT, January 6, 1954, p. 31.

50. Edward A. Vandeventer, "Death Valley Scotty," *Sunset* 56 (March 1926), pp. 22-72. NYT, May 16, 1934, p. 11; March 15, 1936, pt. 4, p. 2; October 3, 1936, p. 3; December 23, 1937, p. 2.

51. *Salt Lake Mining Review,* December 15, 1916, p. 27; October 30, 1918, p. 28. *Denver Post,* July 2, 1923, p. 7.

52. NYT, February 8, 1930, p. 5; March 15, 1936, pt. 4, p. 2.

53. NYT, February 5, 1930, p. 1; January 8, 1937, p. 14; January 15, 1937, p. 6; January 25, 1937, p. 8. LAT, January 8, 1937, p. 1.

54. NYT, March 14, 1940, p. 25.

55. NYT, March 17, 1939, p. 23.

56. NYT, August 1, 1940, p. 12.

57. NYT, March 14, 1941, p. 23; March 15, 1941, p. 19.

58. RMN, April 24, 1943, p. 4. "Scotty and His Gold..." *Desert* 12 (January 1949), p. 33. NYT, September 22, 1951, p. 19.

59. NYT, January 8, 1954, p. 22.

Chapter 20. The Literary Life: Julian Hawthorne, Mining Swindler

1. Maurice Basson, *Hawthorne's Son* (Columbus: Ohio State University Press, 1970). Vernon Loggins, *The Hawthornes* (N.Y.: Columbia University Press, 1951).

2. Julian Hawthorne, *The Memoirs of Julian Hawthorne,* ed. E. G. Hawthorne (N.Y.: Macmillan, 1938), 51.

3. Charles Honce, *More Julian Hawthorne Fruits* (N.Y.: 1941), 22-23.

4. Morton to Hawthorne, July 10, 1908, Hawthorne Family Collection, BANC MSS 72/236z, Box 5, The Bancroft Library, University of California, Berkeley. Ontario Bureau of Mines, *17th Annual Report,* 1908, p. 34.

5. Morton to Hawthorne, September 18, 1908, Hawthorne Family Collection, BANC MSS 72/236z, Box 5, The Bancroft Library, University of California, Berkeley.

6. Julian Hawthorne, "Journalism the Destroyer of Literature," *Critic* 48 (February 1906), pp. 166-70.

7. NYT, December 11, 1912, p. 8.

8. EMJ, October 17, 1908. NYT, December 7, 1912, p. 9.

9. L. D. Huntoon, "Tamagamie," *Mines and Minerals* 29, no. 8 (March 1909), p. 367.

10. EMJ, September 19, 1908; October 17, 1908.

11. NYT, October 24, 1911, p. 22; March 13, 1912, p. 3.

12. NYT, December 10, 1912, p. 10. EMJ, December 14, 1912.

13. *Canadian Mining Journal,* April 15, 1909, p. 227.

14. *Canadian Mining Journal,* December 1, 1910, pp. 710-11.

15. *Annual Report Geological Survey of Canada* 7 (1902): 46-52. Ontario Bureau of Mines, *20th Annual Report,* 1911, p. 108. Ontario Bureau of Mines, *21st Annual Report,* 1912, pp. 27, 159. E. Lindeman and L. L. Bolton, *Iron Ore Occurrences in Canada* 1 (Canada Department of Mines, no. 217), 136.

16. *Canadian Mining Journal,* April 15, 1909.

17. MJ, November 5, 1910. TL, November 5, 1910. FT, July 22, 1910, p. 6; August 13, 1910, p. 6; November 5, 1910, pp. 6, 10.

EMJ, October 28, 1911.

19. Julian Hawthorne, *The Subterranean Brotherhood* (N.Y.: McBride Nast, 1914), 36.

20. EMJ, December 30, 1916, p. 1149.

21. Hawthorne, *Subterranean Brotherhood.* Unidentified to Hawthorne, April 19, 1920, Hawthorne Family Collection, BANC MSS 72/236z, Box 6, The Bancroft Library, University of California, Berkeley.

22. Morton to Hawthorne, April 5, 1917, and September 23, 1917, Hawthorne Family Collection, BANC MSS 72/236z, Box 5, The Bancroft Library, University of California, Berkeley.

23. Morton to Hawthorne, May 14, 1920, August 30, 1928, and n.d., Hawthorne Family Collection, BANC MSS 72/236z, Box 5, The Bancroft Library, University of California, Berkeley.

24. Philip Smith, *Harvest from the Rock* (Toronto: Macmillan, 1986), 185-89.

Chapter 21. Selling the Pure Blue Sky

1. RMN, December 25, 1901, p. 1; December 27, 1901, p. 1. SFCh, December 25, 1901, p. 2. WSJ, March 5, 1904, p. 1.

2. MSP, February 7, 1920, pp. 181-82.

3. *Pacific Coast Miner,* March 14, 1903, p. 185.

4. EMJ, December 2, 1905, p. 1031; May 5, 1906, p. 851; April 21, 1906, p. 773; February 23, 1907, p. 400. H. C. Beeler, "The Recent Agitation," *Mines and Minerals* 27, no. 7 (February 1907), p. 304. GFTr March 31, 1909, p. 2. TS, April 19, 1909, p. 4.

5. *Saturday Evening Post,* December 2, 1911. MSP, March 16, 1912, p. 397; March 10, 1917, p. 323.

6. "The Basis of Blue Skyism," *Mining & Oil Bulletin* 2 (February 1916), pp. 67-68. MSP, May 14, 1921, pp. 673-76. J. E. Davis, 1999, "Corporate Disclosure through the Stock Exchanges," unpublished undergraduate paper for Harvard Law School, p. 41, http://cyber. law.harvard.edu/rfi/papers.

7. *National Petroleum News,* March 10, 1920, p. 28.

Bibliography

Bauer, Georg (Georgius Agricola). *De Re Metallica.* Translated by H. C. and L. H. Hoover. 1556. Reprint, New York: Dover, 1950.

De Quille, Dan. *The Big Bonanza.* 1876. Reprint, Las Vegas: Nevada Publications, 1974.

Goodwin, Nat C. *Nat Goodwin's Book.* Boston: Gorham, 1914.

Harpending, Asbury. *The Great Diamond Hoax and Other Stirring Incidents in the Life of Asbury Harpending.* 1915. Reprint, Tulsa: University of Oklahoma Press, 1958. Harpending's charm warms this entertaining autobiography.

Johnson, Hank. *Death Valley Scotty: "The Fastest Con in the West."* Corona del Mar, Calif.: Trans-Anglo, 1974. Wherever other sources are not cited in the chapter on Death Valley Scotty, readers may assume that the source is Johnson's thorough biography of Scott.

Lingenfelter, Richard E. *Death Valley & the Amargosa: A Land of Illusion.* Berkeley and Los Angeles: University California Press, 1988.

Lord, Elliot. *Comstock Mining and Miners.* U.S. Geological Survey Monograph 4. Washington, DC: Government Printing Office, 1883.

Rice, George Graham. *My Adventures with Your Money.* 1913. Reprint, Las Vegas: Nevada Publications, 1986.

Smith, Grant H. *The History of the Comstock Lode.* Bulletin Number 3. Reno: University of Nevada, 1943.

Spence, Clark C. *British Investments and the American Mining Frontier, 1860-1901.* Moscow: University of Idaho Press, 1995.

Stevenson, Robert Louis. *From Scotland to Silverado.* Edited by J. D. Hart. Cambridge: Harvard University Press, 1966.

Twain, Mark. *Mark Twain's Letters.* Edited by E. M. Branch et al. Berkeley and Los Angeles: University of California Press, 1988.

Woodward, Bruce A. *Diamonds in the Salt.* Boulder, Colo.: Pruett, 1967.

Index

Page numbers in italics refer to figures